D1234712

Graph Databases in Action

EXAMPLES IN GREMLIN

DAVE BECHBERGER
JOSH PERRYMAN
FOREWORD BY TED WILMES

MANNING
SHELTER ISLAND

For online information and ordering of this and other Manning books, please visit www.manning.com. The publisher offers discounts on this book when ordered in quantity. For more information, please contact

Special Sales Department
Manning Publications Co.
20 Baldwin Road
PO Box 761
Shelter Island, NY 11964
Email: orders@manning.com

Manning Publications Co.
20 Baldwin Road
PO Box 761
Shelter Island, NY 11964

Development editor: Frances Lefkowitz
Technical development editor: Nick Watts
Review editor: Aleks Dragosavljević
Production editor: Lori Weidert
Copy editor: Frances Buran
Proofreader: Melody Dolab
Technical proofreader: Alex Ott
Typesetter: Dennis Dalinnik
Cover designer: Marija Tudor

ISBN: 9781617296376
Printed in the United States of America

contents

foreword

At the dawn of a new decade, developers are confronted with a myriad of database options when beginning a new project. The stalwart relational database still rules the roost, maintaining popularity in both legacy and greenfield projects. This is for good reason; flexibility and forty plus years of cumulative engineering history are hard to argue with. Despite the success of relational databases, the last decade saw an explosion of new commercial and open-source database systems that were designed around alternative models and query languages. Some tackle traditional RDBMS workloads with a new twist, perhaps focusing horizontal scale out or high performance via the embrace of in-memory optimization that have become available due to decreases in RAM prices. Many other systems diverged from the relational model altogether. Out of this set, we find a variety of focus areas and modeling paradigms. This book focuses on one of the more expressive and powerful developments, the graph model, and the property graph in particular.

Graph databases aren't a new thing. Hierarchical and navigational databases have existed since the 60s, but these have recently experienced an increase in developer popularity. I think this is largely due to the intuitiveness of the property graph data model. People are already wired to think in graphs. If you draw a graph on a whiteboard, technical and non-technical folks get it. Consequently, after you overlay the graph model onto your software tasks at hand, everything starts to look like a graph problem.

With all that said, we're still dealing with technology, and the available property graph databases are the newer technology at that, so there isn't any magic. This is where Dave and Josh come in. I can't imagine a better pair to help lay out the signposts and guide you on the journey to graph understanding. Both are accomplished graph architects and developers that have been involved in this junior space since before its recent uptick in popularity. Having worked in graph-based product development and consulting, they've racked up years of real-world experience.

This experience has influenced their pragmatic approach to the problems of graph application development, and though both proponents of graphs, they're proponents with a healthy dose of skepticism and are not overly fascinated with the technology. After all, as mentioned, one of the first and most important questions new developers have is, "Is this a graph problem?" As you make your way through this book, you'll hone an intuition for translating real world problems into graph data models and build up your Gremlin query chops, a popular and powerful property graph query language. The rubber meets the road in chapter 6 where you use this knowledge to build your first graph application. By the time you've finished, you'll have the knowledge to evaluate if a graph database is a good fit for your next project, and if so, to execute on that vision having already built an example graph database application.

TED WILMES
Data Architect & JanusGraph Technical Steering Committee Member
Expero Inc.

preface

Two complementary trends started in the mid to late 2000s. First, companies began using and collecting more data on their customers, competition, and users than ever before. Second, the information companies wanted from this data became more complex, often containing hidden connections. These two trends drove the need for an easier exploration of expansive, yet highly connected data. Graph databases met that need.

Both the authors have gotten an up-close and personal view of this market as the technology, usage, and adoption of graph technology has matured. We both started using graph databases in the mid 2010s while working for a niche software consulting company. Independently, we each worked on projects that used graph databases to solve specific types of complex data problems. At that time, graph databases were new and very rough. Despite the challenges of working with new technologies, we both recognized the power of this tool and were hooked.

Since then, we have spent countless hours banging our heads against a proverbial wall to understand all the intricacies and nuances of building graph-backed applications. This book is the distillation of those countless hours of struggle. It is our hope that the hands-on nature of this book will provide a solid, foundational understanding of the skills needed to build graph-backed applications and, in the process, help you to avoid some of the pitfalls that we encountered.

acknowledgments

This book has been a labor of love, and sometimes frustration, so we first and foremost need to thank our wives (Melody and Meredith), and then acknowledge family and friends for their endless patience and for indulging us as we shared our latest esoteric discoveries while working with graph databases. Without their support we never could have made it through the countless hours it took to create this book.

A big thank you goes out to Dr. Denise Gosnell, Kelly Mondor, Ted Wilmes, and Daniel Farrell for all the specific insights, interviews, and support you provided, which helped us immensely in creating this book.

We would also like to thank the team at Manning Publications for allowing us the time and opportunity to publish this book. We would like to thank the entire Manning staff and specifically our publishers Marjan Bace and Michael Stephens, as well as our editors Frances Lefkowitz, Nick Watts, Alex Ott, Lori Weidert, and Frances Buran for all the amazing feedback and endless patience you have shown. Our appreciation also goes out to all the reviewers whose comments and reviews were invaluable in solidifying the organization and in clarifying the focus of this book: Scott Bartram, Andrew Blair, Alain Couniot, Douglas Duncan, Mike Erickson, John Guthrie, Mike Haller, Milorad Imbra, Ramaninder Singh Jhajj, Mike Jensen, Nicholas Robert Keers, Mladen Knežić, Miguel Montalvo, Luis Moux, Nick Rakochy, Ron Sher, Deshuang Tang, Richard Vaughan, and Matthew Welke.

We would also like to thank the team at Expero Inc., without whom Josh and Dave would never have met, nor would have ever started their exploration of graph databases. Our many years of working side by side with the exceptionally talented Experonauts were a fruitful starting point that eventually led to writing this book.

about this book

This book is written for anyone building applications using graph databases. It is designed to provide a foundational understanding of graphs and graph databases, as well as to provide a framework for building applications using common graph database patterns. To teach this framework, this book follows the development lifecycle of a fictitious application called DiningByFriends. We use this application throughout the book to provide a realistic grounding of graph principles and examples of the concepts and content we teach. In many areas throughout this book, we compare and contrast the differences between building a graph-backed application and using the more traditional relational database model. By the end of this book, you will not only have the skills needed to build your own graph-backed application, but you will have built your first application, DiningByFriends.

Who should read this book

This book is for application developers, data engineers, and database developers who want to use graph databases as the backing data store for their applications. Throughout this book, we do not expect the reader to have any prior experience using graph databases, but you should be familiar with data modeling concepts, specifically with relational database development, as these are used heavily throughout as a common point of reference. Although all the application code is written in Java, any developer with object-oriented application development experience should be able to follow along with the concepts and content.

How this book is organized: A roadmap

This book is organized into 3 parts, comprising of 11 chapters. In part 1, "Getting started with graph databases," we establish the foundation for our DiningByFriends application:

- Chapter 1 begins with an introduction to graphs and graph terminology. We discuss how graph databases differ from relational databases and how you can use graph databases to solve highly connected data problems. We finish this chapter by discussing what makes a problem a good candidate for using a graph database.
- Chapter 2 is where we hit the ground running by building an initial data model for our DiningByFriends application. We start with the types of information needed to begin the data modeling process. We then show how to turn this information into a conceptual data model. Finally, we walk through a framework for taking our business needs and our conceptual data model and turn that into our initial data model using the elements of a graph database: vertices, edges, and properties.
- Chapter 3 begins a set of three chapters focused on learning the process of querying a graph database, known as traversing. We begin by teaching you how to retrieve and filter data from our graph. We follow this with learning how to navigate the structure of our graph and how that differs from working with a relational database. Then we finish up this chapter by demonstrating the ease with which you can recursively traverse through a graph to retrieve complex, interconnected data.
- Chapter 4 continues our exploration of graph traversals with data mutation use cases. We then show how you can traverse the graph to find the entities and relationships that connect two items, known as the path. Finally, we look at how to leverage properties on relationships to filter the traversals and increase their performance.
- Chapter 5 finishes our initial focus on graph traversals with a discussion of ways to format the results of our traversal into a desired output. Additionally, you learn how to perform common operations such as sorting, filtering, and limiting the results returned.
- Chapter 6 begins the process of building our DiningByFriends application by taking the traversals we developed in chapters 3, 4, and 5 and walking through incorporating these into a Java application. Then we'll process the results to complete this first part.

In part 2, "Building an application with graph databases," we extend the concepts introduced in part 1:

- Chapter 7 uses the foundations of data modeling from chapter 2, as well as what you learned about traversing a graph, to extend the data model for more complex use cases, such as recommendation engines and personalization.

- Chapter 8 leverages a recommendation engine use case to demonstrate the power of using a known-walk pattern to create a robust recommendation application pattern.
- Chapter 9 uses our personalization use case to demonstrate how to use a subgraph access pattern within a graph-backed application.

In part 3, "Beyond the basics," we move past the DiningByFriends application to discuss our next steps in the application development process.

- Chapter 10 discusses how to debug and troubleshoot common performance problems with traversals. We then investigate exactly what supernodes are and why they cause issues in graph-backed applications. We follow up these common performance problems with common application and traversal pitfalls and anti-patterns, as well as how to recognize and avoid them.
- Chapter 11 takes a forward-looking view and discusses some of the next steps you might want to take with your graph-backed application. We also discuss some of the most common graph analytics algorithms and how you can apply these to solve a specific problem. Finally, we wrap up this chapter with a brief overview of how to leverage graphs in machine learning (ML) application.

About the code

This book contains many examples of source code, both in numbered listings and in line with normal text. In both cases, source code is formatted in a `fixed-width font like this` to separate it from ordinary text.

In many cases, the original source code has been reformatted; we've added line breaks and reworked indentation to accommodate the available page size in the book. In rare cases, even this was not enough and code listings include line-continuation markers (➥). Additionally, code annotations accompany many of the listings, highlighting important concepts.

The code for the examples in this book is available for download from the Manning website at https://www.manning.com/books/graph-databases-in-action, and from GitHub at https://github.com/bechbd/graph-databases-in-action.

About the technologies

Our goal throughout this book is to equip the reader with the conceptual knowledge needed to build graph-backed applications. However, in order to provide practical examples of these concepts, we had to make decisions regarding the technologies used for demonstration.

Our first decision was to pick the type of database. We decided to use a labeled property graph database, instead of, for example, an RDF store or triplestore database. Labeled property graph databases are the most common type we have seen in production use and seem to be the ones with the most momentum behind them. Additionally, these are the closest to the familiar concepts of relational databases, so labeled property graph databases are quite effective for comparisons.

This lead us to our next decision: the traversal language to use, openCypher or Gremlin.

While there's a strong case for using openCypher, the goal of this book is to remain as vendor-agnostic as possible. It is important to us that these concepts and techniques are easily transferable to many popular databases when you start to build your applications. In the end, we decided to use the Apache TinkerPop version 3.4.*x* framework because it currently has the most database vendors with compatible implementations.

We have been questioned multiple times during the proposal and review processes as to why we chose this stack over a Neo4j/Cypher stack. Given the popularity of the Neo4j ecosystem this is a fair question which deserves fuller comment. There are three reasons we chose TinkerPop's Gremlin for the illustrations throughout this book:

- Gremlin is a better tool for teaching how a traversal works.
- Gremlin is a common language of choice for enterprise applications.
- Gremlin is the most portable language between property graph databases.

As for the first reason, we believe that the imperative design of Gremlin provides a better teaching tool for learning how a graph traversal works compared to the declarative approach of Cypher/openCypher. The syntax of Gremlin requires that we think about how we are moving through our graph in order to determine where we will move next. While we do appreciate the simplicity of Cypher/openCypher, it can also obfuscate critical technical matters, especially when dealing with issues of performance or scale. So while Cypher/openCypher is a great starting point for learning how to work with connected data, we feel that Gremlin is better suited for building high performing, scalable data applications.

Because Gremlin is the common language of choice for enterprise applications, many of these applications were built using TinkerPop-enabled databases. This means that Gremlin is the query language of choice. Some organizations have both Cypher/openCypher and Gremlin applications. But in our experience, the bigger, more complex enterprise-level projects seem to have chosen one of the many TinkerPop-enabled databases or cloud services.

As for our third choice, at this time, it is easy to say that Gremlin is the most widely available query language across graph database engines. Nearly all of the major cloud vendors (Amazon Web Services, Microsoft Azure, IBM, Huawei, and so forth) offer graph databases or services compatible with Gremlin. The lone exception is the Google Cloud Platform, which offers Neo4j as a service.

Our goal is not to advocate for one database or language over another. We seek to provide you with a solid foundation for how to use a graph database when building applications with highly connected data and to illustrate how graph databases work under the cover. We think that Gremlin provides the best path to accomplish this.

With the decision to use TinkerPop's Gremlin made, we had to pick a specific TinkerPop-enabled database to use. In the spirit of remaining vendor agnostic, we've decided to use TinkerGraph for the examples. TinkerGraph is the graph implementation used in the Gremlin Server and Gremlin Console, the reference software provided as part of the Apache Software Foundation's TinkerPop project.

Finally, we had to decide on an application programming language to build our example application, DiningByFriends. As Java is the most common language we have used with graph databases, we chose that as our application language. We should note that it is possible to build the same application with other languages such as C#, JavaScript and Python. Not only is it possible, we have done so ourselves. But all the traversals provided in this book are written in Gremlin and any application code is written in Java.

While almost all the concepts presented throughout this book are not specific to TinkerPop-enabled databases, there are a few we discuss that are unique to TinkerPop. When this is the case, we'll note where a TinkerPop-specific feature is used so that you're aware that a particular feature might not be available in your graph database of choice. If no such note is given, it is safe to assume that the concept we discuss is applicable to other labeled property graph databases as well.

liveBook discussion forum

Purchase of *Graph Databases in Action* includes free access to a private web forum run by Manning Publications where you can make comments about the book, ask technical questions, and receive help from the authors and from other users. To access the forum, go to https://livebook.manning.com/#!/book/graph-databases-in-action/discussion. You can also learn more about Manning's forums and the rules of conduct at https://livebook.manning.com/#!/discussion.

Manning's commitment to our readers is to provide a venue where a meaningful dialogue between individual readers and between readers and the authors can take place. It is not a commitment to any specific amount of participation on the part of the authors, whose contribution to the forum remains voluntary (and unpaid). We suggest you try asking the authors some challenging questions lest their interest stray! The forum and the archives of previous discussions will be accessible from the publisher's website as long as the book is in print.

about the authors

DAVE BECHBERGER is a data architect and developer with over two decades of experience. He uses his extensive knowledge of graph and other big data technologies to build highly performant and scalable data platforms in complex data domains such as bioinformatics, oil and gas, and supply chain management. Since the mid-2010s, Dave has worked with graph databases as a consultant, consumer, and vendor. He is an active member of the graph community and has presented on a wide range of graph-related topics at national and international conferences.

JOSH PERRYMAN also has over two decades of experience building and maintaining complex systems. Since 2014, he has focused on graph databases, especially in distributed or big data environments, and he regularly blogs and speaks at conferences about graph databases. Josh has worked with a variety of industries, including enterprise software, financial services, consumer products, and government intelligence agencies. In addition to consulting and product work, he has designed Gremlin training courses that have been delivered all over the world.

about the cover illustration

The figure on the cover of *Graph Databases in Action* is captioned "Femme de la Foret Noire," or a woman from the Black Forest, in Southwest Germany. The illustration is taken from a collection of dress costumes from various countries by Jacques Grasset de Saint-Sauveur (1757–1810), titled *Costumes civils actuels de tous les peoples connus*, published in France in 1788. Each illustration is finely drawn and colored by hand. The rich variety of Grasset de Saint-Sauveur's collection reminds us vividly of how culturally apart the world's towns and regions were just 200 years ago. Isolated from each other, people spoke different dialects and languages. In the streets or in the countryside, it was easy to identify where they lived and what their trade or station in life was just by their dress.

The way we dress has changed since then and the diversity by region, so rich at the time, has faded away. It is now hard to tell apart the inhabitants of different continents, let alone different towns, regions, or countries. Perhaps we have traded cultural diversity for a more varied personal life—certainly for a more varied and fast-paced technological life.

At a time when it is hard to tell one computer book from another, Manning celebrates the inventiveness and initiative of the computer business with book covers based on the rich diversity of regional life of two centuries ago, brought back to life by Grasset de Saint-Sauveur's pictures.

Part 1

Getting started with graph databases

Journeys into new technologies take work, and in this book, our journey will extend your current knowledge of building relational database applications to demonstrate how you can solve complex data problems by building graph databases and graph-backed applications. In this first part, we ease into your journey by establishing concepts, terms, and processes, while highlighting the critical differences required when approaching a problem with a graph mindset.

Chapter 1 introduces the core concepts of graphs and discusses the types of problems that are well suited for these models. In chapter 2, we establish a data modeling methodology and build a simple data model for a social network that we'll use in our example application, DiningByFriends. The next three chapters introduce the most common operations that you'll use to find and manipulate data in graph databases. We approach these operations in three stages, starting with the basics of moving around a graph in chapter 3. Chapter 4 then covers how to perform basic CRUD (Create/Read/Update/Delete) operations before extending the work we did in chapter 3 to perform more complex recursive and pathfinding traversals. In chapter 5, we close our introduction by using simple graph operations to examine ways to organize your results. Chapter 6 completes this part by synthesizing the work from chapters 2 through 5 into our working Java application, DiningByFriends.

1 Introduction to graphs

This chapter covers

- An introduction to graphs and graph terminology
- How graph databases help solve highly connected data problems
- The advantages of graph databases over relational databases
- Identifying problems that make good candidates for using a graph database

Modern applications are built on data—data that is ever increasing in both size and complexity. Even as the complexity of our data grows, so do our expectations of what insight our applications can derive from that data. If you are old enough, you likely remember when applications took a long time to load data and had limited features. Today's reality is different; applications provide powerful, flexible, and immediate insight into data. But for every 100 questions modern applications answer, the most common data tool these use (namely, a relational database) handles only about 88 of those questions well. That leaves 12 types of questions where relational databases struggle. These remaining questions deal with the links and connections within the data, those aspects of the data that can generate powerful and unique insights. This puts us at a crossroad: we can use the relational database "hammer" to pound away at those questions and make this work well enough, or we can take a step back and look at what other tools can answer these questions better, faster, and with less effort.

By reading this book, you decided to take a step back from your relational database hammer and investigate a road less traveled: graph databases. This book is written for developers, engineers, and architects who are interested in other ways to solve problems specific to working with highly connected data. We assume you are already familiar with relational databases but are interested in learning when, where, and how graph databases are a better tool.

Our goal with this book is to equip you with the techniques needed to add graph databases as another tool in your toolbelt. We like to think of this book as the guide that we wish we had when we started building graph-backed applications. Throughout this book, we'll demonstrate common graph patterns that highlight how graph databases enable navigation and exploration of data in ways not easily accomplished with a traditional relational database.

Our primary approach is through an example of building a fictitious restaurant review and recommendation application we call "DiningByFriends." As we move through the software development life cycle from planning, to analysis, to design, and on to implementation, this application demonstrates how to think about and work with graph data. Each chapter builds on the previous chapter, and by the end of this book, we'll have created a functioning application on a graph database. We believe that putting the concepts immediately to work by solving a realistic set of problems, even if they are somewhat simplistic, is the best way to get comfortable using a new technology. Let's begin our journey with an introduction to what graphs and graph databases are and how they compare with traditional tools such as relational databases.

1.1 What is a graph?

When you look at a road map, examine an organizational chart, or use social networks such as Facebook, LinkedIn, or Twitter, you use a graph. Graphs are a nearly ubiquitous way to think about real-world scenarios as these abstract out the items and the relationships being represented, and this abstraction allows for quick and efficient processing of the connections within the data.

Let's demonstrate with a common task: going to the supermarket. Take out a piece of paper and draw out a plan for getting from your house to your supermarket. Chances are it looks something like figure 1.1.

Figure 1.1 shows a graph where the key items and relationships are represented by abstractions. First, we abstracted key locations, like intersections, and represented these as circles. We then designated the connections between these key intersections as lines, showing how the key intersections are related. This is just one example of how we naturally represent real-world problems as graphs.

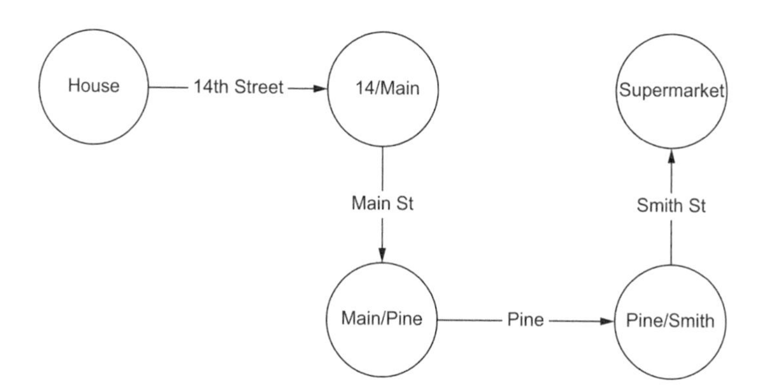

Figure 1.1 A graph representing directions to the supermarket

It is human nature to abstract real-world entities and their relationships, and the mathematical name for this abstract construct is a *graph*. When thinking about a set of data that contains a vast array of highly interconnected items, we might also describe this data set as a web of interconnected things, which is just another way of saying a graph.

On maps, cities are frequently represented by circles, and the roads that connect these are represented by lines. On an organizational chart (org chart), a circle usually represents a person, normally with an associated title, and lines that connect these people together show the employer-employee relationship. In a social network, people connect to one another via friending or following. This process of generalizing entities and the connections between them is the fundamental basis for graphs and graph theory. Because graphs have been defined and studied by mathematicians for centuries, we can offer these definitions used in graph theory as our starting terms:

- *Graph*—A set of vertices (singular, vertex) and edges
- *Vertex*—A point in a graph where zero or more edges meet, also referred to as a node or an entity
- *Edge*—A relationship between two vertices within a graph, sometimes called a relationship, link, or connection

Euler and origins of graph theory

The origins of graph theory are generally attributed to a paper published by Leonhard Euler (pronounced "Ol-ler") in 1736, concerning the Seven Bridges of Königsberg. Königsberg (now known as Kaliningrad) was a Prussian city located on the Pregel river. The river contained two islands and was traversed, or connected, by seven bridges. The experiment was to devise a path that would allow citizens of the town to cross all seven bridges exactly once. Euler approached this problem by creating an abstract representation of the land masses (as the vertices) and the bridges (as the connections or edges) between these. Based on this abstraction, Euler stated that it was not the items specifically that mattered, but the topology of how these items were connected that played the most significant role.

(continued)

In his "Seven Bridges of Königsberg" paper, Euler stated that for the problem to be resolved, the graph needed either zero or two nodes with an odd number of connections. Nowadays, any graph meeting this condition is known as an *Eulerian graph*. If the path visits each edge exactly once, then it contains an Eulerian path. If the start and end vertex are the same, then it has an Eulerian circuit, which is also known as an Eulerian cycle. We share this as an interesting bit of historical context, but in our combined experience, we have never used these academic facts or Eulerian definitions in any real-world problems.

While definitions are nice, graphs have the advantage of being simple to illustrate. When working with graphs, diagrams usually consist of circles representing vertices and lines representing edges, as figure 1.2 shows.

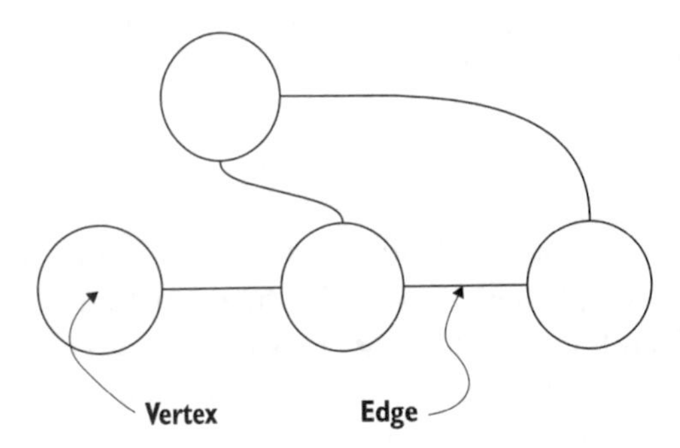

Figure 1.2 A graph is easily illustrated with circles for the vertices and lines for the edges.

NOTE We use the terms *vertex* and *edge* throughout this book. Some graph databases use the term *node* instead of vertex and *relationship* instead of edge, but these are conceptually the same.

Graphs are not new concepts to software developers. These are the basis of many common data structures that we use in software development all the time, likely without even realizing it. Common data structures such as linked lists and trees are simply types of graphs with specific rules applied to them. While these data structures are well known to developers, the actual implementation details specific to graphs are usually abstracted away.

1.1.1 What is a graph database?

A graph database is a data-storage engine that combines the basic graph structures of vertices and edges with a persistence technology and a traversal (query) language to create a database optimized for storage and fast retrieval of highly connected data. Unlike other database technologies, graph databases are built on the concept that the relationships between entities are as or more important than the entities within the data. Because entities and relationships are treated with equal importance in a graph database, we can represent and reason over real-world relationships more accurately and easily, especially when compared to other database technologies. As we'll show in this book, graph databases are better tools for both representing the rich and varied relationships between things, and recognizing patterns based on these relationships.

Let's briefly look at some of the challenges of representing multiple varying types of relationships with relational databases. Relational databases (in a fit of naming irony) are rather poor at representing rich relationships. The relationships in relational databases are foreign keys, which are pointers to primary keys in other tables. These pointers are not things we can observe and manipulate easily. Instead, the foreign keys are followed (at query time) from one row to another row. (Though possible, it is often expensive to follow these in the reverse direction.) Lookup or linking tables move away from the query-time-only-pointer construct to allow for storing attributes about the relationship, similar to the edge-construct in graph databases.

On the other hand, graph databases provide excellent tools for moving through relationships in our data. By making the connections (edges) as important as the items, the edges connect to (vertices), graph databases represent these associations as full-fledged constructs of the database that can be easily observed and manipulated. This ability to store rich relationships is one of the main reasons that graph databases are better suited to handling complex linked-data use cases. In developer parlance, we might say that edges are "first-class citizens" just like the vertices. That is, the relationships are as critical and useful in the data model as the things or entities.

As a final point, graph databases enhance developer productivity for certain problems in ways that other technologies cannot. Storing data in a manner that better represents its real-world counterpart can make it easier for developers to reason over and understand the domain in which they are working. This allows new team members to get up to speed more quickly on the domain. They learn the domain and its database representation simultaneously.

1.1.2 Comparison with other types of databases

Though this book is focused on graph databases, and it uses relational databases as the primary foil for comparison, we should note that the database world is not limited to these two types of data stores. In the broadest of terms, a database can be categorized as an engine type in one of the five following ways. Figure 1.3 summarizes the relationships between these types of engines:

- *Key-value*—Represents all data by a unique identifier (a key) and an associated data object (the value). Examples include Berkeley DB, RocksDB, Redis, and Memcached.
- *Wide-column (or column-oriented)*—Stores data in rows with a potentially large number or possibly varying numbers of columns in each row. Examples include Apache HBase, Azure Table Storage, Apache Cassandra, and Google Cloud Bigtable.

- *Document*—Stores data in a uniquely keyed document that can have varying schema and that can contain nested data. Examples include MongoDB and Apache CouchDB.
- *Relational*—Stores data in tables containing rows with strict schema. Relationships can be established between tables allowing the joining of rows. Examples include PostgreSQL, Oracle Database, and Microsoft SQL Server.
- *Graph*—Stores data as vertices (nodes, components) and edges (relationships). Examples include Neo4j, Apache TinkerPop's Gremlin Server, JanusGraph, and TigerGraph.

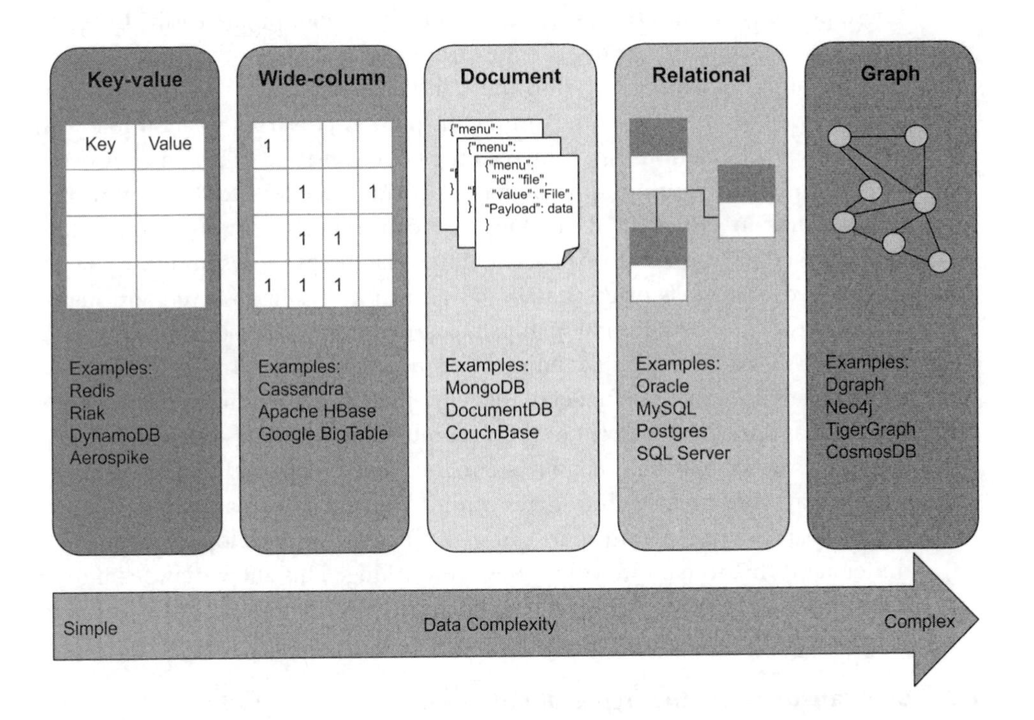

Figure 1.3 Database engine types ordered by data complexity

As you can see from these examples, only the relational databases and graph databases, by default, include the ability to relate entities within the data. It may be possible to do that with specific implementations of key-value, wide column, or document databases, but this is usually an enhancement added by a vendor's specific implementation. Because our focus is on graph databases and only relational databases offer a comparable functionality, the rest of our discussions are exclusive to these two types of engines.

1.1.3 Why can't I use SQL?

As developers, we often choose a familiar tool over an optimal one, especially when dealing with databases. Most development teams have an in-depth knowledge of the ins and outs of relational databases, but few have expertise in other types of databases. Therefore, we often default to the relational database either through convenience or ignorance, while there are better tools in the toolbox to solve certain problems.

We are not trying to say that relational databases are a poor tool. In fact, it's usually the first one that we reach for when working on our own applications. But relational databases have their limitations. While it is possible to use relational databases with highly connected data, in many cases the work can be simplified by using a tool designed for these types of use cases. In this section, we look at three areas where graph databases provide a simpler, more elegant solution than using a relational database:

- Recursive queries (for example, an organization's employee reporting hierarchy, or org chart)
- Different result types (for example, an orders and products reporting example)
- Paths (for example, a river-crossing puzzle)

For this chapter, we chose three different examples to represent these three unique graph database capabilities. Starting with the next chapter, we'll introduce the DiningByFriends problem domain and start the formal data modeling process. At that point, most of the examples will follow with the development of this sample domain. But until then, we'll use a variety of ways to introduce you to the basic concepts of graphs and graph databases.

Recursive queries

Recursive queries are executed multiple times in succession, repeatedly calling themselves until they reach some escape or terminating condition. Relational databases do not handle recursive operations (especially unbounded ones) well, struggling both with syntax and performance. This usually leads to writing and maintaining complex queries, excessive denormalization of our data, or both, all in an effort to return results in a timely fashion.

On the other hand, graph databases use their rich relationship representations to handle these unbounded recursive queries cleanly and efficiently. To see what we're talking about, let's take a look at what a recursive query looks like in both SQL and in a graph database. Given a list of employees and managers in a company, as shown in figure 1.4, let's examine how we determine a person's reporting hierarchy.

To model this hierarchy in a relational database, the following query shows how we would define a table. Then we take this table schema and lay out the data (table 1.1):

```
CREATE TABLE org_chart (
  employee_id           SMALLINT NOT NULL,
  manager_employee_id   SMALLINT NULL,
  employee_name         VARCHAR(20) NOT NULL
);
```

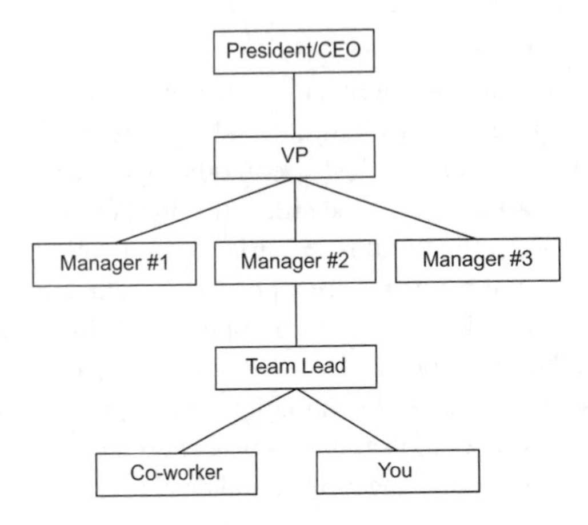

Figure 1.4 Management hierarchy in a company, demonstrating recursive queries

Table 1.1 Example of a company's management hierarchy in a relational database

employee_id	Manager_employee_id	employee_name
1	3	You
2	3	Co-worker
3	4	Team Lead
4	5	Manager #2
5	8	VP
6	5	Manager #1
7	5	Manager #3
8	NULL	President/CEO

We then use a recursive function to query this data to find a user's management hierarchy. The following code snippet show the query:

```
WITH RECURSIVE org AS (
     SELECT employee_id,
            manager_employee_id,
            employee_name,
            1 AS level
     FROM org_chart
  UNION
     SELECT e.employee_id,
            e.manager_employee_id,
            e.employee_name,
            m.level + 1 AS level
     FROM org_chart AS e
       INNER JOIN org AS m ON e.manager_employee_id = m.employee_id
  )
```

```
SELECT employee_id, manager_employee_id, employee_name
FROM org
ORDER BY level ASC;
```

If you've ever written common table expressions (CTEs) in SQL like our management hierarchy query, then you know that these can be complex to write and debug, and are notorious for poor performance. On the other hand, nested and recursive queries like the previous hierarchy example are the types of questions that graph databases are optimized to answer. For example, figure 1.5 shows what the same data looks like as a graph.

To find our user's management chain in our graph, we need to write a query analogous to our SQL query, which in graphs is known as a *traversal*. For our hierarchy example, we would get a traversal like the following one:

```
g.V().
  repeat(
    out('works_for')
  ).path().next()
```

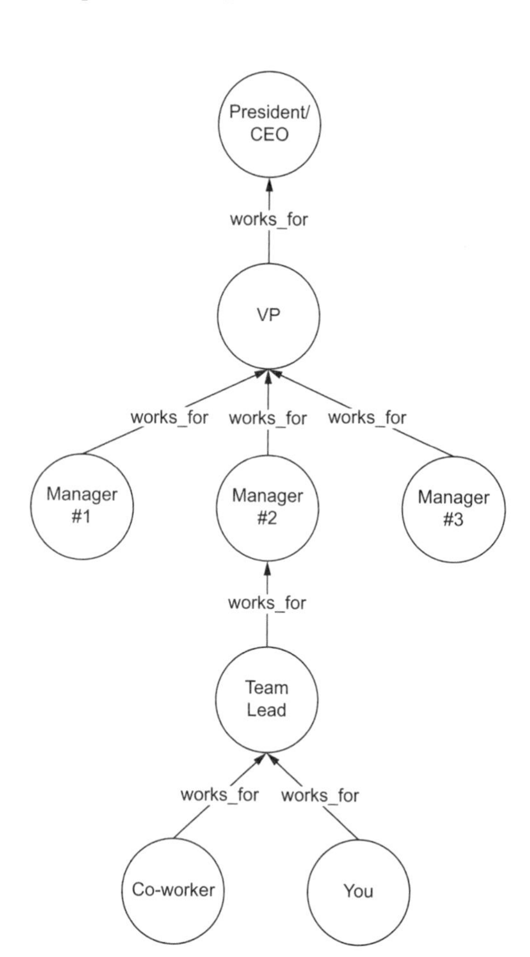

Figure 1.5 Graph representation of organizational hierarchy with the circles as vertices and the arrows as edges

NOTE The traversal is in a graph query language called Gremlin, which we'll use throughout this book. At this point, it isn't necessary to understand precisely how it works. We'll delve into details starting in chapter 3. For now, just notice the relative simplicity of this query compared to the previous SQL query.

This example demonstrates the straightforward nature with which you can recursively ask questions of a graph. If we compare this to figure 1.5, we can see how this traversal naturally maps to our instinct to visually navigate the hierarchy of the data.

DIFFERENT RESULT TYPES

Have you ever needed to return several different data types from a database, all within a single result set? While it is possible to achieve this with a union of all the columns in all of the tables, it tends to yield less than ideal results. One of the strengths of a graph database is the ability to return differing data types in the results. Let's look at how relational and graph databases compare when returning different types.

For instance, let's say that we have an order-processing system and we want to return not only the order information but also the product information. Figure 1.6 represents a traditional implementation with tables in a relational database.

Orders		
id	name	address
1	John Smith	123 Main. St
2	Jane Right	643 Park St.

Products		
id	product_name	cost
123	widget 1	5.95
234	widget 2	10.76

Figure 1.6 Orders and Products tables in a relational database; note the differences in column names.

The following code snippet shows how to write a query to retrieve an order with the associated product information. Table 1.2 shows the result set for this query.

```
SELECT id,
       name,
       address,
       null AS product_name,
       null AS cost,
       'Order' AS object_type
FROM Orders
UNION
SELECT id,
       null AS name,
       null AS address,
       product_name,
       cost,
       'Product' AS object_type
FROM Products;
```

Table 1.2 Results from the SELECT query that retrieves the order and associated product information

id	Name	Address	product_name	cost	object_type
1	John Smith	123 Main St	<null>	<null>	Order
2	Jane Right	234 Park St	<null>	<null>	Order
123	<null>	<null>	widget 1	5.95	Product
234	<null>	<null>	widget 2	10.76	Product

From the results, we see that the union of these two disparate data types dictates that our answer contains a large number of null values (commonly known as *sparse data* or *sparse matrix*). This abundance of null data is caused by the columns between the two tables being inconsistent. A relational database specifies that the returned result set must contain a consistent set of columns. In cases of sparse data, this not only inflates the amount of data returned, but it also reduces the descriptive nature of the data structure. Let's take a look at how that same data appears in a graph database (figure 1.7).

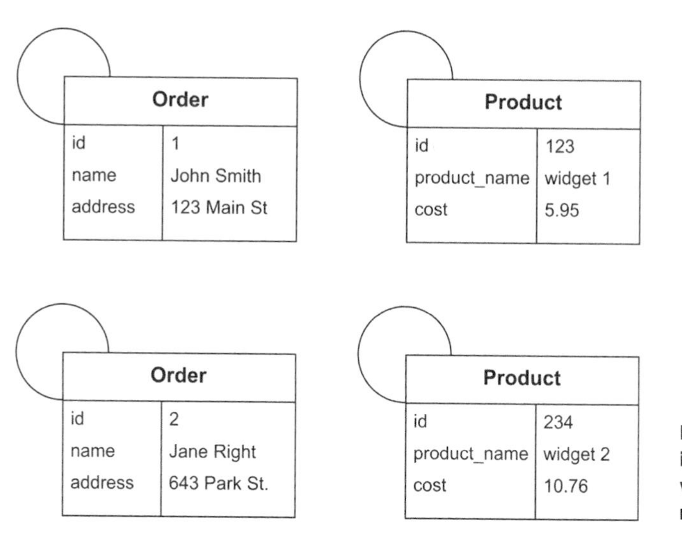

Figure 1.7 Our order product information example shown as vertices in a graph (edges are not modeled)

Using this graph, we can write a graph traversal to return both product and order data. In this example, a graph database returns these results:

```
gremlin> g.V().valueMap(true)
==>[label:order, address:[123 Main St], name:[John Smith], id:1]
==>[label:order, address:[234 Park St], name:[Jane Right], id:2]
==>[label:product, cost:[10.76], id:234, product_name:[widget 2]]
==>[label:product, cost:[5.95], id:123, product_name:[widget 1]]
```

Compared to the earlier SQL results, the data returned from the graph retains the semantic meaning of what the object is and what it represents, without the extraneous null data. Because graph databases provide the flexibility to return disparate data, we can create much cleaner code when working with highly varied data types.

PATHS

A path is the sequence of vertices and edges that describe how the traversal moved through the graph; for example, in Google or Apple Maps, a set of directions between two locations. The ability to return how two objects are connected to each other from within the database is a feature unique to graph databases.

Let's look at a classic puzzle known as the "river crossing puzzle" to illustrate how paths can help solve problems in a novel fashion. In our river crossing puzzle, we have a fox, a goose, and a bag of barley that must be transported across a river by a farmer on a boat. However, this movement is bound by the following constraints:

- The boat can only carry one item in addition to the farmer on each trip.
- The farmer must go on each trip.
- The fox cannot be left alone with the goose or it will eat it.
- The goose cannot be left alone with the barley or it will eat it.

Using a relational database, we can't find a way to solve this riddle without using a brute force method to calculate all possible combinations. However, with a little clever data modeling and the power of a pathfinding algorithm, it's rather straightforward to answer this riddle with a graph.

Let's start by modeling the initial state of our system as a vertex in our graph. We'll call our vertex `TGFB_`, where each character represents part of the problem:

`T` (the boat and the farmer)
`G` (the goose)
`F` (the fox)
`B` (the barley)
`_` (the river)

This `TGFB_` vertex encodes the state of the puzzle by telling us that the boat (`T`), the goose (`G`), the fox (`F`), and the barley (`B`) are all on one side of the river (`_`). Our goal is to achieve a state where these are all on the other side of the river.

With the vertices representing possible states, we use edges to show how we transition from one state to the next. For example, figure 1.8 shows how we can represent the state change of the farmer taking the goose to the other side of the river, leaving the fox and the barley on the initial side. And figure 1.9 shows the result of

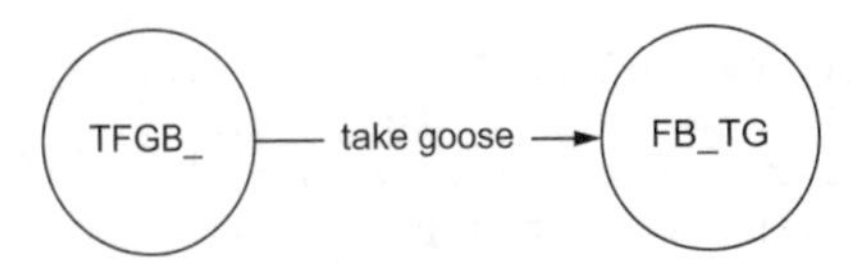

Figure 1.8 Graph representation of the farmer using the boat (`T`) to take the goose (`G`) across the river (`_`), leaving the fox (`F`) with the barley (`B`).

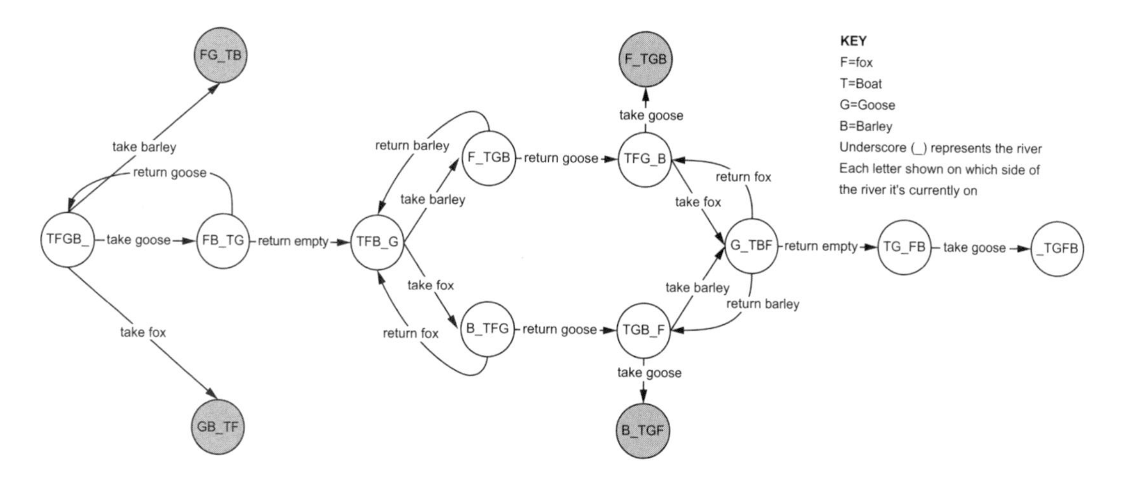

Figure 1.9 The full graph of the river crossing puzzle using a pathfinding algorithm. Notice the clear depiction of the possible solutions with any state that violates the highlighted constraints.

modeling all the potential options as a representation of these states (vertices) and state changes (edges).

Figure 1.10 illustrates what happens if we simplify our graph by removing any state (vertex) that violates a constraint and the adjoining relationships (edges). We can further simplify our graph by removing any edge that connects back to a previous state because this leads us to a previous state (known as a *cycle* in graphs).

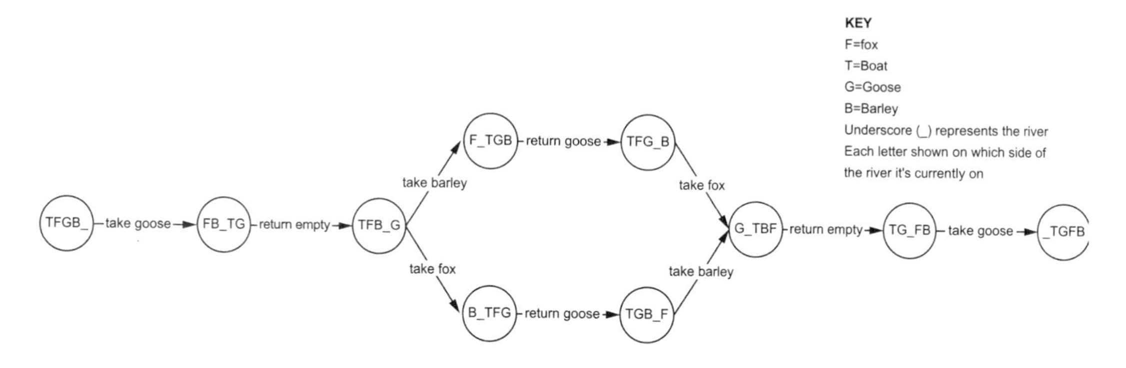

Figure 1.10 The river crossing puzzle using our pathfinding algorithm with only the valid states

By analyzing figure 1.10, we see two separate paths to get to our desired state. To query the graph to return these paths, we simply leverage the pathfinding capabilities of graph databases to return the two appropriate paths as shown by this traversal:

```
g.V('TFGB_').
  repeat(
    out()
  ).until(hasId('_TGFB')).
  path().next()
```

When we run this traversal, it returns not only the first and last vertex visited, but also the entire set of vertices and edges that were visited along the way. The two lists represent two different paths to the solution:

```
TFGB_ -take goose-> FB_TG -take empty-> TFB_G -take barley-> F_TGB -return
    goose-> TFG_B -take fox-> G_TBF -return empty-> TG_FB -take goose-> _TGFB

TFGB_ -take goose-> FB_TG -take empty-> TFB_G -take fox-> B_TFG -return goose->
    TGB_F -take barley-> G_TBF -return empty-> TG_FB -take fox-> _TGFB
```

Although this example is a riddle, it represents the same fundamental problems found in many real-world applications, such as finding a route on a map, finding optimal resource usage in a logistics system, or locating connections between people in a social network. Each of these cases is fundamentally about determining the optimal set of steps to get from one entity to another. The graph data structure allows us to leverage these pathfinding capabilities, which are not a native construct in other database types.

1.2 Is my problem a graph problem?

From social network analysis, recommendation engines, dependency analysis, fraud detection, and master data management, to search problems and research on the internet, you'll quickly encounter a listing of good use cases for graph databases. The difficulty with many of these lists is that unless your problem is one of those specified, it's hard to know how or if it's a good fit for a graph database.

In this section, instead of focusing on specific use cases, we'll look at problems in a more generic way. This is somewhat conceptual, but we find that it can be difficult to generalize from an example to a specific problem domain. We'll start with defining a general problem and then providing some examples to illustrate. We'll then close this section with a general framework for evaluating problems and with a decision tree (which is a form of graph!) to use as a tool for deciding whether to use a graph database or not.

1.2.1 Explore the questions

While reading through the vast array of information on graph databases available on the internet, you might come across the statement that says, ". . . everything is a graph problem." We agree that the real world is easily described in graph terms, but saying that everything is solved by one type of database is a drastic oversimplification. Just because a problem can be represented as a graph doesn't necessarily mean that a graph database is the best technology to choose to solve that problem.

Our process starts with one simple question: "What problem are we trying to solve?" Answering this question provides crucial details about what questions we are going to ask, and this governs the types of data we need to store and how we need to retrieve it. We break down our answers into the following categories of problems:

- Selection/search
- Related or recursive data
- Aggregation
- Pattern matching
- Centrality, clustering, and influence

Let's examine each of these in turn and discuss what makes each a good or bad candidate for using a graph database.

SELECTION/SEARCH

We classify the following types of questions as search or selection problems. These questions narrowly focus on finding a small set of entities that all share a common attribute such as name, location, or employer:

- Give me everyone who works at *X*?
- Who in my system has a first name like John?
- Locate all stores within *X* miles?

These sorts of questions do not require rich relationships within the data. In most databases, answering these questions requires using a single filtering criterion or, potentially, an index. While you can answer these with a graph database, these problems do not use or require graph-specific functionality. Instead, it is advisable to use a relational database such as PostgreSQL (https://www.postgresql.org) or a search technology such as Apache Solr (http://lucene.apache.org/solr) or Elasticsearch (https://www.elastic.co). These databases or tools are either more mature (e.g., RDBMS) or better optimized (e.g., search tools) to answer these sorts of questions. Because these problems don't leverage the relationships in our data, in our experience, it's unlikely that taking on the additional complexities of graph databases is worthwhile.

VERDICT For these types of questions, use an RDBMS or search technology.

RELATED OR RECURSIVE DATA

Questions that explore the relationships between entities add meaning and provide topological value to data, providing a strong use case for a graph database. Some examples of these types of questions include

- What's the easiest way for me to be introduced to an executive at *X*?
- How do John and Paula know each other?
- How's company *X* related to company *Y*?

Graph databases leverage this information better than any other type of data engine, and their query languages are better suited to reasoning over the relationships within the data. Although not impossible in relational databases, these sorts of friends-of-friends queries require complex and difficult to maintain or reason over recursive CTEs or complex joins across many different tables.

VERDICT For these types of questions, use a graph database.

AGGREGATION

Data aggregation queries constitute an excellent use case for a relational database. Relational databases are optimized to perform complex aggregation queries quickly and with a minimal amount of overhead. Example questions might include

- How many companies are in my system?
- What are my average sales for each day over the past month?
- What's the number of transactions processed by my system each day?

These same sorts of queries can be performed in graph databases, but the nature of graph traversals requires that much more of the data is touched. But this causes higher query latency and resource utilization.

VERDICT For these types of questions, use an RDBMS.

PATTERN MATCHING

Pattern matching based on how entities are related is a prime example of how to leverage the power of graph databases. Typical use cases for this sort of query involve things like recommendation engines, fraud detection, or intrusion detection. Some questions might include

- Who in my system has a similar profile to me?
- Does this transaction look like other known fraudulent transactions?
- Is the user J. Smith the same as Johan S.?

Pattern-matching use cases are so commonly done in graph databases that graph query languages have specific, built-in features to handle precisely these sorts of queries.

VERDICT For these types of questions, use a graph database.

CENTRALITY, CLUSTERING, AND INFLUENCE

The relative influence or importance of one entity compared to another is a typical graph database use case. Some example questions might include

- Who's the most influential person I am connected with on LinkedIn?
- What equipment in my network has the most substantial impact if it breaks?
- What parts tend to fail at the same time?

Examples of other problems of this type include finding the most influential person in a Twitter network, identifying critical pieces of infrastructure, or locating groups of entities within your data. Calculating the answers to these sorts of problems requires looking at entities, their relationships, and the incident relationships and adjacent entities. As with pattern-matching use cases, these types of problems often have specific, built-in graph query languages features.

VERDICT For these types of questions, use a graph database.

1.2.2 *I'm still confused. . . . Is this a graph problem?*

The types of problems discussed so far provide a significant first step in deciding if your problem is a good candidate for using a graph, but what if your problem doesn't neatly fit into one of these predefined types? In this section, we use the friends-of-friends problem with a decision framework to help us decide if we have a good problem for a graph.

To illustrate, we use a small social graph that includes Alice, Bob, Ted, and Josh as vertices connected by `follows` edges, as shown in figure 1.11. The question we want to answer is, "Given a person in the graph, of the people that they follow, who do those people follow that the first person might also want to follow?" This question is the same as that answered by sites such as LinkedIn, Twitter, or Facebook to recommend connections to users on a daily basis. Let's break this down into its four basic parts:

- Given a person in the graph . . .
- . . . of the people that they follow . . .
- . . . who do those people follow . . .
- . . . that the first person might also want to follow?

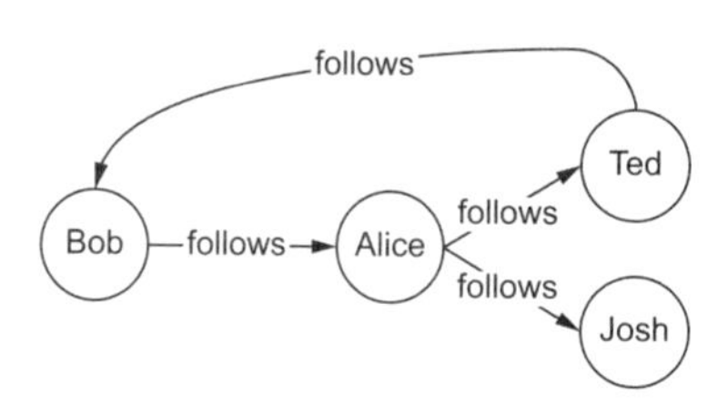

Figure 1.11 A simple social graph illustrates the common friends-of-friends pattern.

Let's take Bob as a place to start (first point). Bob follows Alice (second point). Alice follows both Ted and Josh (third point). Therefore, Bob might want to follow both Ted and Josh (final point).

Take look at the decision tree in figure 1.12, which is designed to answer the question, "Should I use a graph database?" Then we examine each of the questions and analyze why these lead you to using or not using a graph database in your work. We should note at the outset that here we focus on transactional (as in online transactional processing or OLTP) use cases. The decision matrix could be different for analytical use cases (as in online analytical processing or OLAP). We focus almost exclusively on the transactional processing use cases through chapter 10, but in the final chapter, we give some guidance for whole-graph (or graph analytics) processing.

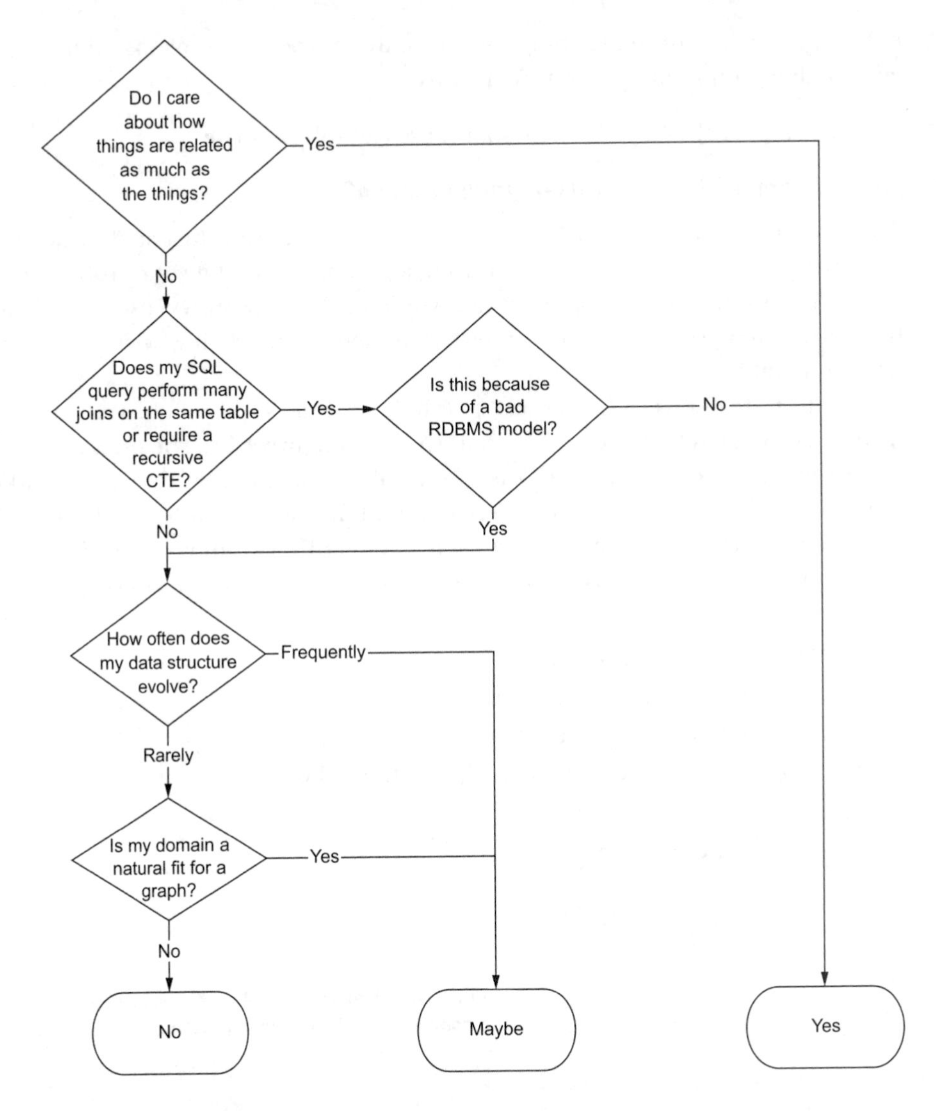

Figure 1.12 "Should I use a graph database?" decision tree. Start at the top and work down to a Yes, No, or Maybe.

DO WE CARE ABOUT THE RELATIONSHIPS BETWEEN ENTITIES AS MUCH OR MORE THAN THE ENTITIES THEMSELVES?

This question is perhaps the most critical clue, which is why we put it first. It speaks to the heart of one of the most powerful features of graph databases: relationships are as meaningful as entities. If our answer to this question is *yes*, then we probably need a data model that allows for sophisticated representations of the relationships—an excellent candidate for using a graph database. But if our answer is *no*, then perhaps another data engine would be a better choice.

In the case of our friends-of-friends problem, the answer to this question is *yes*. After the starting step of our question (Given a person in the graph) each of the remaining steps requires the use of relationships between people to answer.

DOES MY SQL QUERY PERFORM MULTIPLE JOINS ON THE SAME TABLE OR REQUIRE A RECURSIVE CTE?

While a large number of joins in a SQL query can indicate that a graph database might be a good fit, it doesn't make that possibility certain. Large numbers of joins in a SQL query are often a sign of a well-normalized data model. But when those joins are not being used to retrieve reference data (as is done with a third normal form in a relational database) and, instead, are used to link items together (as with a parent-child relationship), then we may want to consider a graph database. Also, recursive query patterns benefit from graph databases when we do not know the number of joins that will be performed.

Taking our friends-of-friends example, let's say that we want to answer the question, "What are the connections to get from Bob to Ted?" Attempting to perform this query in a relational database requires an unknown number of joins, and it might not complete, indicating that no path exists between the two. However, graph databases can recurse efficiently over unbounded hierarchical data such as this. If a recursive approach helps to solve the problem, then a graph database is often preferable.

IS THE STRUCTURE OF MY DATA CONTINUOUSLY EVOLVING?

We won't go so far as to call graph databases *schemaless*, a term indicating that the database engine does not enforce schema on write operations; we know several graph databases that do enforce schema. But we can say that you can design graph databases to be more tolerant of evolving data. Relational databases, on the other hand, have a well-deserved reputation for the strictness of their schema and the complexity associated when making schema changes.

If your problem requires taking in data with different data schemas, such as dependency management, then a graph database may be worth investigating. Flexibility with data schema alone should not be a sufficient reason to choose a graph database, however, but combined with other features, it might be enough to tip the scales in favor of using a graph database.

IS MY DOMAIN A NATURAL FIT FOR A GRAPH?

If you're doing something such as routing, dependency management, social network analysis, or cluster analysis, then your problem revolves around highly interconnected data, so your domain may be well suited for using a graph. A word of caution: although your domain models naturally in a graph, if your questions aren't relying on the relationships in the graph for the answers, then you should consider other options.

In fact, our initial work with graph databases, back in 2014, revealed how the client's data fit very naturally in a graph.[1] We even tried it in three different graph

---

[1] The analysis was redone with a public data set at the "Graph Database Shootout 2.0" talk presented at Graph-Day Seattle in July, 2016 (http://mng.bz/9A7r).

databases. The model presented had built-in inheritance functionality, multi-hop traversals, and a natural requirement for dependency analysis. The two primary data constructs in the customer's application were even called components (an alias for vertices) and relationships (an alias for edges). The fact that it should've been built in a graph database instead of a relational database seemed obvious to all who gave even a cursory look at the data and the domain.

In the end, the right answer for that particular customer wasn't to use one of the three graph engines we evaluated, but to better use their relational database (or rather, use it in a way congruent with their primary access patterns). We then added a read-optimized relational projection, basically a full copy of the legacy model, into the relational database schema designed for performant querying. This is sometimes known as a *command query responsibility segregation* (CQRS) pattern. With this new "fast-read" model in place, we demonstrated a 100-fold performance improvement for some of their most demanding queries.

At first, we were all shocked that the graph databases didn't provide the necessary performance improvement because the data modeling was so naturally a graph. Then we looked more closely at the five queries used to evaluate the performance of each database. Aside from the inheritance modeling, none of the queries required a graph-style access pattern. Because a graph was not required, we used aggressive denormalization to address the inheritance use cases. In fact, the required access patterns were well-suited for relational databases; hence, the outstanding performance improvement when the data was modeled to take advantage of the RDBMS query optimizers' strengths.

Back to the graph database decision tree (figure 1.12); if you can answer *yes* to one or more of those questions, then it's likely that you may have a graph problem. If you are still uncertain—if there is still a perception of risk around the use of a graph database—then execute a small project (between two days and two weeks) to evaluate the graph as a part of a solution. Also, switching to using a graph database does not have to be an all-or-nothing situation. Don't be afraid to experiment with graph databases for solving only a portion of a problem. Multi-model approaches with graph databases are common and, in our experience, tend to be very successful.

As we mentioned at the beginning of this chapter, relational databases solve 88 out of 100 application issues well, so feel free to use them for those problems. The remaining 12 are really the ones where you might want to begin experimenting with graph databases. The rest of this book introduces you to the hows and the whys of building software with a graph database, starting with data modeling in chapter 2.

Summary

- Graph databases are based on the graph theory part of discrete mathematics, which has been around for hundreds of years. This means that mathematicians had centuries of creating nomenclature, not all of which can be considered useful or relevant to building software with graph databases.

- A graph is made up of vertices (also known as nodes and entities) and edges (also called relationships, links, or connections). Edges connect or meet at vertices.
- The five general types of databases are key-value, wide column, document, relational, and graph. Of these five, only relational and graph databases are able to model relationships with any level of sophistication.
- Graph databases are designed with relationships as first-class citizens, making it easier to build software that relies on working with these relationships. When answering questions that heavily rely on the relationships between data, graph databases tend to perform better compared to other types of databases.
- Use cases that require features like recursive queries, returning different result types, or returning paths between things, are easier to code and are better performing in graph databases.
- Due to the power and flexibility of graph databases, a large variety of good and bad graph use cases are cited on the internet. The most important factor in deciding if a use case is good or bad is the knowledge of the desired questions and outcomes from whatever system you choose.

2 Graph data modeling

This chapter covers

- Defining project goals and terminology with business or end users
- Building a conceptual data model for the entities and their relationships
- Translating a conceptual data model into a graph data model
- Comparing graph data modeling concepts to relational data modeling concepts
- Constructing the graph data model for our social network use case

Let's say you want to build a fire pit in your backyard. How would you approach this problem? Would you just start building something and hope that it comes out all right, or would you sit back and draw a picture of what you want to accomplish? When building anything, be it software or a backyard fire pit, it's crucial that you start with a good mental picture of the end result. This picture needs to include the scope that the solution addresses and the requirements to complete the solution. The more details this picture provides, the easier it is to build the solution.

In software, a significant part of the mental picture is the data model. A well-thought-out data model with a helpful level of abstraction and consistent naming conventions is intuitive to work with, maybe even a joy to use. This is as true with graph databases as it is with any other type of database. But graphs add a twist—modeling relationships with greater sophistication. And therein lies our challenge: we need to create a data model that succinctly expresses these relationships, yet with a high level of detail.

This chapter follows a four-step process to graph data modeling. First, we'll start by defining the problem to ensure we understand the details and requirements. Then we'll move on to creating a conceptual data model (a whiteboard model) of our problem from a business point of view, expressing the entities and relationships between these. Third, we'll translate this conceptual data model to a logical data model consisting of vertices, edges, and properties to express the developer's view of the entities and relationships between those. Finally, we'll test our logical data model against our business understanding to ensure that our model is capable of satisfying all the requirements of the problem we need to solve. We'll then conclude the chapter by building a graph data model for the social network use case, DiningByFriends, to learn by doing.

2.1 The data modeling process

Data modeling is the process of translating real-world entities and relationships into equivalent software representations. The extent to which we achieve accurate software representations of these real-world items dictates how well we address the intended problem.

In relational database applications, the process of data modeling is about translating certain real-world problems, understandings, and questions into software, usually focusing on creating a technical implementation involving a database. This includes identifying and understanding the problem, determining the entities and relationships in that problem, and then creating a representation of that problem in the database. The graph data modeling process is largely the same. The main difference is that we must shift from an "entity first" mindset (or perhaps more accurately, an "entity-only" mindset) to an "entity and relationship" mindset.

In this section, we'll demonstrate how to make that mindset shift by executing this process with our DiningByFriends app. Along the way, we'll call attention to specific details that are particular to graph data modeling and show how they differ from other types of data modeling. To start this process, we first go through some terminology.

2.1.1 Data modeling terms

As data modeling is about translating real-world problems, let's begin by defining some generic data terms that we use when discussing the business view of the problem. These will later be translated to graph-specific terms for the technical implementation.

When describing the business view of the problem, we use the following terms. You might not be familiar with these terms as defined here, so we want to be clear about how we use these throughout this process:

- *Entity*—Commonly represented by nouns, entities describe the things or the type of things in the domain (for example, vehicles, users, or geographic locations). As we move from problem definition and conceptual modeling, entities often become vertices in the logical model and technical implementation.
- *Relationship*—Often represented by verbs (or verbal phrases), relationships describe how entities interact with one another. It could be something like *moves* as in "a vehicle *moves* to a location," or *friends* as in the Facebook sense of this word as a verb (for example, "a person *friends* another person"). As we move from problem definition and conceptual modeling, relationships often become edges in the logical model and technical implementation.
- *Attribute*—Like entities, also represented by a noun, but always in the context of an entity or relationship. Attributes describe a quality about the entity or relationship. We limit our use of attributes as we feel that these can distract from the more critical parts of the model development process.
- *Access pattern*—Describes either questions or methods of interaction in the domain. Examples can be questions like, Where is this vehicle going? or Who are this person's friends? As we move from problem definition and conceptual modeling, access patterns often become queries in the logical model and technical implementation.

You'll find some obvious correlations between these data modeling terms (entity, relationship) and the graph elements (vertex, edge) that we introduced in the first chapter. In fact, in some graph database engines, edges are called relationships. This begs the question, why use separate terms when these all mean the same thing?

But these are not the same things. Though there is often a strong correlation between the conceptual model described with entities and relationships and the logical model described with vertices and edges, these are not guaranteed to have a one-to-one correlation. To take an example, it is perfectly normal to have an entity in the conceptual model implemented as a property on a vertex in the logical model.

Let's illustrate this distinction with a preview of an implementation decision we need to make in our DiningByFriends model, as shown in figure 2.1. Consider that restaurants are generally categorized by their type of cuisine. We can implement this as a `cuisine_type` property on the `restaurant` vertex or as a separate `cuisine`

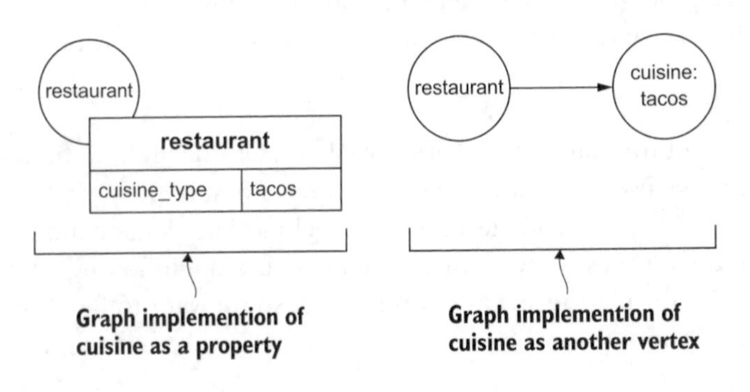

Figure 2.1 Two possible graph implementations for a restaurant's cuisine

vertex that restaurants connect to by an edge. Either could work and both give valid results, but in the end, we usually make our choice based on the predominant access patterns.

Put another way, the physical data model is largely a result of the queries we write. We know that, for some, this feels like putting the cart before the horse. Don't you usually create the data model and then write the queries? Yes, we have done that many times and have painted ourselves into a corner with design mistakes more often than we would like to admit. The approach we take is designed to reduce that risk and to minimize the pain of data model changes.

Going back to the use of different terms for different parts of the process, the other reason for this goes back to our main point as stated at beginning of this section: we translate a real-world problem into a technical domain. We use the technical terms *vertex* and *edge* when working with a specific type of data engine, a graph database in this case. If we use a relational database, the technical terms become *table* and *column*. But data modeling starts with engagement with the business, with the users and their perspectives. The business and end users do not think in terms of vertices and edges, nor do they use these terms in their normal day-to-day tasks, and they shouldn't. The process we describe here uses different terms, like entity and relationship, to remind us that the conceptual model is a tool for communicating requirements between the end users and the developers.

2.1.2 Four-step process for data modeling

Having defined our data modeling terms, this leads us to the data modeling process itself. The process uses the following four steps, and each step is covered in detail in its own section:

1. *Understand the problem.* We start in section 2.2 with a focus on the business or domain terms and language to ensure a clear understanding of the end user's perspective. We'll explore project goals to make certain that the domain and scope of the problem is clear. At the end of this step, we'll have specified the common terms and the core access patterns of users.
2. *Create a whiteboard or conceptual model.* After understanding the problem and the language used to describe it, in section 2.3, we'll move from text to a picture, focusing on drawing a diagram that makes sense to the business users and one that's useful to the technical developers. We'll define the conceptual model, which includes codifying the main entities and relationships between those entities. After completing this step, we'll have a high-level picture of the problem domain from the business perspective.
3. *Create a logical data model.* In this step, covered in section 2.4, the technical implementers (that's you!) combine the domain defined in the first step with the conceptual model from the second step to create the physical description of the graph data model. This includes defining the vertices and the edges, as well as specifying the properties on those.

Most graph databases are schemaless, so once we define this logical model, we're ready to begin working on our queries. If your chosen graph database requires explicit schema definitions (similar to defining tables and keys in an RDBMS), we'll also look at those at this time. In either scenario, once we finish this section, we'll have completed the data model for our social network use case.

4. *Test the model.* In section 2.5, we'll verify that our developed model satisfies the defined problem, that the entities and relationships needed to answer our user's questions exist, and that these are properly named. This step focuses on validating coherence in the three previous steps, where we moved from a textual description of the domain to a simple picture of the entities and relationships and, finally, constructed our model with vertices and edges. This is largely a matter of asking, "Does this make sense given the other steps?" and "Did we leave anything out which we established before?"

The first two steps in this process are a partnership between the business users and the technical development staff, first defining a common set of terms and then illustrating how those terms relate to one another in a simple diagram. For the last two steps, a technical team member takes the diagram and builds and tests the logical model that becomes the basis for implementation, as shown in table 2.1.

Table 2.1 Summary of the design process for developing a logical data model

Data modeling step	Participants	Tools	Output
1. Problem definition	Business, Developers	Domain, scope, business entities, functionality	Textual description of problem
2. Conceptual data model	Business, Developers	Entities, relationships, access patterns	Picture of entities and their relationships
3. Logical data model	Developers only	Vertices, edges, properties	Diagram of graph elements
4. Test the model	Developers only	Preceding steps	Coherence of the outputs of prior steps

For some of us with a highly structured mindset, we know that providing a four-step process to guide both this chapter and all future data development deeply resonates with our desire for a well-ordered world (and development process). But we also know that some "code-first cowboys" out there feel that four enumerated steps are three steps too many. Sure, we know that in many cases "code wins," and that an 80% implementation is often preferable to a 100% design. But this book is not targeted at building toys. Instead, we aim to build production-level applications with highly connected data in complex domains.

We are not suggesting spending endless hours/days/weeks agonizing over the perfect data design before writing any code. Designs change just as quickly as the business can create new requirements, so yesterday's perfection is tomorrow's functionally incomplete application.

We know from experience that mistakes in the data design phase cause problems during implementation, and that these problems are significantly harder to fix at that stage. Don't be deceived by the apparent simplicity of a graph or by the schema-lite/schemaless nature of some graph databases. Any implementation implies some level of written and tested code, and data that is loaded. As with any relational database project, design changes usually mean schema changes, which leads to code change and, likely, a data migration of some sort. All of these additional downstream effects have to be dealt with, and often with less mature tooling.

2.2 Understand the problem

Whether working in a large enterprise, a small company, or just on a side project, the first step in data modeling is understanding the problem, the domain, and the scope of the work we are addressing. In a large enterprise, this work may have already been done for us through some requirements document. In a small company or with personal projects, that is unlikely to be the case. In the end, it doesn't matter if our project has a requirements document or not; it is up to us to have a sufficient understanding of the problem before beginning work on our data model.

In this section, we examine several types of questions we need to answer before we can develop our data model. Ideally, these sorts of questions are already identified in functional and business requirements before beginning the project. The goal of these questions is to define how users interact with the system so we can develop a logical data model that supports the users' preferred access patterns.

> **NOTE** If you already have a strong background in data modeling, feel free to skip this section and move on to the next. If not, then read on and learn about what you need to know to ensure that you understand the problem.

We've found that users are very clever. Even if your model doesn't directly support their preferred access patterns, they find a way to make it work. But if they can't find a workaround, then they stop using the tool. Think about that! Fail here and your application is abandoned. Now isn't that a cheery thought?

While the questions vary by project, there are different categories of questions that help us gain a clear view of the problem. These categories include the following:

- Domain and scope
- Business entity
- Functionality

In the next few sections, we explore each of these types of questions, discuss why these are important, and provide some examples from our DiningByFriends application.

2.2.1 Domain and scope questions

Every problem can expand in infinite directions, so the more precisely we define the scope, the more likely we are to succeed. Domain and scope questions define the boundaries of the problem. If we make the domain too broad, then we risk not understanding its boundaries and may never complete the application. If we make the domain too narrow, then we may miss out on critical features and not provide sufficient functionality to our users. Properly defining the domain and scope of the problem you work on is therefore crucial to building a complete and functional application. The following sections provide example questions and answers to narrow the scope of the problem for our DiningByFriends app.

WHAT WILL DININGBYFRIENDS DO FOR ITS USERS?

DiningByFriends provides users with personalized restaurant recommendations. When using DiningByFriends, users have three main needs that the application must satisfy:

- *Social network*—Users want to connect with friends who are also using the application. This functionality is similar to the way people connect with friends on any social network such as Twitter, LinkedIn, or Facebook.
- *Restaurant recommendations*—Users want to create and look at reviews of restaurants and then get recommendations for a restaurant based on these reviews. This is the central service the app provides.
- *Personalization*—Users want to rate the reviews of restaurants to indicate whether the review was helpful or not. Then they want to combine these reviews with their friends' ratings to receive personalized recommendations based on the restaurants their friends also like.

WHAT TYPES OF INFORMATION DOES THE APPLICATION NEED TO RECORD TO PERFORM THESE TASKS?

To answer this question, DiningByFriends should include at least the following information:

- All the basic identifying information about users, such as a name and a unique ID, so people can find and connect with them on the social network. (In a real-world scenario, this would likely include many additional attributes, but we keep it limited for this example application.)
- Restaurant identifiers and details, such as the name, address, and cuisine, to provide location-specific recommendations.
- The text of the review, along with the rating and a timestamp of the rating in order to get personalized recommendations.
- Reviews need to include ratings of its helpfulness (for example, up/down thumbs) so that friends know if a user agreed or disagreed with those reviews.

WHO ARE THE USERS OF OUR APPLICATION?

We have one type of user for our application. This includes users of the application who connect with friends, enter reviews, and receive recommendations.

> **NOTE** We know that nearly all complex applications have internal or system users of some sort. These can include system administrators, customer service personnel, and others responsible for the maintenance and operation of a complex technical solution. We have elected to ignore such requirements in an effort to streamline the design of the use case. We therefore only focus on the traditionally understood end user.

From the questions in this section, we now have a fairly clear picture of the problem domain and its scope, as well as a set of terms that make sense to the business or end users we want to serve. And we now know the critical items needed to construct a personalized restaurant recommendation: people (users), restaurants, restaurant reviews, and ratings of those reviews.

2.2.2 Business entity questions

This type of question identifies the business entities and relationships within our problem domain. Looking at artifacts such as a relational database schema, entity relationship diagrams (ERDs), or other architectural documentation often helps us obtain a sense of the structure, language, and terminology already in use. Our goal is to identify the fundamental building blocks of our application and how these are related to one another. The following sections provide a few examples of the business entity questions we might ask.

WHAT SORT OF ITEMS OR THINGS DOES THE APPLICATION UTILIZE?

The application works with people, reviews, and restaurants.

HOW DO THESE ITEMS INTERACT WITH ONE ANOTHER?

- People *write* reviews.
- Reviews *discuss* restaurants.
- Restaurants *serve* one or more types of food.
- A person *is friends with* another person.
- People *rate* reviews.

WHAT ARE THE CRITICAL PIECES OF DATA YOU NEED TO KNOW ABOUT EACH ENTITY?

While not an exhaustive list, here are some items we need to store:

- *User data*—First and last name to help identify users
- *Restaurant data*—Details such as names, addresses, and types of foods served
- *Reviews*—Descriptions of the users' experiences.
- *Ratings*—Rankings of a review so that friends know if it is helpful or not

Graphs and graph data models derive much of their power from having well-defined relationships between entities, which is a change for those of us used to relational data modeling. A well-defined relationship in a graph requires not only a name for the relationship but also an understanding of how that relationship connects entities, as well as any potential attributes required to define the relationship. Therefore, it is critical to spend extra time exploring the relationships between entities, looking for potentially important interactions that are not immediately obvious. Looking at the answers, we often find a relationship and entity not called out specifically but hidden in the replies.

EXERCISE Do you see any hidden entities or relationships in the previous lists of questions about the business entities for our DiningByFriends app?

As we examine the list, we see a hidden relationship between a restaurant and the type or types of food served. In our restaurant recommendations application, it is highly likely that a user wants to search a specific type of food or cuisine to get recommendations. This desire means that it's likely beneficial to make cuisine (Pizza, Chinese, Indian, etc.) an entity itself and to add a corresponding relationship between a restaurant and its cuisine.

2.2.3 Functionality questions

Questions concerning functionality reveal how our business entities interact, which represents the relationships between these entities. These questions start by exploring what the user might ask of the system or what problems users have that they want the system to solve for them. These problems determine both the questions that the user asks and, sometimes, the order in which they ask those.

When we get to the conceptual model in the next phase (section 2.3), we codify the functionality as access patterns. Later, we test our logical model (section 2.5) to see if it can provide the described functionality, or put another way, support the identified access patterns. The final use of functionality is in the actual implementation, when we build the queries for the system in chapter 5. The definition work we do in this step becomes the bedrock on which we build our application. In slightly more practical and graph-oriented terms, functionality definitions lead directly to the edges we define in our logical model and, likely, some of the properties for the edges as well. Let's look at a few functionality questions for our use case.

HOW ARE PEOPLE GOING TO USE THE SYSTEM?

Users create friendships with people they know, provide reviews, rate restaurants, and read and rate reviews submitted by their friends.

WHAT QUESTIONS DOES DININGBYFRIENDS NEED TO ANSWER FOR THE USER?

These questions about functionality fill in the details of how a user is going to interact with the system:

- Who are my friends?
- Who are the friends of my friends?
- How is user *X* associated with user *Y*?

- What restaurant near me with a specific cuisine is the highest rated?
- Which restaurants are the ten highest-rated restaurants near me?
- What are the newest reviews for this restaurant?
- What restaurants do my friends recommend?
- Based on my friends' review ratings, what are the best restaurants for me?
- What restaurants have my friends reviewed or rated in the past *X* days?

We now know what our users are going to do with the app and also what they are going to ask of it. In other words, this is a first pass at our queries using natural language. (Remember that this process is done with the business or end user and should be completely understandable by them.) As we mentioned at the top of this section, gaining this understanding of this information ensures that we model our data in a way that matches the users' desired access patterns.

2.3 Developing the whiteboard model

The second step in our modeling process is to develop the conceptual or whiteboard model. We need to get a high-level diagram of what the schema for DiningByFriends looks like from a business perspective. This is our first tangible picture of the system, and it must be driven by the business view of the problem.

As builders, it is in our nature to solve problems, usually right away. But it is vital to take time to understand and define the business perspective of a domain. This accelerates our development process in the long run. Isolating what is most important to the business is crucial to making informed decisions and preventing unnecessary complexity and excessive rework.

2.3.1 Identifying and grouping entities

We develop our conceptual data model by first extracting the entities in our domain. As you'll recall, entities refer to the things in our application domain and represent either physical items, such as people and places, or logical items, such as reviews and ratings.

TIP Start by looking for the nouns.

Once we locate the entities, we need to identify items that can be easily grouped into a single entity. When making these groupings, it pays to listen to the way business and other non-technical users discuss the problem. These users live this problem almost daily; if they use nouns interchangeably, signaling that the nouns are synonyms, then it is likely these can be combined into a single entity. For example, if we are working on an internal application and the business users mention user, employee, or client interchangeably, then we could probably group these nouns into a single entity within our conceptual model.

As a best practice, we should make all the names of our entities *singular* because each entity represents a single instance of that item. We know that there are those who prefer to use plural nouns for their entity-naming schemes, but we have found that singular names tend to be a better fit for graph data modeling.

> **EXERCISE** Looking back at the answers in section 2.2, identify what you think the entities for DiningByFriends should be.

Looking at the answers, we find four entities for DiningByFriends:

- *Restaurant*—Represents a restaurant, which includes name and location.
- *Cuisine*—Describes the type of food served. This entity was not explicitly defined as one of the nouns but was found by listening to how the business described its needs.
- *Person*—Represents a system user, which includes the first and last name.
- *Review*—Actual review content, which includes the full review text and rating.

How does this compare to your list? If you have more, less, or a different set of entities, don't worry. There is no one correct answer, and different people often come up with different solutions. We chose the entities in the list because we thought these were the highest-level items that could be derived from the information available, while still providing context to our problem.

Now that we have our entities, let's put that "whiteboard" to use. We start our diagram with a box for each identified entity, as shown in figure 2.2.

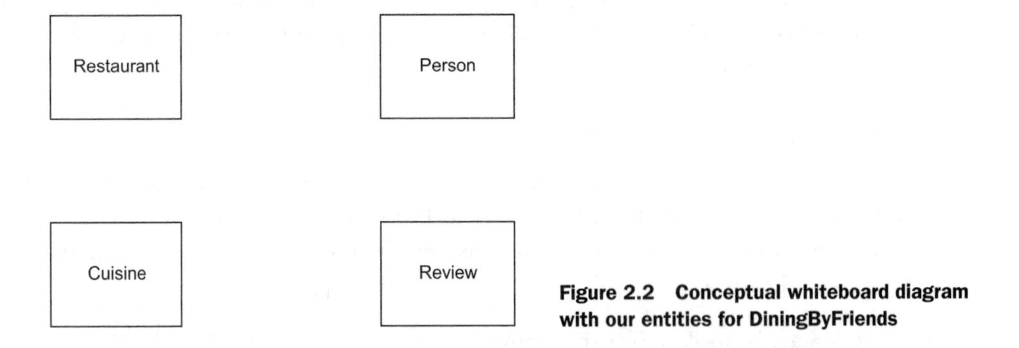

Figure 2.2 Conceptual whiteboard diagram with our entities for DiningByFriends

2.3.2 *Identifying relationships between entities*

The entities we just identified represent the *what* in our data model. The next step is to determine the relationships or the *how*. Extracting relationships is similar to locating the entities, except instead of looking for the nouns in our answers to the functionality questions, we look for the verbs. The verbs describe how any two of our entities interact. Once we have identified the verbs, we need to provide names to describe the relationships. To name these, we take each verb and combine it with the entity names in the form noun–verb–noun or entity–relationship–entity to form a short, understandable phrase (Restaurant Serves Cuisine, for example).

EXERCISE Look for the verbs in our functionality questions and locate what you think the relationships in DiningByFriends should be.

Were you able to generate a list of relationships? It is a bit harder than finding the entities, isn't it? Then again, this step is complicated by not having a user to interview. Here is our list along with the functional questions each relationship addresses (questions can appear more than once):

- *Person–Friends–Person*—This relationship helps construct the social network component of our app by allowing the application to answer user questions such as
  - Who are my friends?
  - Who are the friends of my friends?
  - How is user *X* associated with user *Y*?
- *Person–Writes–Review*—This relationship enables us to construct the recommendation engine for DiningByFriends as the reviews serve as the underlying data to provide recommendations. This relationship allows the app to answer user questions such as
  - Which restaurant near me with a specific cuisine is the highest rated?
  - Which restaurants are the ten highest-rated restaurants near me?
  - What are the newest reviews for this restaurant?
  - What restaurants do my friends recommend?
  - Based on my friends review ratings, what are the best restaurants for me?
  - What restaurants have my friends reviewed or rated in the past *X* days?
- *Reviews–Are About–Restaurant*—This relationship allows our application to formulate our recommendation engine as these reviews need to be associated with a restaurant that the app recommends. Here's a list of questions to consider:
  - Which restaurant near me with a specific cuisine is the highest rated?
  - Which restaurants are the ten highest-rated restaurants near me?
  - What are the newest reviews for this restaurant?
  - What restaurants do my friends recommend?
  - Based on my friends review ratings, what are the best restaurants for me?
  - What restaurants have my friends reviewed or rated in the past *X* days?
- *Restaurant–Serves–Cuisine*—This relationship allows us to provide some of the filtering capabilities for the recommendation engine, especially the ability to recommend a restaurant based on cuisine:
  - Which restaurant near me with a specific cuisine is the highest rated?
- *Person–Rates–Reviews*—This relationship enables our app's personalization component by providing the information needed to tailor a recommendation to a specific user based on their friends' reviews:
  - Based on my friends review ratings, what are the best restaurants for me?
  - What restaurants have my friends reviewed or rated in the past *X* days?

How does our relationship list compare to yours? As with entities, there is not one correct list. Many people go a step further and describe the attributes, or properties, of the entities and relationships. We shy away from this whenever we can because defining attributes this soon devolves into an exercise in bikeshedding (focusing on the comparatively trivial aspects of a problem instead of targeting the goal, which is to develop a high-level understanding of our application).

Although there is certainly a danger of "tyranny of the trivial" by spending too much time on properties, there is also the potential that useful details can come up in such a discussion. To mitigate the risk of losing those helpful details—details that can be used later, particularly in building the filtering and search components—we time-box the exercise by limiting ourselves to 10 or 15 minutes of discussion. During this time, we capture key points on Post-It notes, on a dry-erase board, or in a Google document, and drop these as a task in the backlog for later review and action.

> **TIP** Don't underestimate the power of a virtual "idea parking lot" for staying focused on the primary tasks at hand.

When it is time to put pen to paper to the conceptual model, we prefer a much simpler whiteboard—a box-and-arrow conceptual data model built with flowchart software such as Microsoft Visio or Lucidchart, or even with simple presentation tools such as PowerPoint or Google Slides—to a more defined methodology, such as Unified Modeling Language (UML). We build these flowchart models by representing the entities as labeled boxes and the relationships as labeled arrows. Doing this for DiningByFriends, we get the chart depicted in figure 2.3.

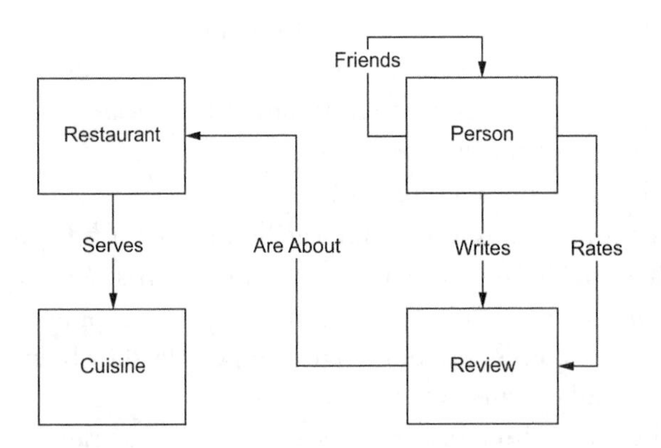

Figure 2.3 Our conceptual data model shows the entities (boxes) and relationships (arrows) for DiningByFriends.

We've found that both technical and non-technical users easily understand whiteboard models, which are intuitive. At this point, the target audience for this model is a business user, not a developer, so we want to stay away from thinking about the physical implementation. That will come soon enough.

2.4 Constructing the logical data model

Now we are ready to build our logical data model and translate those entities and relationships to the graph concepts of vertices, edges, and properties. The outcome of this process is another diagram, but this one contains sufficient detail to provide the necessary schema information for a developer to begin coding an implementation. In this section, we only work on the first use case in our application: the social network functionality. Later in the book we extend our data model for the other features. We start with social networking for several reasons:

- Our social network is the basis for how we eventually extend DiningByFriends for our recommendation engine and personalization features.
- The number of questions the social network answers is small enough to allow us to acquaint ourselves with the patterns and processes of graph data modeling.
- This network is the most straightforward, but it still retains several features, such as recursion and self-referencing edges, that enable us to highlight some of the unique abilities of graph databases.

To refresh our memory, the requirements for our social network provide the ability to connect with friends to see what they are reviewing. It addresses these questions for the Person–Friends–Person relationship:

- Who are my friends?
- Who are the friends of my friends?
- How is user *X* associated with user *Y*?

Figure 2.4 shows the portion of our conceptual data model that we work on in this section. Because we use this part of the model throughout the rest of the chapter, we repeat this same diagram so that it is always readily accessible for you.

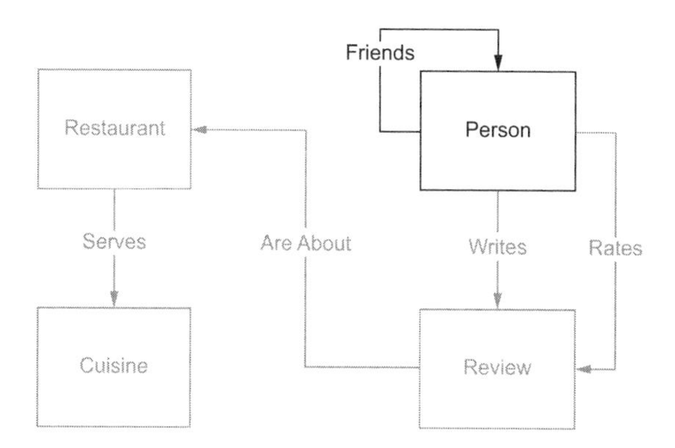

Figure 2.4 Conceptual data model for DiningByFriends with the relevant parts required for the social network highlighted

We start by showing the completed data model and work backward to show the patterns and processes we used to come up with this model. To begin, let's examine the final graph data model for our social network, as shown in figure 2.5.

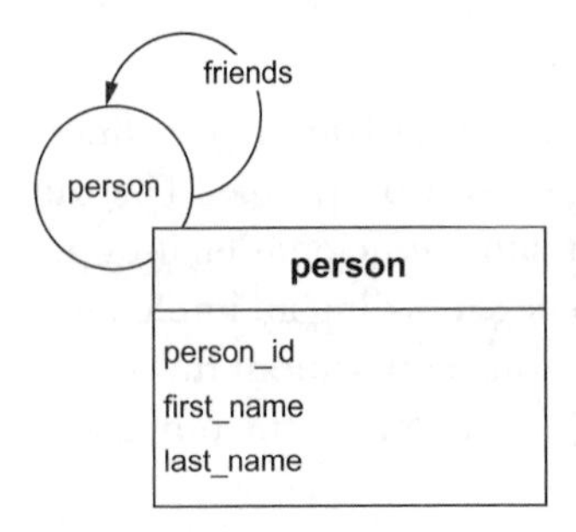

Figure 2.5 The logical graph data model for the social network for DiningByFriends

As figure 2.5 shows, the graph data model for this use case is one vertex with the label `person`, one edge with the label `friends`, and three vertex properties with the keys `person_id`, `first_name`, and `last_name`.

> **NOTE** In some environments, keys and key-value pairs are called property names and name-value pairs.

Now that we know our end state, let's see how we got there. There are four steps to building a graph data model from a conceptual model:

1. Translate entities to vertices.
2. Translate relationships to edges.
3. Find and assign properties to vertices and edges.
4. Check your model.

Wait…what does that say… "properties to vertices *and* edges?" Yes, you read that correctly. This ability for both vertices and edges to have properties highlights another fundamental difference between a relational database and a graph database. Because relationships are first-class citizens in a graph data model, both vertices and edges can have properties associated with those. While this addition might seem trivial, it is one of the more powerful aspects of a graph database because it opens up several useful data modeling options that we explore throughout this book.

2.4.1 Translating entities to vertices

The first step in creating our graph model is to identify all our vertices. Much of this work was done when we developed our conceptual model because the entities in a conceptual data model map almost directly to the vertices in a logical graph model. The creation of the vertices in our graph model requires two things:

- Identify all the relevant entities from our conceptual model
- Give the vertex a name in the form of a label to uniquely identify that type of entity in our graph model

To begin these two tasks, let's glance at the social network section of the conceptual data model. Figure 2.6 presents this section, and then we examine each point in more detail.

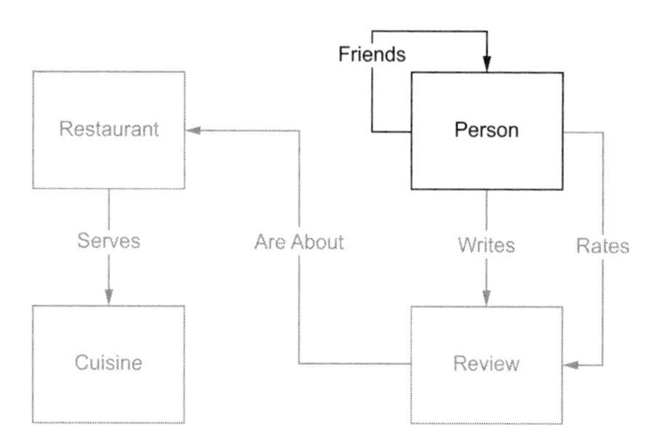

Figure 2.6 The conceptual data model with the relevant parts required for the social network highlighted

FINDING THE CONCEPTUAL ENTITIES

In our conceptual model, we located the entities Person, Restaurant, Review, and Cuisine. Now we need to narrow those down to only the entities required to answer the questions from section 2.3 for our social network functionality:

- Who are my friends?
- Who are the friends of my friends?
- How is user *X* associated with user *Y*?

In this case, the questions refer to only one entity, Person, because "my friends" and "the friends" are both people. While there are other logical entities in the model (Review, Restaurant, Cuisine), these are not required for the app's social networking, so we can ignore these for now. Although this is a simple example, this step will be more involved when we get to more-complex use cases in chapter 7.

As a general rule, we look for the entities in our application by looking for the nouns in our list of functionality questions. Because nouns represent physical or logical items, these frequently are the best indicator of which entities are required to solve the questions in the application.

NAMING THE VERTEX LABELS

Now that we have identified our entities, we need to assign each a label. A label in a property model graph categorizes or groups vertices that represent similar concepts. As has been observed by many prominent coders, there are only two hard things in computer science: cache invalidation, naming things, and off-by-one errors.[1]

Deciding on label names is not a trivial undertaking. A good label name is short, descriptive, and precise. As with properly naming variables in an application, if you do it well, the names add significant clarity and provide additive value when reading and working in the code. If you do it poorly, names can obfuscate their purpose and can

[1] The exact source for this bit of developer humor errors is not available, but it's generally attributed to Tim Bray, Phil Karlton, and Leon Bambrick.

be misleading. In software development terms, a label in a graph database is analogous to a class in object-oriented languages such as C++, C#, or Java: both contain a definition to explain how an object is structured and both can be used to classify like items together.

For the Person entity, the conceptual model also uses the entity name `Person`. Going back to our problem definition, let's pretend that we often talk with the business and end users and find that there is an existing implementation using a relational database. In this feature-discovery fantasy, we see that the business users refer to that entity as both people and users. With different terms being used for the same entity, we decide that Person is the most descriptive name for this entity.

IMPORTANT It is a best practice to make vertex labels singular because each vertex only refers to a single instance of an item.

We could have gone with the name User, but this is specific to one type of potential person within the application. While we currently do not have this requirement, we might need to represent other types of people, such as employees or owners, in the future. By choosing the more generic label of `person`, we can represent these potential future entities more easily without losing the type information in our current system.

IMPORTANT It is also a best practice to make labels as generic as practical. While we will go into this in greater detail in chapter 7, a rule of thumb is that if we expect that we might need to represent other, similar concepts in the future, then it's worthwhile to use a more generic term.

As with other databases, a consistent naming convention for label and property names is critical to the maintainability of an application. Consistency provides predictable behavior for developers and administrators. As developers, we find nothing more frustrating than inconsistent naming conventions in databases, and graph databases are no exception to this rule. For this book, we use lower_snake_case names and make all label names singular. Applying these best practices, we settle on the label `person`, as shown in figure 2.7.

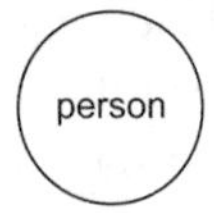

Figure 2.7 Example vertex with the label `person`

NOTE It is generally a safe bet that each vertex in a graph database can only be associated with a single label. That is the approach of Apache TinkerPop, and it is the approach we take in this book. There are situations, such as modeling inheritance, where having multiple labels per vertex is appropriate. And some graph databases, such as Neo4j and Amazon Neptune, do support multiple labels per vertex. Be sure to understand your vendor's capabilities before starting the data modeling process.

2.4.2 Translating relationships to edges

Now that we have identified and labeled our `person` vertex, it is time to define our edges. Edges are based on the relationships from our conceptual model. Defining edges takes a little more effort than finding the vertices. The edges in graph databases include features like directionality and uniqueness, which do not have direct counterparts in relational databases. Therefore, defining these relationships is more involved than just applying a name as you would do in a relational database. The four steps to defining an edge are

1. Identifying the relevant relationships from the conceptual data model
2. Naming the edge in the form of a label to uniquely identify that relationship in our graph data model
3. Assigning a direction to the edge by defining the start and end vertex types
4. Specifying the uniqueness of the edge by deciding on the number of times this edge can exist between two specific instance vertices

FINDING THE RELATIONSHIPS

Recall from the conceptual model that the social network component includes a single entity, Person, but this particular entity has three relationships associated with it: Friends, Rates, and Writes. (Friends here is a verb.) Let's look again at the functionality questions for the social networking we discovered while building our conceptual model:

- Who are my friends?
- Who are the friends of my friends?
- How is user *X* associated with user *Y*?

All the questions revolve around how one person is connected to another as a friend. The Friends relationship is the only link from a `person` to a `person` vertex that's required for our social network. The Rates and Writes relationships are not required because these reference an entity (Review) that is not required for this use case. Let's see what our conceptual model looks like with the relevant sections highlighted in figure 2.8.

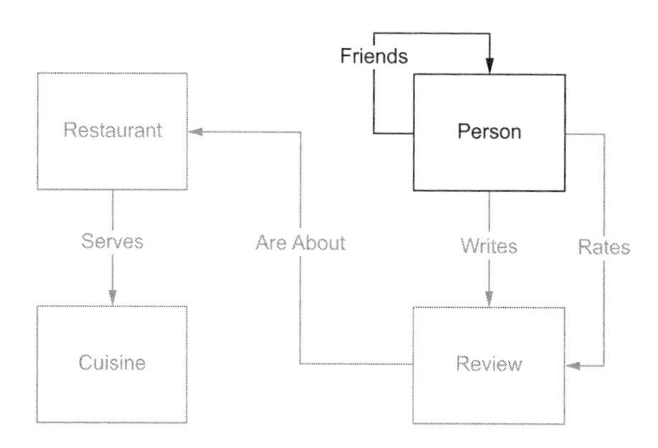

Figure 2.8 Conceptual data model with the relevant parts required for the social network highlighted

Because we were thorough when creating the conceptual model, this part of the translation to the logical model is almost trivial. If we missed a relationship and a corresponding edge, then that begins to surface as we evaluate the access patterns against the logical model in our test phase (more on testing in section 2.5).

NAMING THE EDGE LABELS

Now that we know that we need to represent the `Friends` relationship as an edge in our model, it is time to name it (step 2), just like we did with vertices. To decide on the edge label for our data model, we apply the same best practice naming rules: be concise, descriptive, and generic. By applying these rules, we get an edge with a `friends` label that starts at a `person` vertex and ends at a `person` vertex, as shown in figure 2.9.

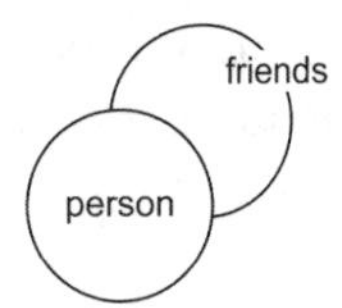

Figure 2.9 Adding a `friends` label connecting a `person` vertex to a `person` vertex in a loop

Don't be alarmed that the edge is connecting back to the same vertex type. It is acceptable, even common in some models, for an edge to have the same vertex type at both ends. This is known as a *loop*. A loop edge is similar to a foreign key referencing the same table or a linking table that connects back to the original table, as in figure 2.10.

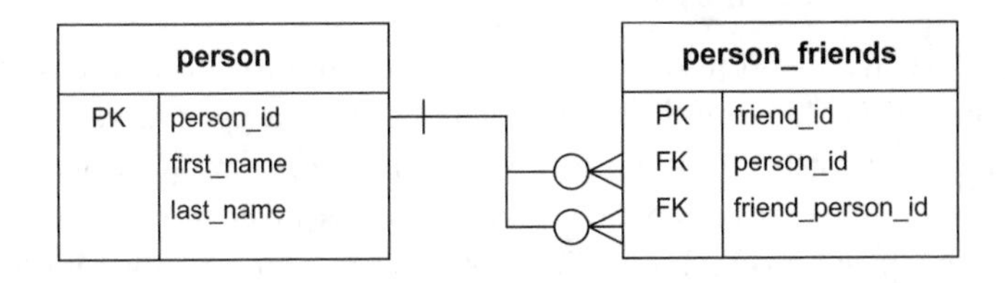

Figure 2.10 A loop in a graph data model is similar to a linking table in a relational data model. It references back to the original table.

GIVING THE EDGE A DIRECTION

Once we have a label for our edge, the next step is to give the edge a direction. By convention, the direction of an edge is described as being from one vertex, the *out* vertex, to another vertex, the *in* vertex. In figure 2.11, we see that the `Bill` vertex is the out vertex and the `Ted` vertex is the in vertex.

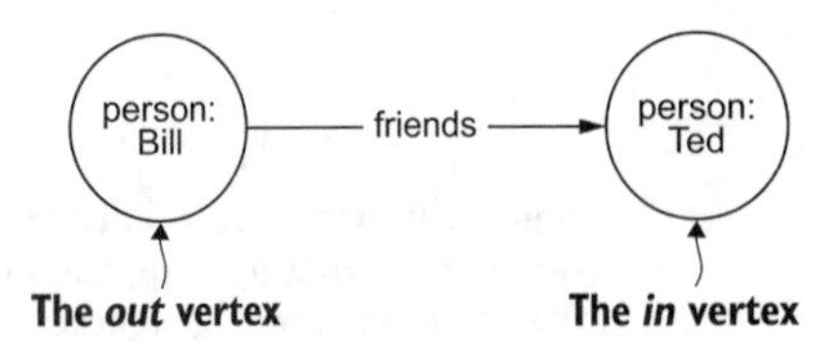

Figure 2.11 Example of data in a graph with an out vertex for Bill and an in vertex for Ted

In a good graph model, the vertex–edge-vertex combinations read as a sentence. In figure 2.11, we read the vertex–edge-vertex as Bill friends Ted. When looking at your label names and edge directions, don't be afraid to reword the label or switch the direction of the edge to make the data model more understandable. The direction of an edge should complement the edge label to make a sentence that sounds natural (or mostly natural) and that fits the functional needs of the use case.

Consider the example of a simple graph, shown in figure 2.12. It tracks the cities people live in. There are two vertex labels, `person` and `city`, and one edge label, `lives_in`, between the vertices.

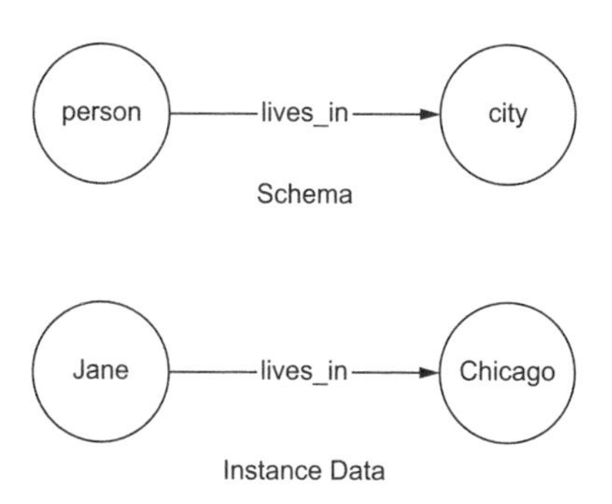

Figure 2.12 A sample graph with the schema, `person-lives_in-city`, and the instance data, `Jane-lives_in-Chicago`

In figure 2.12, we see both the schema of our graph and the instance data (the graph of the data stored). Looking at the instance data graph, we see that Jane lives in Chicago. Looking at this data, it makes logical sense and reads fluently. If we reverse the direction of the edge so that the in vertex is a person and the out vertex is a city, then the instance would read as Chicago lives in Jane. As cities don't live in people, this sentence no longer makes sense. So, what can we do to make this sentence make sense? The simplest solution is to reword the edge label to something that makes the sentence more understandable, such as `is_residence_of`, as shown in figure 2.13.

Now if we read the instance of the reworded graph, we see that the sentence, "Chicago is residence of Jane," makes sense. Returning back to our DiningByFriends model,

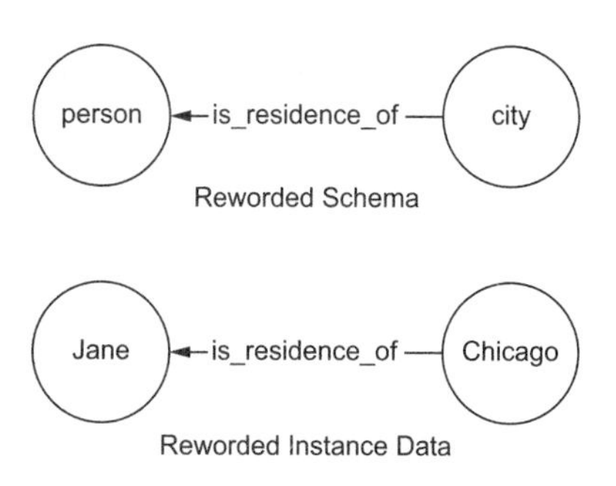

Figure 2.13 Reworded graph with the edges `is_residence_of` for both schema and instance data

because our `friends` edge connects a `person` vertex to a `person` vertex, the direction is irrelevant. Figure 2.14 shows that the start and end vertex labels are identical.

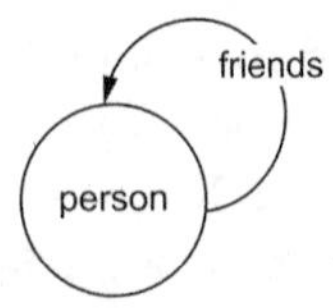

Figure 2.14 The `friends` edge, now with an added direction pointing from a `person` vertex to a `person` vertex

It certainly simplifies things in social networking queries when the edge direction, while specified, is not consequential, but don't expect to see this often! This situation is very uncommon.

DETERMINING EDGE UNIQUENESS

The final step to address when defining edges is *uniqueness.* Edge uniqueness describes the number of times an instance of a vertex is related to another instance of a vertex with an edge having the same label. Whew, that definition is a mouthful, so here's another way to define this concept: uniqueness describes what is an allowable number of edges of a given label between two vertices. That is still a bit abstract, so let's take a look at an example that demonstrates what we mean by edge uniqueness.

In figure 2.15, vertex *A* is connected to vertex *B* by edge *Y* more than once, so edge *Y* is a *multiple uniqueness* edge. Vertex *A* is also connected to vertex *C* by edge *X*, but it is only connected one time; thus, edge *X* is a *single uniqueness* edge.

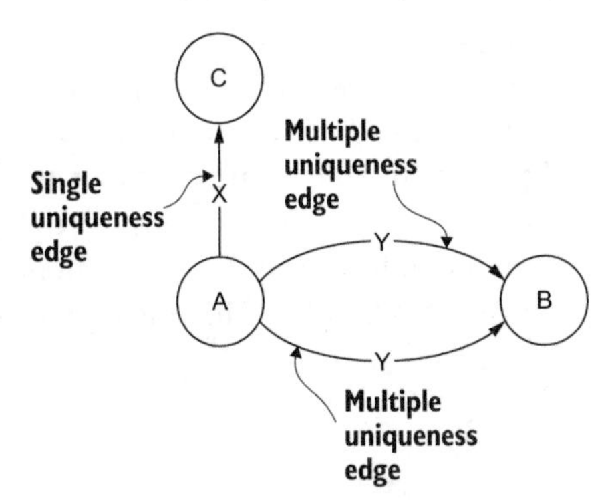

Figure 2.15 Entities A and C have one single uniqueness edge or relationship with each other. Entities A and B are connected more than once, having multiple uniqueness edges.

> **Why uniqueness and not cardinality or multiplicity?**
>
> If you come from a relational database background, you are likely wondering why we use the term *uniqueness*. This was actually a topic of quite a lively discussion among the authors and several of our peers.

The term *cardinality* is often used (wrongly) to refer to a many-to-many or a one-to-many relationship. As Martin Fowler explained, these descriptions show multiplicity, not cardinality (http://mng.bz/QxY4). We fully agree with Fowler's definition of cardinality and multiplicity as

- *Cardinality*—The number of elements in a set (for example, figure 2.15 has two Y edges between A and B).
- *Multiplicity*—A specification of the minimum and maximum cardinality that set can have (for example, one-to-many, zero-to-many, many-to-many).

With these definitions, why don't we use these terms to describe our edge schema? Because we are not describing the characteristics of a single edge; instead, we are describing the characteristics of a group of edges.

Because cardinality describes a quantifiable number, it can be used to define the number of edges in the instance data of our graph (the cardinality of Y edges between A and B in figure 2.15 is two). However, because cardinality represents a single number, it cannot describe the range of potential options needed by the schema for the same reasons Martin Fowler points to with relational databases. So why not use multiplicity?

Using the term multiplicity to describe the characteristics of a group of *edges* causes a problem. In traditional *UML®*/ERD terminology, multiplicity constrains the number of related entities. Based on that understanding, the multiplicity of a graph database would always be many-to many because, by design, graphs connect vertices (the simplest analog to entity) to multiple other vertices. Because, traditionally, we would only ever have one multiplicity, this term isn't suitable for graph databases as it does not add any descriptive value to our data model. We could alter the definition of multiplicity for the context of graph databases, but this would only cause additional confusion to those familiar with the traditional usage.

This led us to describe the edge schema with a different term—*uniqueness*. Data uniqueness describes the measure of duplication of identified data items within a data set. In this case, we define uniqueness as *the allowable number of edges of a given label between two vertices*. So single uniqueness refers to zero or one edge, and multiple uniqueness means more than one possible edge. This is much the same as how a unique constraint on a SQL column represents single uniqueness.

In data structure terms, single uniqueness is a set of edges: there can be only one of a given edge label between two instance vertices. Multiple uniqueness is like a collection: there can be one or more of the given edge label between two instance vertices.

Why does edge uniqueness matter? Let's use a simple movie example as shown in figure 2.16 to see why. This graph consists of three people (entities)—Bob, Joe Dante, and Phoebe Cates—and shows the relationships (edges) to the movie *Gremlins* (also an entity).

In our *Gremlins* graph in figure 2.16, we have four vertices (three `person`'s and one `movie`) and three edges (`watched`, `acted_in`, and `directed`). Each person in the

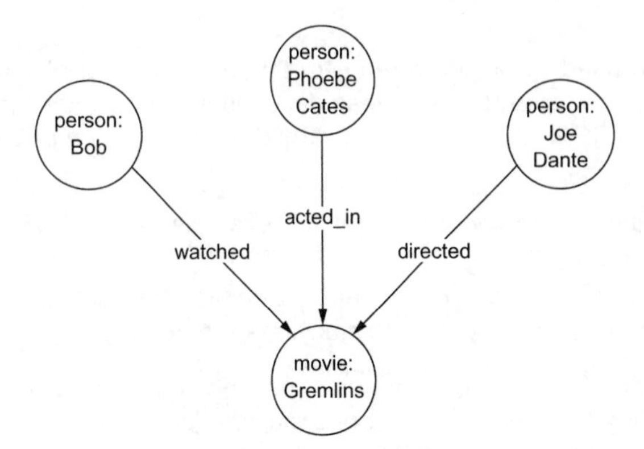

Figure 2.16 A simple graph with four entities, or vertices, and three edges

graph is connected to the `movie` vertex using one of the three edges. In this example, the following relationships exist:

- Bob watched *Gremlins.*
- Phoebe Cates acted in *Gremlins.*
- Joe Dante directed *Gremlins.*

Let's begin by examining the `directed` edge in our graph. Remember, we are looking for what is the allowable number of edges of a given label between two vertices.

We can all agree that a person (Joe Dante) can only direct a movie (*Gremlins*) once, so the `directed` edge has single uniqueness. This does not preclude Joe Dante from having multiple `directed` edges because he also directed *Gremlins 2*, or mean that *Gremlins* could not have multiple `directed` edges going into it. It merely enforces that there can be only a single `directed` edge from `Joe Dante` to `Gremlins`.

Single uniqueness is, in fact, significantly more common than multiple uniqueness. As a rule, assume single uniqueness and think about multiple uniqueness only when there is a specific requirement dictating multiple instances of the same edge between two vertex instances.

When does an edge require multiple uniqueness? Take the example of the `watched` edge; we can all agree that it is possible, and certainly likely, that a person would watch the movie more than once. As a result, there would be multiple `watched` edges between `Bob` and `Gremlins` as shown in figure 2.17.

Multiple uniqueness is less common than single uniqueness. But it is useful when the same relationship can exist between the same two distinct items multiple times, such as tracking the times a person has ordered a product or documenting connections between items on the internet.

Now, let's review the last edge in our movie graph, the `acted_in` edge, and see what its uniqueness should be. Based on the definition of uniqueness (What is the allowable number of edges of a given label between two vertices?), what do you think its uniqueness should be?

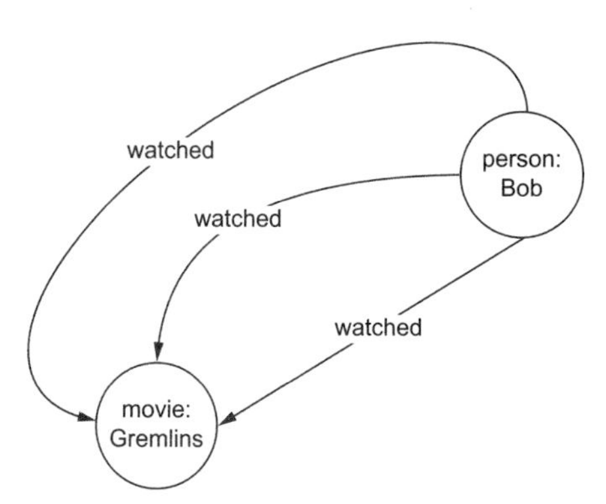

Figure 2.17 Multiple uniqueness demonstrated with multiple `watched` edges between `Bob` and `Gremlins`.

We would probably only ever have a single `acted_in` edge between a person and a movie, so it would have single uniqueness. But if we think about people acting in movies, it is possible that a user acts in the same movie multiple times in different roles. Examples with Eddie Murphy or Tyler Perry come readily to mind. How would we go about modeling the requirement to store the fact that a person was in a movie in more than one role?

In figure 2.16, we could allow for multiple `acted_in` edges between a person and a movie, but there is no way to distinguish the role associated with that `acted_in` edge from the role associated with another `acted_in` edge. If we were to change the `acted_in` edge to have a `role` property, then we could see that Eddie Murphy `acted_in` *The Nutty Professor* in two separate roles, Sherman Klump and Buddy Love. We would also be able to identify the role that is associated with each edge, as shown in figure 2.18.

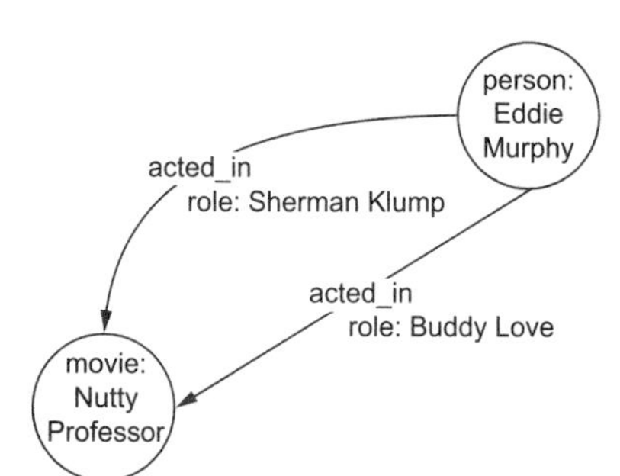

Figure 2.18 Adding a `role` property to the `acted_in` edge creates a mutiple uniqueness edge to express the fact that Eddie Murphy acted in multiple roles in the movie *Nutty Professor*.

If we add the `role` property as shown in figure 2.18, then the `acted_in` edge would be considered a multiple uniqueness edge. This is just one example of how you can use the uniqueness of your edges to represent information in your domain.

Returning to our graph data model for DiningByFriends, what would the correct uniqueness of the `friends` edge be? We could say that it is a single uniqueness edge because a `person` can only be friends with a `person` once. How do we know if we have the right edge uniqueness? To answer this question, let's first look at how incorrect uniqueness can affect an application. Improper uniqueness usually appears in one of three ways:

- Too little data returned
- Duplicated data returned
- Poor query performance

IMPORTANT Incorrect edge uniqueness is one of the most common problems in graph data modeling, and it is frequently a root cause of query issues.

The first symptom of incorrect uniqueness is that too little data is returned from a query. This occurs when we have an edge with single uniqueness that really should be multiple. In this scenario, the query returns only the first edge saved or the last edge saved. The exact response returned depends on how the database handles data concurrency. But either way, it's incomplete.

The second symptom, having duplicated data returned, occurs when we have multiple uniqueness edges but we should only have one. In this scenario, our application incorrectly retrieves data for multiple edges on each query because multiple edges exist between two instance vertices. Because we can have multiple edges with the same label between any two vertices, each time our application saves an edge between vertex *A* and vertex *B*, for example, a new edge is created. Over time, this means that we end up with a collection of many edges when we might have intended for only a single edge between vertex *A* and *B* to ever exist.

The third symptom of incorrect edge uniqueness is harder to diagnose because it appears as poor query performance. This can, of course, be caused by many things. But most often having a multiple uniqueness edge instead of a single uniqueness edge is what causes poor query performance, because the database has to do more work to return the data for a query with multiple edges. We discuss more about debugging and troubleshooting query performance in chapter 10, but let's quickly look at why incorrect uniqueness causes a problem.

In the earlier *Gremlins* movie example, if the `directed` edge had multiple uniqueness, it would be possible that an incorrect traversal could create five `directed` edges between `Joe Dante` and `Gremlins`. If we run a query to return all the movies that Joe Dante has directed, which it does using *all* of the `directed` edges connected to Joe Dante, it requires the database to do five times the work because there are five times the edges to traverse between the two vertices shown in figure 2.19. And this can cause a noticeable performance impact to both your application and the database!

If we continue with the extra edges example and imagine that we move out further in our graph, from the movies to the actors which Joe Dante directed, we can quickly see how this seemingly simple oversight exponentially increases the number of vertices and edges we traverse. The error in the `directed` edges would result both in repeated movies and in repeated actors. The good news is that there are a few ways to

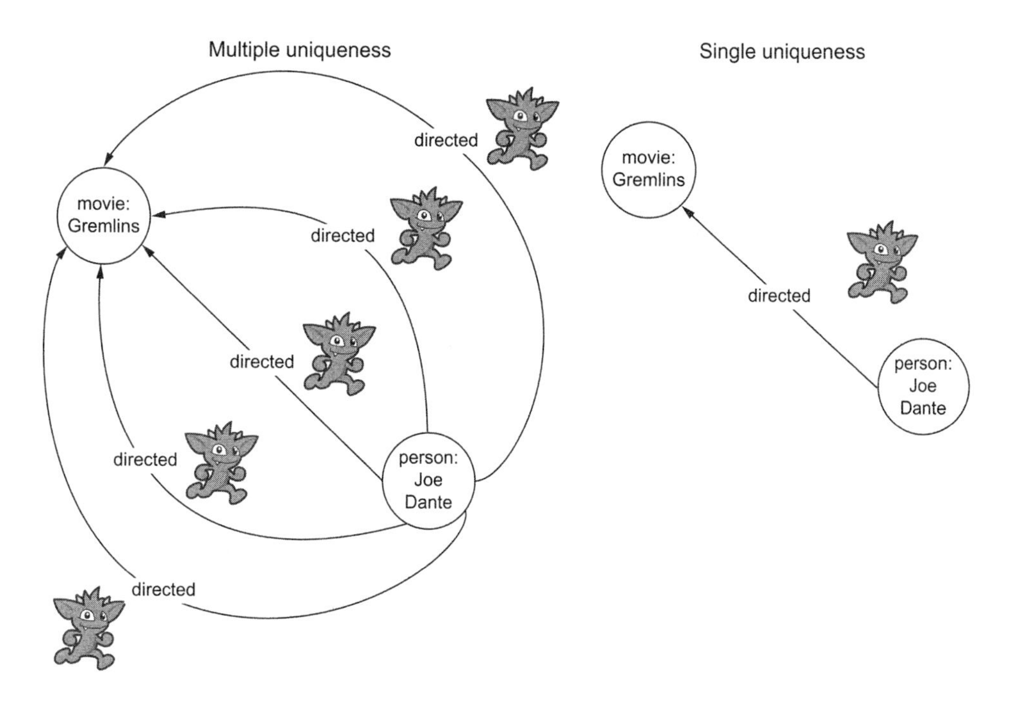

Figure 2.19 The comparative amount of work required for multiple uniqueness edges versus single uniqueness edges

mitigate this problem at query time, which we cover when we start writing queries in chapter 3. But the best way to prevent this problem is to properly design the data model to reflect the correct uniqueness of the edges.

Some graph databases, such as DataStax Enterprise Graph and JanusGraph, set the uniqueness of an edge explicitly as part of a schema definition. But many other graph databases do not define schema explicitly, so there is no way for the database to enforce a uniqueness constraint. This schemaless approach of some graph databases means that we must write application logic to enforce this uniqueness within the application.

2.4.3 Finding and assigning properties

Now that we've created the structure of our graph model, composed of the vertices and edges, it is time to define the properties and assign these to the vertices and edges. Properties in a graph data model are key-value pairs that describe a specific attribute of a vertex or edge.

> **Default and null values in graph databases?**
>
> Properties in a graph database are similar to the columns in a row in a database table: these store the relevant data about a specific entity. Unlike columns, an application does not insert default values or null values into properties in graph databases.

(continued)

Due to their rigid structure, relational databases require that data appears in every column for every row. However, graph databases store data similar to the way a key-value database stores its data, where the data either exists or does not. This means that storing null values or populating properties with default values is not needed, saving space and reducing the payload sent to the client. But it also means that it is possible that some of the properties on a vertex simply might not exist, requiring more defensive coding in the application.

Before we can assign the properties to edges and vertices, we first have to decide

- What properties are required?
- How we are going to name them?
- What is their data type?

To answer these questions, we need to consider what we know about the domain as well as our conceptual data model to decide what information needs to be stored.

NOTE When migrating an existing system, it is beneficial to reference the data model for that system as a blueprint. Data models have matured over the years and so provide a rich perspective on the necessary data requirements for an application. If you are developing a greenfield application unconstrained by parameters set by prior work (such as we are with DiningByFriends), then now is the time to sit with the technical and non-technical people to determine what the specific data fields, names, and data types should be.

The first place we look for the properties is, once again, in our list of functionality questions for our social networking use case. If you remember, these are

- Who are my friends?
- Who are the friends of my friends?
- How is user *X* associated with user *Y*?

Based on these questions, we can see that we need to store the `first_name` and `last_name` for a person to identify who our `friend` is. We can also assume that we are going to need a unique identifier of each person (`person_id`) to differentiate between people with the same name. Without this additional attribute, we are unable to discern one John Smith from another John Smith. Adding our properties to our current data model, we get the model in figure 2.20.

In this case, we did not add properties to an edge because all the required attributes (`person_id`, `first_name`, `last_name`) describe the `person`, not the `friends` relationships between people. However, adding properties to edges is common, and we will have a chance to do so in our DiningByFriends model in chapter 7.

At this point in the process, if you chose to use any of the schemaless graph databases that exist on the market, then you are complete. But if you are using a database

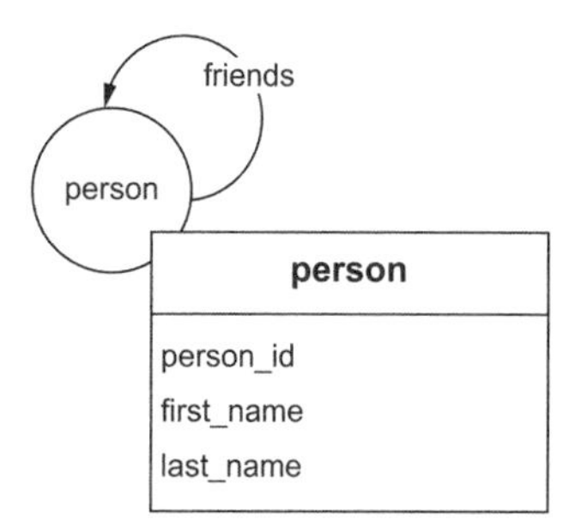

Figure 2.20 Our social network graph data model with the properties added (shown in the box)

that requires the schema to be explicitly specified, then you have one additional step: translating your logical data model into the physical data model required by your chosen database. The actual mechanics for how this is done are specific to each database as each tool has its own unique definition language to describe the physical graph data model. Due to this lack of standardization across graph database vendors, we recommend reading the documentation for your chosen tool to complete this step.

2.5 Checking our model

The last step in creating our logical data model is to validate that we can answer the questions for our social network use case and that the model we built, shown in figure 2.21, follows best practices for graph data modeling.

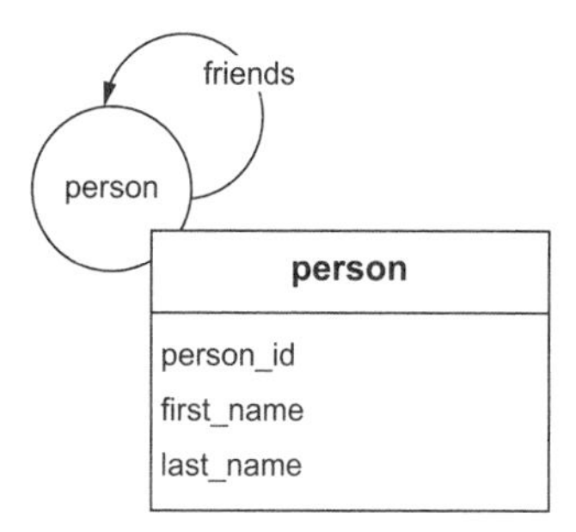

Figure 2.21 The final logical graph data model for the social network DiningByFriends use case

Looking at our questions, can we answer these using our graph data model?

- *Who are my friends?* We can answer this question by starting at a specific `person` found by `person_id` and traversing the `friends` edge to see all of their friends.
- *Who are the friends of my friends?* We can answer this question by starting at a specific `person` found by `person_id` and traversing the `friends` edge, and then traversing the `friends` edge again to see all of the friends of my friends.
- *How is user X associated with user Y?* We can answer this question by starting at a specific `person` found by `person_id` and traversing the `friends` edge until we

either have no more `friends` edges to traverse or we traverse to the destination person.

Later in this book, we discuss precisely how to achieve this sort of query, but be aware that this sort of unbounded recursive query is one of the most compelling use cases for graph databases.

Now that we have a validated model, there are a few additional best practice checks to make sure that our data model provides a solid graph model:

- *Do the vertices and edges read like a sentence? Yes.* While this is not an absolute requirement, it is an excellent general check to verify that vertex labels represent the nouns in your model and the edge labels describe the actions or verbs in your model.
- *Do I have different vertex or edge labels with the same properties? No.* In this case we only have a single vertex and a single edge label. In a more complex model, as we see later in this book, this check is a helpful way to validate that you have made your labels generic enough.
- *Does my model make sense? Yes.* While this step can seem like an obvious check, it does pay to take time to step back and double-check that your graph data model has not strayed too far from your conceptual data model and that it makes sense for the problem that you are solving.

In this chapter, we built our data model for our social networking use case and validated that it makes sense for our problem. In the next chapter, we are going to start querying our database to answer the questions for our social network use case.

Summary

- Strong, early investment in understanding the problem, use cases, and common domain terminology are the foundation of building a good data model. This also reduces the risk that you'll need to radically change the design later.
- A conceptual data model provides an overarching view of the scope, entities, and application functionality from the point of view of a business user.
- Translating a conceptual data model to a logical data model requires four steps: translate entities to vertices, translate relationships to edges, assign properties, and check the model.
- Translating entities to vertices involves identifying the required conceptual entities, creating corresponding vertices, and providing those vertices with a label that is concise, descriptive, and generic.
- Translating relationships to edges consists of identifying the required conceptual relationships, creating corresponding edges, labeling each edge, assigning a direction to each edge, and determining the edge uniqueness.
- Edge uniqueness defines the number of times an instance of a vertex can relate to another instance of a vertex with an edge with the same label. Incorrectly

identifying edge uniqueness is a common problem in graph data modeling, which causes data and performance issues.

- To validate a graph data model, verify it against the requirements and conceptual model, check that the vertex and edge labels read like a sentence, and ensure that the model does not have duplicated edge or vertex types. Finally, perform a "gut-check" to see that the model makes sense.

3 Running basic and recursive traversals

This chapter covers

- Navigating the structure of a graph
- Performing filtering operations with traversals
- Using recursive traversals

With our graph data model in hand, the next three chapters focus on how to navigate through our graph and how to return data. In this chapter, we'll start by filtering and navigating edges, the fundamental building blocks of graph traversals. We then extend these concepts to cover a powerful feature of graphs: the ability to easily write recursive queries. Additionally, we examine how to leverage these techniques to answer common graph questions, such as how people are connected in social networks.

Throughout these next three chapters, we'll use the social network use case defined in chapter 2. As we move through this process, we'll use Gremlin as our query language and introduce its syntax, known as *steps*. Don't worry if you don't know Gremlin or are using a different language; we'll thoroughly explain each step as we encounter it. Although we cover quite a few Gremlin steps throughout this chapter, you can refer to the Apache TinkerPop official documentation on Gremlin for a more thorough explanation (http://tinkerpop.apache.org/docs/current/reference/).

NOTE Source code for this chapter is available here https://github.com/bechbd/graph-databases-in-action. For simplicity's sake, we use an environment variable named `$BASE_DIR`, which should be set to the local path where you placed the source code you downloaded. We use this environment variable throughout this and later chapters to simplify our script commands.

3.1 Setting up your environment

Now is the time to set up your local environment for development. This section discusses the minimum steps to get Gremlin up and running. There's a more detailed treatment of the TinkerPop project, associated artifacts, and how to set up Gremlin (including a manual launch) in appendix A. In this section, we show you how to do three things to get you started:

- Get a Gremlin Server running and available to receive connections
- Connect a Gremlin Console to your Gremlin Server with a session
- Load test data into the server

If you already have a Gremlin Console running locally, and it's connected to a Gremlin Server, then you can skip this section. If you need to download the Gremlin Console and Server, you can do that at http://tinkerpop.apache.org/.

3.1.1 Starting the Gremlin Server

Using a terminal window, navigate to the directory where you have unzipped the Gremlin Server download. On MacOS or Linux systems, start the Gremlin Server with `bin/gremlin-server.sh start` (on Windows, use `bin\gremlin-server.bat`). Here's the syntax for MacOS or Linux:

```
$ cd apache-tinkerpop-gremlin-server-3.4.6
$ bin/gremlin-server.sh start
Server started 10066.
```

Running this command yields the process ID of the server that was started, which is a helpful indication that everything worked as expected. To stop the server on a MacOS or Linux system, use

```
bin/gremlin-server.sh stop
```

On a Windows system, use

```
bin/gremlin-server.bat stop
```

WARNING When you stop the server, you lose all the data stored in the database because the Gremlin Server is *in-memory* only.

To restart the server on MacOS or Linux systems, use

```
bin\gremlin-server.sh restart
```

On Windows systems, use

```
bin\gremlin-server.bat restart
```

This performs a stop and then a restart on the server. Now that we have an instance of the Gremlin Server running, let's move on to our next task.

3.1.2 Starting the Gremlin Console, connecting to the Gremlin Server, and loading the data

In this section, we launch Gremlin Console, connect it to our server, and load some data. Because we want to get up and running as quickly as possible, we provide the scripts to accomplish these tasks.

Using a terminal window, navigate to the directory where you have the unzipped version of the Gremlin Console download. On MacOS or Linux systems, run the Gremlin Console with this script argument:

```
bin/gremlin.sh -i $BASE_DIR/chapter03/scripts/3.1-simple-social-network.groovy
```

On Windows, use this command:

```
bin\gremlin.bat -i $BASE_DIR\chapter03\scripts\3.1-simple-social-network.groovy
```

Running this command or script first launches the Gremlin Console and then connects it to the Gremlin Server. Finally, it loads a small data set into our database.

> **IMPORTANT** These commands return your database to this known state, overwriting any data that you may have added.

You can verify that all of this was successful by entering the single character g in the Gremlin Console. Your terminal screen should display an output like the following:

```
$ bin/gremlin.sh -i $BASE_DIR/chapter03/scripts/3.1-simple-social-network.groovy

         \,,,/
         (o o)
-----oOOo-(3)-oOOo-----
plugin activated: tinkerpop.server
plugin activated: tinkerpop.utilities
plugin activated: tinkerpop.tinkergraph
gremlin> g
==>graphtraversalsource[tinkergraph[vertices:4 edges:5], standard]
gremlin>
```

If you don't see output like that, then close your terminal and start again with these instructions or with the instructions in appendix A. Now you are connected to a graph database with four vertices and five edges. This is the social network data as shown in figure 3.1. We use this data for all examples throughout this chapter.

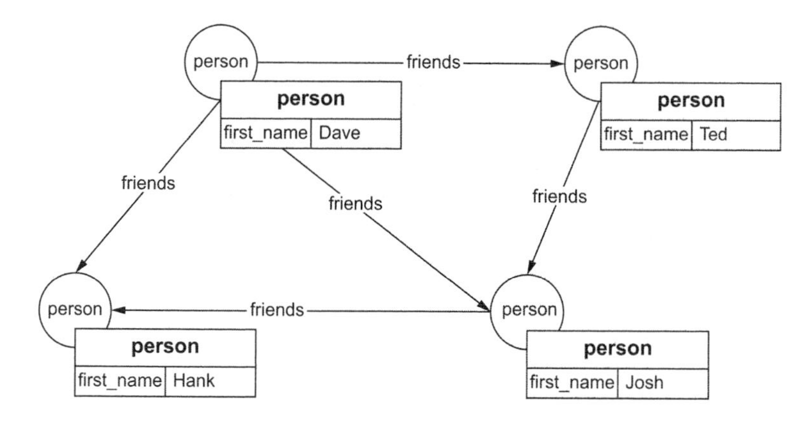

Figure 3.1 The social network graph. Setting up the environment as detailed in section 3.1 loads the data for this use case, which we use throughout the chapter.

3.2 Traversing a graph

Now that we've added data for our graph, let's begin by writing our first graph traversal. Let's say we want to look at our social network to answer the question, "Who are Ted's friends?" In a relational database, we would use a query to answer this question, but in a graph, we perform a traversal. This process of moving through the graph is known as *traversing*. The definition of the set of steps and actions we perform to retrieve this data is known as the *traversal*, analogous to our SQL query.

> **Traverse, traversal, traversal source, and traverser**
>
> Throughout this book, we use several similar sounding terms to describe the process of moving through a graph. To avoid any misunderstandings, here is a summary of these terms, all in one place:
>
> - *Traverse*—The process of moving from vertex to edge or edge to vertex as we navigate through a graph. Traversing a graph is analogous to the act of querying in a relational database.
> - *Traversal*—A specification of one or more steps or actions to perform on a graph, which either returns data or makes changes, or in some cases, does both. In a relational world, this would be the actual SQL query. In the graph world, this is the set of operations, called steps, that are sent to the server to be executed.
> - *Traversal source*—The traversal source is a concept specific to TinkerPop. It represents the base or starting point from which steps traverse the graph. By convention, this is usually represented with the variable `g` and is required to begin any traversal.

(continued)

- *Traverser*—The computing process associated with a specific branch of a traversal's execution. A traverser maintains all the metadata about the current branch of the graph it's moving through (e.g., current object, loop information, historical path data, etc.). A unique traverser represents each branch through the data.

Another way to think about these terms is that a traversal begins at a traversal source by sending one traverser per branch to traverse a graph. The traverser can either be removed or returned with the results.

3.2.1 Using a logical data model (schema) to plan traversals

Traversing a graph database focuses on how to traverse from one element to another. To do this effectively, we leverage our logical data model to understand what the relevant schema elements are for each element in the graph. Let's look at what elements are in the graph and what is most relevant to think about when writing a traversal, as shown in table 3.1.

Table 3.1 Summary of graph elements and relevant schema elements

Graph element	Relevant schema elements
Vertices	Vertex label, vertex properties, and the connected edge labels
Edges	Edge label, edge properties, edge direction, and connected vertex labels

NOTE Notice that edges in our data model are part of each row in table 3.1. This demonstrates the importance of relationships in graph databases and their value when working with highly connected data.

We have included as figure 3.2 the final logical data model from chapter 2. Let's determine the most relevant schema elements for each of our graph elements, starting with our vertices. From our model, we can see that we only have one vertex label, `person`, to consider. The `person` vertices contain `person_id`, `first_name`, and `last_name` properties, representing attributes of people in the data. The final portion to consider

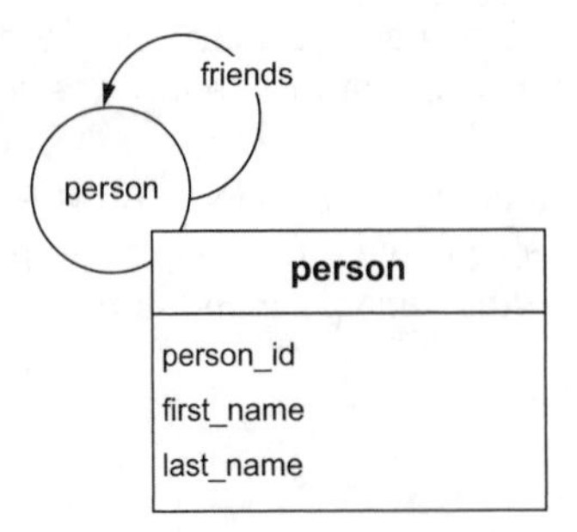

Figure 3.2 The final logical graph data model from chapter 2 for the social network DiningByFriends use case

is the edge labels that are connected (i.e., incident) to the `person` vertex. In this model, there is only one label, `friends`.

Next, we'll determine the edge direction, which helps us know how to move through our directed graph. Although this is important in most scenarios, because edges go from one label to a different label, in this example, the direction is not relevant. That's because the `friends` edge goes from a `person` vertex to `person` vertex (a loop edge).

These are the relevant schema elements that we'll use throughout this chapter to help us plan and write our traversals. We'll use these identified schema elements to guide our work, but when demonstrating the concepts, we'll use instance data.

All of our work at this stage is built on the foundation of our data modeling process. When we clearly state the business questions and thoroughly understand the use cases, we should find that our logical model and identified relevant schema elements aid in writing traversals. If we find that it is difficult to write traversals to address the use cases, then we likely missed something in our data modeling process.

3.2.2 Planning the steps through the graph data

Having walked through the logical data model, let's leave the abstract behind and take a look at our social network graph data from figure 3.1. Let's decide what we need to do to answer the question, "Who are Ted's friends?" The first step in answering this question is to establish a starting point in the graph. For this question, we need to find the `Ted` vertex, as shown in figure 3.3.

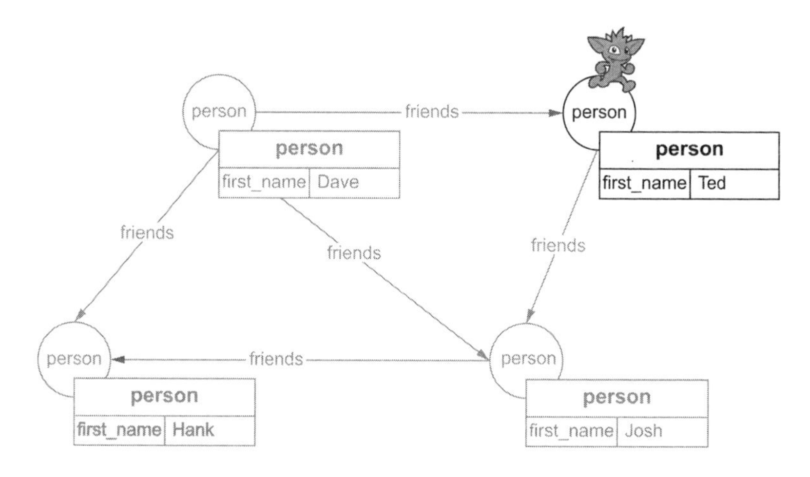

Figure 3.3 Our social network graph highlighting our starting location, Ted

With the `Ted` vertex as the starting point of our traversal, our next step is to find Ted's friends. Looking at our graph, we notice a descriptively named `friends` edge connected to our `Ted` vertex. Let's traverse the edge between Ted and Josh as depicted in figure 3.4.

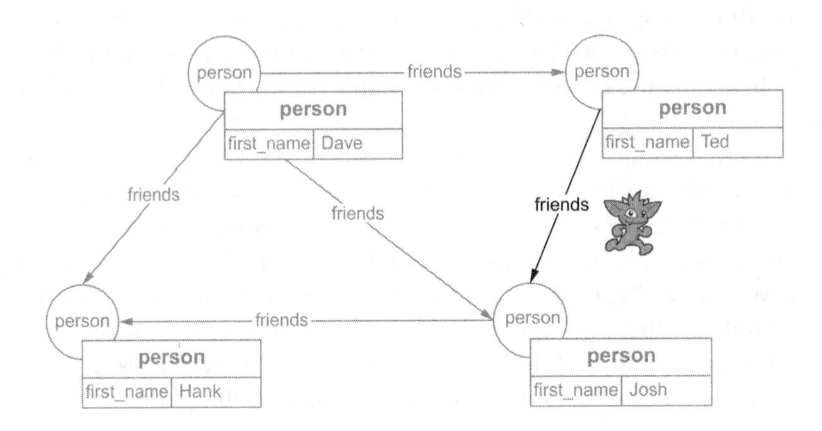

Figure 3.4 The social network graph highlighting our location after we traverse the outgoing `friends` edge

Examining this graph, it's reasonable to wonder why we didn't follow the other `friends` edge (from Ted to Dave). This is where the directed nature of edges in a graph comes into play.

Remember that our data model is similar to Twitter when making friends. In Twitter, when you follow someone that does not mean they follow you. In DiningByFriends, we have also chosen a model where a person can friend someone who doesn't connect back as a friend. This isn't the style of reciprocated friending found on Facebook.

By "Who are Ted's friends," we mean who are the people that Ted friended, not the people who friended Ted. This means we only follow the `friends` edges that begin at the `Ted` vertex, not those that end at `Ted`. To find people that friended Ted, we need to locate edges that end at the `Ted` vertex, not those that begin at his vertex. This directed nature of edges is a key distinguishing capability in graph databases and is useful for filtering or deciding which edges to traverse.

> **NOTE** In a graph, edges are represented by lines where the start or source of the edge *from* the vertex is represented without an arrowhead, and the destination or target *to* the vertex is represented with the arrowhead. See figure 3.4 for an example.

For simplicity's sake, we only traverse a single edge in this example. However, if there were multiple outgoing `friends` edges, such as if we were to start at the `Dave` vertex, then multiple parallel processes would be traversing around our graph. Each of these parallel processes is called a *traverser*. Now that we're located on the `friends` edge for

Ted, the last step is to complete the traversing of the edge to the `person` vertex at the other end, as shown in figure 3.5.

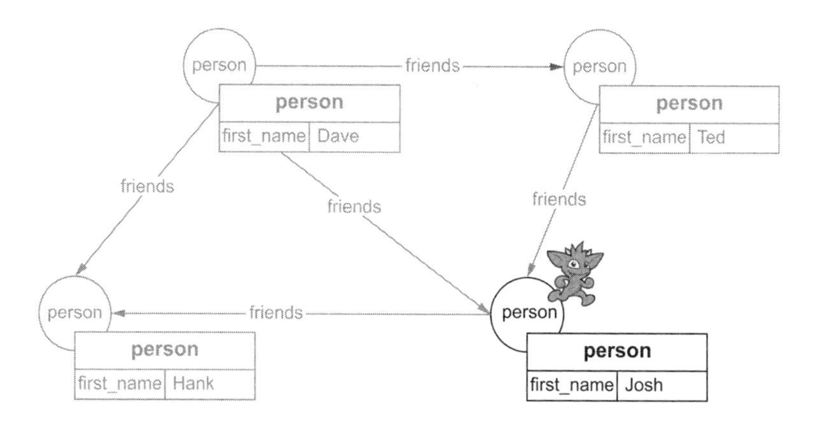

Figure 3.5 The social network graph highlighting our location after we traverse to the adjacent `Josh` vertex

As figure 3.5 shows, we reached the conclusion that Ted is friends with Josh. This simplified example demonstrates how traversals use the structure of the graph data model to move from one vertex or edge to another.

3.2.3 *Fundamental concepts of traversing a graph*

The process of traversing a graph can be broken down into these basic operations: find a starting vertex, identify an edge to traverse, traverse that edge, and finally, complete the traversal by arriving at the destination vertex. We'll use this same series of steps when we traverse through graph data in all of the graph traversals both in this book and in our own applications; although for this book, we won't usually break it down to this level of detail. With that in mind, let's flesh out our understanding of traversing by examining the four critical characteristics of the process.

TRAVERSING IS A SERIES OF STEPS

Traversing a graph entails defining the series of steps for moving through the graph. These steps can also include various operations for manipulating the graph data, such as filtering data, as well as the traversing of edges. Our example traversal only consisted of three individual steps, but the number of steps can grow quite drastically as our traversals become increasingly complex. The important tip to remember is that each step in a traversal starts at one location and (almost always) ends at a different location. This leads us to the second fundamental concept of graph traversals.

TRAVERSING REQUIRES KNOWING WHERE WE ARE

Traversals require knowing our location in a graph. This concept is foreign for those of us coming from a relational database background. In a relational database, our SQL queries are capable of joining any two tables at any point in the query. In a graph, we're limited to using the edges or vertices next to our current location in a graph. To navigate efficiently throughout the graph, we have to keep track of where we are within the structure of our graph data model. In our experience, this is the most difficult skill to master for people new to graphs.

> **A graph "escape room" with drawers and doors**
>
> One technique we use to help people conceptualize how to work with a graph is the metaphor of a graph as a series of rooms connected by hallways. Imagine that you're a little Gremlin, sitting on a vertex. What we actually see through the eyes of the Gremlin is a closed room. Our perception is limited to what's visible within the room. (Our room is always a plain white, but you can pick any color you like.) As we look around the room, we observe the following:
>
> - *A chest of drawers with a label on each drawer.* Each drawer is a vertex property, and the label is the property name.
> - *A series of doors, also with labels and with IN and OUT plaques.* Each door is an incident edge, and the plaques represent the edge direction.
> - *The doors themselves also have drawers.* These drawers are the edge properties.
>
> We have immediate access to everything visible in the room (vertex properties, edges, and edge properties). But access to anything beyond the drawers and the doors requires additional effort; in the case of our traversal, this means an additional step. That's right—our mental model of a graph's vertex is basically an "escape room."

EDGE DIRECTION MATTERS

As we saw in our example, the direction of an edge matters. In this case, even though Ted has two incident `friends` edges, only one of those represents a person he friended, while the other one signifies someone who friended him. Again, these `friends` edges work more like Twitter's "follows" relationships than real-life friendships.

This directionality of a relationship is different than in a relational database, where all relationships are bidirectional. In a graph, it is up to us to determine not only the direction of the edge but also how we want to traverse that edge. This ability to control whether we traverse only incoming, outgoing, or both edge directions provides us with a powerful tool to customize our traversals.

TRAVERSALS DON'T HAVE HISTORY

As we traverse the graph, we only have knowledge of where we currently are, not where we've previously been. (Remember the graph escape room mental model.) This is a fundamental difference from the way a relational database works and another common frustration for new graph developers as this concept catches them by surprise. In a relational database, if we write a query similar to the graph traversal depicted in figures 3.3 and 3.4, it might be something like this:

```
SELECT *
FROM person AS Ted
     JOIN friends ON friends.person_id = Ted.id
     JOIN person AS Friend ON Friend.id = friends.friend_id
WHERE Ted.first_name = 'Ted';
```

For this query, we expect to get back all the columns joined together from the `person AS Ted`, `friends`, and `person AS Friend` tables. By contrast, in a graph, the only value returned from the traversal is the ending vertex (or room). While there are ways to retrieve data from the other steps in our traversal, or even to bring the full history of our traversal along with us at each step, these require that we explicitly ask for that data or use specific steps to indicate that the history must be retained.

3.2.4 Writing traversals in Gremlin

Now that we have covered the core operations involved in traversing a single edge, let's write the code for our first traversal, or series of steps, by finding all of Ted's friends. We start by outlining the steps we need to take through the graph:

- Given all the vertices in my graph
- Find all the `person` vertices with a `first_name` of `Ted`
- Walk the outgoing `friends` edges to the incident vertex
- Return the `first_name`

Next, we map these plain English steps to the corresponding steps in Gremlin. Figure 3.6 shows the mapping.

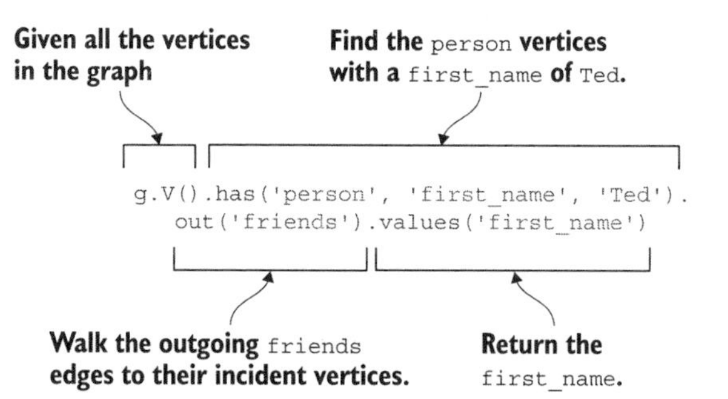

Figure 3.6 Mapping the plain text steps to the corresponding Gremlin steps for the question, "Who are Ted's friends?"

Now we take this traversal and run it in our Gremlin Console. We see that we get the correct answer, Josh, back from our database:

```
g.V().has('person', 'first_name', 'Ted').
   out('friends').values('first_name')
==>Josh
```

We realize that we haven't yet introduced even the most basic Gremlin syntax before showing the entire answer. We did this on purpose. Our experience is that people new to Gremlin tend to understand it better if they first see the answer, and then we break down what each step does. This approach provides a better mental model of how to move through our graph.

TRAVERSAL SOURCE

The `g` step in figure 3.6 is always the first step in every Gremlin traversal. The `g` represents the traversal source for our graph and is the base on which all traversals are written. This variable could be called anything, but the convention with a TinkerPop graph in transactional mode is to use `g`.

> **Gremlin key concepts: `g != graph`**
>
> Throughout this chapter, we refer to `g` as the traversal source and not as a graph. This is another important, and potentially confusing, aspect of TinkerPop: there are two APIs!
>
> The predominant API is the Traversal API that starts, by convention, with a variable defined by `g = graph.traversal()`. This is the API used throughout this book. It's a process that knows how to efficiently navigate its associated graph structure.
>
> The other API is an internal API designed for use by developers creating graph database engines. It's (confusingly) called the Graph API. It's an interface that defines a container object for the collection of `Vertex`, `Edge`, `VertexProperty`, and `Property` objects. It's also a data structure, and it doesn't provide an efficient means for navigation or for anything beyond the most basic ability to locate individual data elements in the graph.
>
> Those with a strong relational database background might think, "Two APIs—that must be like DDL (data definition language) and DML (data manipulation language) in SQL." But that would be wrong. DDL focuses on schema, and there is no corresponding set of language features in TinkerPop. The TinkerPop project does not specify how vendors should declare schema in their graph, so different vendors have different APIs for the graph schema definition. TinkerPop itself avoids the question of schema definition altogether by allowing any schema if it is used in the Gremlin code.
>
> Instead, the Graph API is like a relational database publishing an API in C/C++, C#, or Java for directly manipulating database files below the SQL language abstractions. Imagine handling data operations at the file level, including changes in the transaction logs and other low-level files. This is what using the Graph API is like.
>
> We only mention the Graph API here and in a few other places, such as in chapter 10 when covering anti-patterns. Throughout the book, all examples use the Traversal API.

GLOBAL STEPS

The second step in our traversals is the `V()` step (figure 3.6). The `V()` step returns an iterator that contains every vertex in the graph. It's one of two global graph steps. The other global graph step is `E()`, which returns an iterator that contains every edge in the graph. With few exceptions, one of these two steps is always the second step of our traversals. Our specific choice depends on whether we want to start our traversal on an edge or on a vertex.

Using `V()` to start on a vertex is by far the most common. In fact, except for some very exceptional operations, usually for maintenance or data integrity, we rarely use an `E()` step. Although we work with highly connected data and traverse or reason over the edges, we still operate in an entity-focused world, and we almost exclusively start and end with vertices.

Consider our "Find Ted's friends" example. The natural starting place is the `Ted` vertex. In fact, we venture to say that nearly every traversal you write for transactional operations starts on a vertex or set of vertices. Even in the DiningByFriends domain, we always start with a vertex of some sort, be it a person, restaurant, city, or review. This is both normal and good, and so in our traversals, we'll pretty much always start with `V()`.

FILTERING STEPS

The next step in our traversal is the `has()` step (figure 3.6), the first filtering step we introduce in this book. This is one of the most common Gremlin steps because it only passes through any vertex or edge that

- Matches the label specified, if a label is specified
- Has a key-value pair that matches the specified key-value pair

This filter step is the primary one for filtering traversals in Gremlin. Check the TinkerPop documentation for all forms of the `has()` step. The most commonly used forms include

- `hasLabel(label)`—Yields all vertices or edges of the specified `label` type
- `has(key, value)`—Yields all vertices and edges with a property matching the specified `key` and `value`
- `has(label, key, value)`—Yields all vertices and edges with both the specified `label` and with a property matching the specified `key` and `value`. This performs the same function as this combination:

```
g.V().hasLabel('person').has('first_name', 'Ted')
```

The `has()` step, as with most of the Gremlin steps, can be chained together to perform more complex filtering operations. This is much like using the `AND` in a `WHERE` clause in SQL. For example, we could find all people named Ted who are 40 years old simply by adding on an additional `has()` step to the previous traversal like this:

```
g.V().hasLabel('person').has('first_name', 'Ted').has('age', 40)
```

Our sample graph doesn't include an age property, however, so you can't test that traversal without first updating the data. As written, it doesn't return any results because no vertex can match the second `has()` step and Ted isn't 40. Not that there's a real Ted.

When working with a transactional graph, it's vital to narrow down the number of starting traversers as quickly as possible. This is done for reasons of load and performance. Fewer starting places usually means less work overall in traversing the graph. Therefore, it's quite common for the first step in any traversal to filter the possible vertices to a small subset with one or more `has()` steps. This is similar to filtering the base table in a SQL join.

TRAVERSAL STEPS

The `out(label)` step (figure 3.6) traverses all outgoing edges to the incident vertex with the specified `label`, if a label is provided. If a label isn't provided, then it just traverses all outgoing edges. This is one of the two most common traversal steps we use to navigate from one vertex to another. The other common traversal step is `in(label)`, which traverses all incoming edges to the incident vertex with the specified `label`, if a label is provided.

> **NOTE** Remember, an outgoing vertex is the vertex where an edge starts, and an incoming vertex is a vertex where an edge ends.

The `out(label)` traverses from a vertex (the `Ted` vertex, in our example), to the adjacent vertex along outgoing edges. We specified the label `friends` so our traversal only traverses `friends` edges. Figure 3.7 illustrates traversing from the `Ted` vertex to the adjacent `Josh` vertex on the single outgoing `friends` edge.

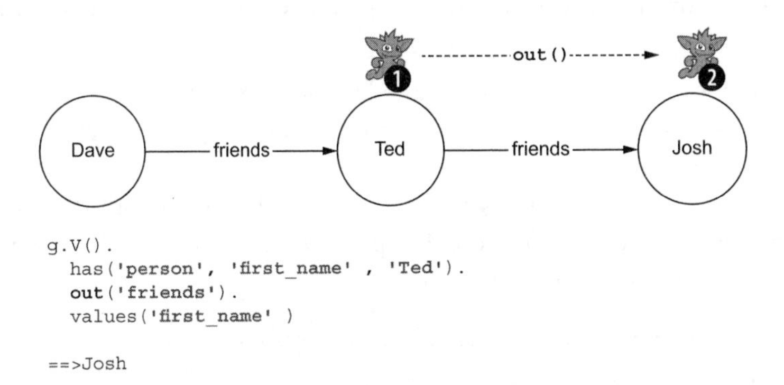


```
g.V().
  has('person', 'first_name' , 'Ted').
  out('friends').
  values('first_name' )

==>Josh
```

Figure 3.7 Traversing the outbound `friends` edge from Ted to Josh

Let's say we want to do the opposite and find people who have friended Ted instead of people whom Ted has friended. This is essentially the same query we described in figure 3.7 except we traverse the incoming `friends` edges instead of the outgoing ones. We switch our query to use the `in()` step, as shown in figure 3.8.

With this simple change, we can now traverse in the opposite direction. This flexibility to traverse relationships in either direction is a fundamental capability of graph

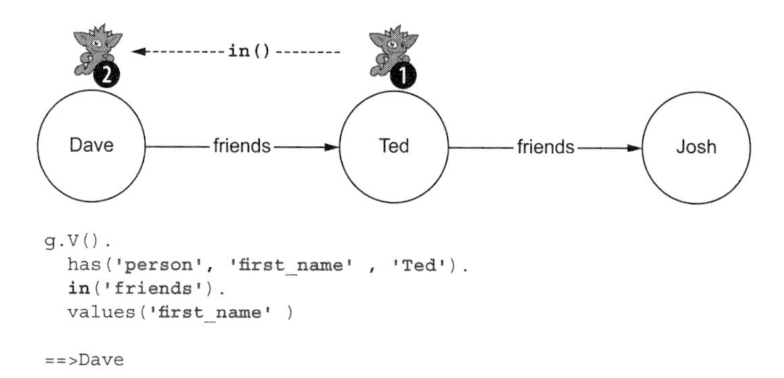

Figure 3.8 Traversing the inbound `friends` edge from Ted to Dave

databases. But it can be a double-edged sword (pun unintended). This directionality filters our traversals, which aids in both readability and performance, but it carries limitations as well. We might not know the direction of the edges we want to traverse, or we might not care in what direction we traverse. Whether we suffer from directional ignorance or directional apathy, Gremlin has a step to help us with that.

What if we want to find two sets of people at the same time? To answer this, we traverse the `friends` edge in both the incoming and outgoing directions *simultaneously*. Let's introduce another Gremlin step: `both(label)`. This step traverses from a vertex to the adjacent vertex along edges with the given `label`. Using this step, we write our traversal to find everyone who has friended Ted, as well as everyone whom he has friended, as shown in figure 3.9.

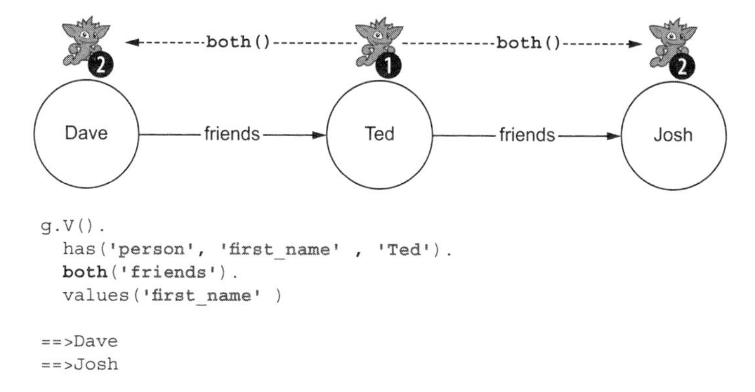

Figure 3.9 Traversing both the inbound and outbound `friends` edge from Ted

3.2.5 Retrieving properties with values steps

The final step in our traversal (figure 3.6) is the `values(keys...)` step, which returns the values of the element's properties. A separate line displays each resulting value. If the element has *N* properties, then the output contains *N* lines. If one or more `keys` are specified, then only properties with that key or keys are returned.

This is one of several different ways to return the property values of an element in our graph. The other commonly used step is `valueMap(keys...)`, in which both the keys and values for the properties matching those keys are returned. We'll go into much more depth on how to use these and other methods to format results in chapter 5.

Using this relatively simple query (figure 3.6), we begin to see how the syntax of Gremlin requires us to think about how we move around our graph in order to retrieve data. While we demonstrate this with Gremlin, the need to understand filtering and edge directionality to move around a graph is common to all graph query languages. We need to understand the direction we're walking an edge in order to understand what data we're getting. Once we make this mind shift to thinking about our traversals in terms of where in the graph we are currently located, we have the necessary mindset to leverage the relationships in the data.

3.3 Recursive traversals

Up to now we have found a specific vertex or traversed to adjacent vertices, but this is only scratching the surface. Stopping here would be like buying a sports car only to drive it around the neighborhood. It's time we put the pedal to the metal and see what we can do on the proverbial open road. In this section, we'll start writing and running one of a graph database's most powerful features: *recursive traversals*, which are also sometimes known as *looping traversals*.

3.3.1 Using recursive logic

We use *recursive traversals* for problems where some portion of the traversal needs to be executed multiple times in succession. There are many problems that require recursive traversals. Here are a few examples:

- *Bill of materials*—A standard bill of materials is made up of pieces, each of which is made up of more pieces, each of which is made up of still more pieces, and so on for an unknown number of levels.

  Example query: given the ID of a piece of equipment, walk the bill of materials to find all of the individual items required to build the equipment.
- *Map directions*—This is one that's familiar to many of us and that most of us use frequently, if not daily. Given two locations on the map, provide a listing of streets and turns to get from a starting location to an ending location. Though

two locations are connected, we cannot predict ahead of time the number of turns required.

Example query: given two locations, provide turn-by-turn directions to get from point *A* to point *B*.

- *Task dependency*—Let's say we're building a software application. Because we're all good developers, we begin by listing the different work items required for completion. For each of these items, we can link these to any dependent work items, which are then connected to their dependent items, and so on.

  Example query: provide an ordered list of items needed to remove the dependency to another item.

Each of these problems requires traversing an unknown number of links. In the case of the bill of materials, the links represent the hierarchy of pieces. For the map directions, the links serve as the connections between intersections. For the task dependencies, the links serve as correlations on the dependency tree.

When we have a problem that requires traversing an unknown number of edges to find the answer, we use a recursive traversal. In a relational database, this is likely handled by a recursive common table expression (CTE), which can be difficult to code and to maintain. However, because graph databases are optimized to handle highly interconnected data, their query languages and underlying data structures are also optimized for quickly executing recursive queries.

Let's extend the traversal from the last section to see recursive queries in action. Instead of trying to find Ted's friends, let's locate all of the friends of Ted's friends. This friends-of-friends-type question is a common pattern in social networks and is similar to what Facebook, Twitter, or LinkedIn do to recommend potential connections. If we want to accomplish this in our social network graph, we would need to execute the following steps. (Note: because *every* traversal starts in the context of all vertices in the graph, we don't explicitly state that when planning our traversing through the graph.)

1. Find all the `Ted` vertices.
2. Traverse the outgoing `friends` edges to the incident edge.
3. Traverse to the incoming vertex (at this point, we're at Ted's friends).
4. Traverse the outgoing `friends` edges to the incident edge.
5. Traverse to the incoming vertex (at this point, we're at the friends of Ted's friends).
6. Return the `first_name` property value.

To refresh our memories of what the process looks like for steps 1–3, covered in section 3.2.2, let's look at figures 3.10, 3.11, and 3.12.

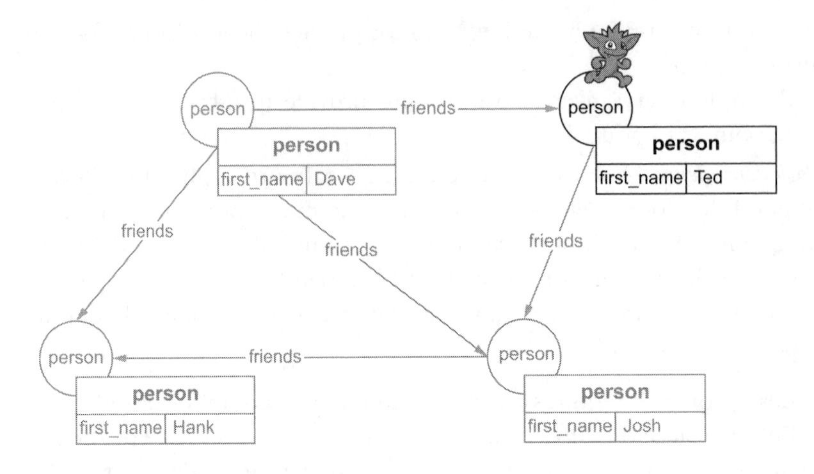

Figure 3.10 Step 1: Locating the `Ted` vertex

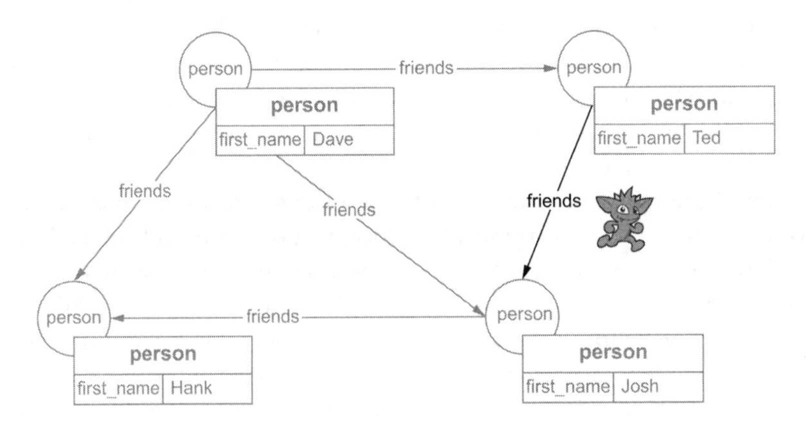

Figure 3.11 Step 2: Traversing from the `Ted` vertex to the incident outgoing `friends` edges

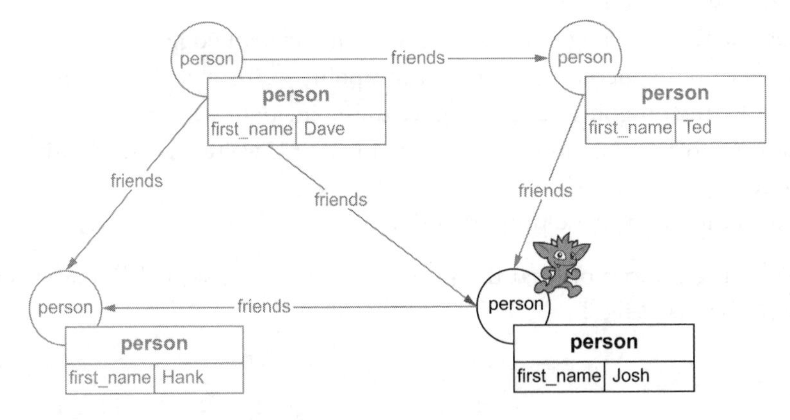

Figure 3.12 Step 3: Moving from the incident `friends` edges to the adjacent vertex

Now that we are located at Ted's friend Josh in our graph, let's examine how these two additional steps take shape. Figures 3.13 and 3.14 show this transformation.

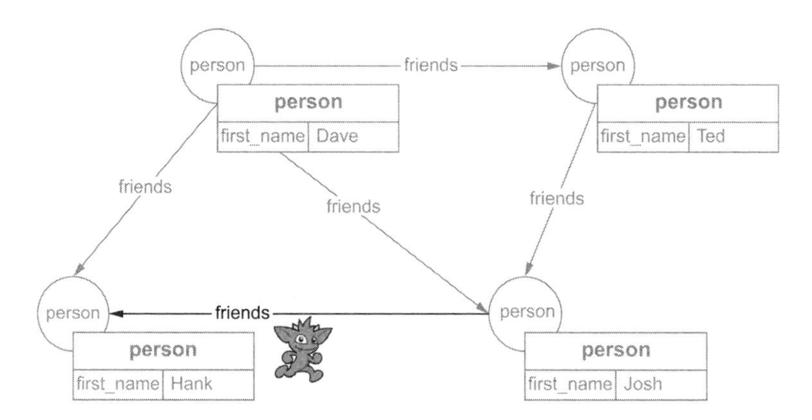

Figure 3.13 Step 4: Continuing from the vertex of Ted's friend Josh and traversing out Josh's outgoing `friends` edge

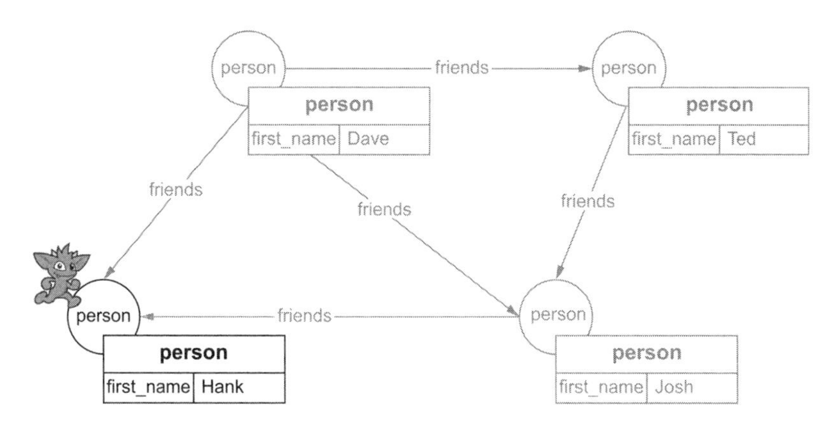

Figure 3.14 Step 5: Continuing from the vertex of Ted's friend Josh, traversing Josh's outgoing `friends` edge, and then moving to Josh's friends

Examining both the text and the diagrams, we see that steps 4 and 5 are just repeats of steps 2 and 3. This repetition of actions is what we're referring to when we talk about recursive queries. In this example, our actions only repeat one time, but many real-world use cases require that a section of a query be repeated many times, sometimes until a specific condition is reached, leading to an indeterminate number of repetitions.

3.3.2 Writing recursive traversals in Gremlin

Now that we have a picture in our mind of how recursive traversals operate, let's take a look at how to use Gremlin to find the friends of Ted's friends. Looking back at the steps we need to do to find the friends-of-friends for Ted, we noticed that we need to repeatedly go out the `friends` edge. We could extend our traversal from before by adding another `out('friends')` step like the following:

```
g.V().has('person', 'first_name', 'Ted').
  out('friends').
  out('friends').
  values('first_name')
==>Hank
```

This works and provides the correct answer, but it only works because we knew that we needed two repetitions. In many cases, we don't know how many repetitions we'll need. We need a way to loop through a defined series of Gremlin steps until a condition is met and the loop exits. In our case, we want to loop two times through the `out('friends')` step from our earlier traversal. To accomplish this in Gremlin, we need to introduce a few new steps:

- `repeat(traversal)`—Repeatedly loops thorough the steps until instructed to stop. The `traversal` parameter represents the set of Gremlin steps to be repeated within the loop.
- `times(integer)`—A modifier for a `repeat()` loop. The `integer` parameter represents the number of operations for the loop to execute.
- `until(traversal)`—A modifier for a `repeat()` loop. The `traversal` parameter represents the set of Gremlin steps that evaluate for each loop. When the `traversal` evaluates to `true`, the `repeat()` step exits.

Gremlin traversal parameters

One important difference to note about the `repeat()` and `until()` steps is that unlike the previous steps, which took a label, a string, or an integer, these steps expect a `traversal` as a parameter. What does it mean to take a `traversal` as a parameter?

While the `repeat()` and `until()` steps are our first introduction to the `traversal` parameter type, this is a common pattern that we'll see again and again as we add Gremlin steps to our toolbelt. When we see `traversal` as a parameter, what we pass into the step is one or more steps that are performed in the context of the step. In the case of the `repeat()` step, the traversal parameter is a set of steps to repeat in a loop. For the `until()` step, the traversal parameter is the stopping condition of the `repeat()` step.

Traversal parameters are similar to lambda expressions in Java. A Java *lambda* expression allows us to provide method arguments to perform complex tasks in the context of that function. A traversal argument in Gremlin enables us to provide a series of Gremlin steps that can execute complex movements within our graph in the context of the step it was passed. For example, if we want to continuously traverse `friends` edges until we come to a person with the first_name of Dave, we could do this by passing in a traversal has('person', 'first_name', 'Dave') to the until() step as shown:

```
g.V().has('person','first_name','Ted').
    repeat(
      out()
    ).until(has('person','first_name','Dave')).
    values('first_name')
```

Note that the example traversal from Ted to Dave is for illustration only. It does not return a result with the sample data used in this chapter. The curious reader can look at the previous diagrams of the data to see why; as for the lazy ones, well, we'll just point out that it had to do with the directions of the edges in the graph.

With these new steps, let's explore what our traversal looks like to find the friends-of-friends for Ted:

```
g.V().has('person', 'first_name', 'Ted').
   repeat(
     out('friends')
   ).times(2).
   values('first_name')
==>Hank
```

Written in English, the traversal reads as

- Given all the vertices in the graph
- Find all the person vertices with a first_name of Ted.
- Repeat the following step or steps.
- (Repeated block) Walk the outgoing friends edges to the adjacent vertex.
- Execute the repeated step(s) two times.
- Return the first_name.

Review figure 3.15 to see how these text steps correlate with the corresponding steps in Gremlin.

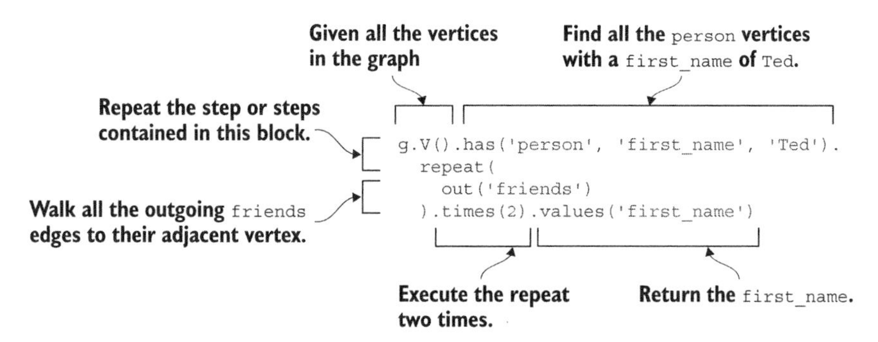

Figure 3.15 Mapping of the plain text steps to the corresponding Gremlin steps for our friends-of-friends query

Let's walk through each step of our traversal on our data model and see how we arrived at our answer of Hank. The first step of our traversal filters our query to just a single traverser on the `Ted` vertex, as seen in figure 3.16.

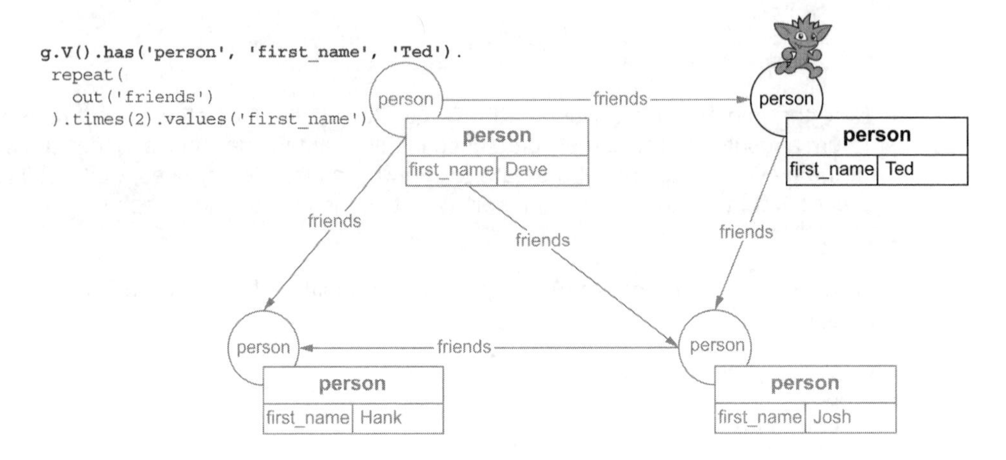

Figure 3.16 The first steps of our Gremlin traversal for our friends-of-friends query for Ted

In the second step of our traversal, we enter the `repeat()` loop and process the interior traversal for the first time; in this case, `out('friends')`. Examining our graph shows us that there's only a single outgoing `friends` edge that, when we traverse, brings us to the `Josh` vertex, as shown in figure 3.17.

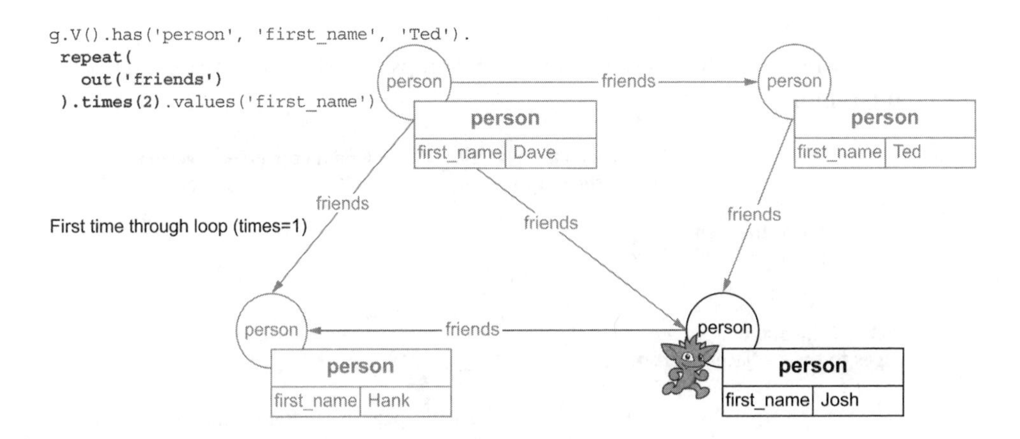

Figure 3.17 The second step of our Gremlin traversal for our friends-of-friends query for Ted

In third step in our traversal, we remain in the `repeat()` loop once again. We know that we need to repeat this loop because we specified, using the `times(2)` step, that we want to repeat the loop twice. We again traverse all outgoing `friends` edges to the adjacent vertex, of which there is only one, to arrive at the `Hank` vertex, as shown in figure 3.18.

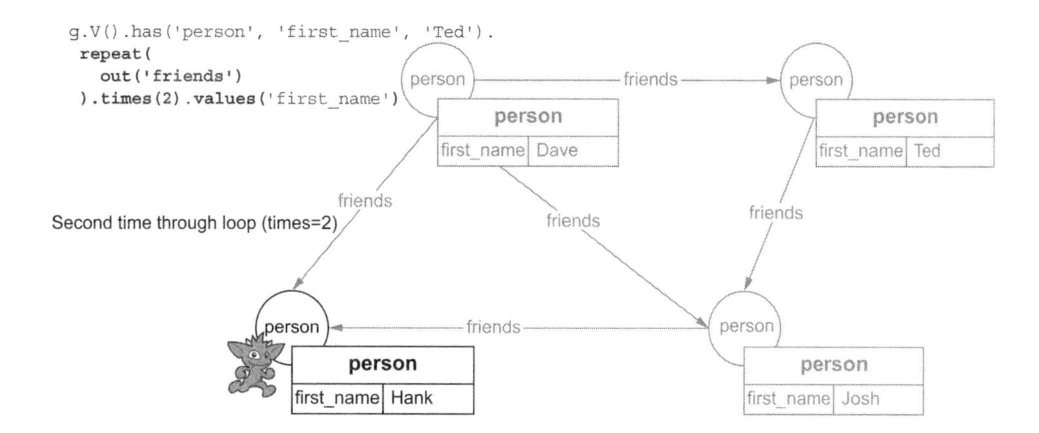

Figure 3.18 The third step of our Gremlin traversal for our friends-of-friends query for Ted

In our traversal, we specified to perform two iterations through our `repeat()` loop. After we iterate through the interior traversal twice, we exit the loop. The last portion of our traversal specifies that we return the `first_name` value of any vertex that we're on; in this case, `Hank`.

What if we don't know how many times we need to repeat our traversal to find Hank from Ted? Imagine we want to continue looping until we find an element that matches a specific set of criteria. For the situations where we don't know the number of times we need to recurse, we use the `until()` step. The `until()` step allows us to loop continuously until a specified condition is met.

> **IMPORTANT** Queries that use the `until()` step can create performance issues because the traversal runs until the condition is met. If the condition is never met, then it continues until it exhausts every potential path in the graph. This scenario is known as an *unbounded traversal.* When using the `until()` step, we recommend providing a maximum number of iterations using the `times()` step or using a time limit with the `timeLimit()` step.

If the `until()` step comes before the `repeat()` step, then the loop operates as a while-do loop. If it appears after the `repeat()`, then it functions as a do-while loop.

Do-while versus while-do loops

Both do-while and while-do (or while) loops are programming constructs used in many programming languages. Both options execute a block of statements continuously while the specified expression is true. In Java, do-while loops look like this:

```
do {
  //Execute some statements
} while (expression)
```

And while loops look like this:

```
while(expression) {
  //Execute some statements
}
```

The fundamental difference is that a do-while loop checks the expression at the end of the loop, but a while-do loop verifies the expression at the beginning. This means that if the `expression` evaluates to `false`, a do-while only executes one time, but a while-do loop won't execute. In other words, a do-while loop always executes at least once, whereas a while loop may not execute at all.

If we want to write a traversal to find a `Hank` vertex from the `Ted` vertex, and we don't know the number of loops it takes to get there, we can do this using the `until()` step as shown here:

```
g.V().has('person', 'first_name', 'Ted').
  until(has('person', 'first_name', 'Hank')).
  repeat(
    out('friends')
  ).
  values('first_name')
==>Hank
```

Although this traversal returns the data we expected, it only provides information about Hank, not how we traversed from Ted to Hank. It only reveals that the two are connected. What if we want to see how Ted and Hank are connected?

To determine the intermediate steps, we need to introduce a modifier step to the `repeat()` step, known as `emit()`. The `emit()` step informs the `repeat()` step to emit the value at the current location in the loop. Let's add `emit()` to our traversal and examine our results:

```
g.V().has('person', 'first_name', 'Ted').
  until(has('person', 'first_name', 'Hank')).
  repeat(
    out('friends')
  ).emit().
  values('first_name')
```

```
==>Josh
==>Hank
==>Hank
```

Interestingly, we got two Hank responses in the results. Remember that when we use the until() step before the repeat, it's a while-do approach. When we execute this traversal, the following occurs:

- Given all the vertices in our graph
- Finds the person vertex where the first_name is Ted
- Evaluates the until() statement to see if we're on a person vertex with a first_name of Hank (in this iteration, it is false)
- Traverses the outgoing friends edges to the adjacent vertex
- Emits the current vertex (in this iteration, it's Josh)
- Evaluates the until() statement to see if we're on a person vertex with a first_name of Hank (again, in this iteration, it is false)
- Traverses the outgoing friends edges to the adjacent vertex
- Emits the current vertex (in this iteration, it's Hank)
- Evaluates the until() statement to see if we're on a person vertex with a first_name of Hank (in this iteration, it is true)
- Emits the first_name property of the current vertex (Hank)

By emitting the vertices as we traverse through our graph, we end up with three vertices, one of which is a duplicate. This duplicated vertex, once from being emitted in the loop and once for being the current vertex at the end, is why the results contain a duplicate Hank. But what if we want to see Ted as part of the results? And maybe also get rid of that extra Hank?

> **NOTE** The emit() step is similar to the until() step, whether it's placed before or after the repeat() step, that impacts how it behaves. If the emit() is placed before the repeat(), it includes the starting vertex. If it's placed after the repeat(), it only emits the vertices traversed as part of the loop.

To ensure that Ted is part of our results, we need to move the emit()step before the repeat() step as illustrated here:

```
g.V().has('person', 'first_name', 'Ted').
  until(has('person', 'first_name', 'Hank')).
  emit().
  repeat(
    out('friends')
  ).
  values('first_name')
==>Ted
==>Josh
==>Hank
```

Why does this traversal add Ted and remove the duplicated Hank? Let's take a closer look at how this works and see if we can spot why moving the `emit()` fixes both problems:

- Given all the vertices in our graph
- Finds the `person` vertex where the `first_name` is `Ted`
- Evaluates the `until()` statement to examine if we're on a `person` vertex with a `first_name` of `Hank` (in this iteration, it is `false`)
- Emits the current vertex (in this case, it's `Ted`)
- Traverses the outgoing `friends` edges to the adjacent vertex
- Evaluates the `until()` statement to see if we're on a `person` with a `first_name` of `Hank` (again, in this iteration, it is `false`)
- Emits the current vertex (in this case, it's `Josh`)
- Traverses the outgoing `friends` edges to the adjacent vertex
- Evaluates the `until()` statement to determine if we're on a `person` with a `first_name` of `Hank` (in this iteration, it is `true`)
- Emits the `first_name` property of the current vertex (`Hank`) and all previously emitted vertices

By placing `emit()` before we `repeat()`, our traversal yields not only our initial vertex (`Ted`) but also avoids duplicating our final vertex (`Hank`). This ability to combine the do-while and while-do capabilities based on the location of the `emit()` step provides us tremendous flexibility for defining the results returned from our recursive queries. However, this flexibility comes at the cost of added complexity. As we saw in this example, simply changing the location of the `emit()` step also modifies the results of this recursive loop.

The graph traversal languages, and Gremlin in particular, provide a rich set of tools for traversing and looping through the structure of our graph within a single traversal. If we compare the simplicity of writing recursive queries in a graph to the complexity of answering the same types of questions in SQL, you'll start to notice why graph databases excel at answering these sorts of problems.

In this chapter, we learned many of the basic building blocks of how traversals work. In the next chapter, we'll extend these building blocks with another powerful tool available to us in graphs—*paths*. Paths allow us to return a result that includes not just the end location of our traversal, but also the route traversed through a graph to get there.

Summary

- The process of moving through a graph is known as traversing a graph. The set of steps and actions that define how we traverse a graph is called a traversal.
- Traversing a graph is done via a series of steps. Each step continues from the location where the last step ended.

- Traversing a graph requires that we understand the structure of the graph, where we're located in the graph at any time, and what the incident edges, adjacent vertices, and available properties are at each location.
- Knowing the direction of the edges we want to traverse from a specific location is crucial because it's required when writing a traversal.
- Graph traversal languages are optimized to process recursive traversals with either a known or an unknown number of loops.

4 Pathfinding traversals and mutating graphs

This chapter covers

- Writing traversals to add, modify, and delete vertices, edges, and properties
- Finding the paths that connect two vertices
- Refining pathfinding traversals using edges and edge properties

Getting lost has itself become a "lost art" since the advent of GPS devices and smartphones. Long gone are the days of stopping at a local service station to ask for directions. But while we may fear that our direction-finding skills are atrophying in the real world, in the data domain, graph's pathfinding algorithms come to the rescue—or maybe these precipitated the digitization of this real-world skill.

We emphasized in the last chapter how critical it is to know your location in the graph at all times in the traversal-writing process. In this chapter, we take that concept a step further with pathfinding algorithms. A *path* is a listing of the vertices and edges visited from the beginning vertex to the ending vertex of a traversal. Paths tell us not only that two vertices are connected, but also show us all of the intermediate elements in between. Paths are the turn-by-turn directions between two points in a graph. But because there aren't a lot of pathfinding options in a graph with four vertices, we'll begin by mutating our graph to add some more data. *Mutating* simply means changing the graph by adding, modifying, or deleting vertices, edges, and/or properties.

After we enlarge our graph in this chapter, we'll extend our recursive traversal knowledge from the last chapter using pathfinding algorithms. We'll close by refining our pathfinding traversals through filtering on edges. By the end of this chapter, you'll know how to add, edit, and delete elements within the graph, how to extend recursive traversals with pathfinding algorithms, and how to refine these algorithms by filtering on edges.

If you haven't done so already, download the corresponding source code for this chapter: https://github.com/bechbd/graph-databases-in-action. The code for this chapter is located in the chapter04 folder. All examples begin with the assumption that our simple social network data set is loaded. This is effectively the same data we used in chapter 3, but we'll add more vertices and edges to it as we work through this chapter.

Using a terminal window, navigate to the directory where you have the unzipped version of the Gremlin Console download. On MacOS or Linux systems, run the Gremlin Console with this script argument:

```
bin/gremlin.sh -i $BASE_DIR/chapter04/scripts/4.1-simple-social-network.groovy
```

On Windows, use

```
bin\gremlin.bat -i $BASE_DIR\chapter04\scripts\4.1-simple-social-network.groovy
```

4.1 Mutating a graph

Until now, we've worked with a small prepared set of data. Figure 4.1 shows our current social network data.

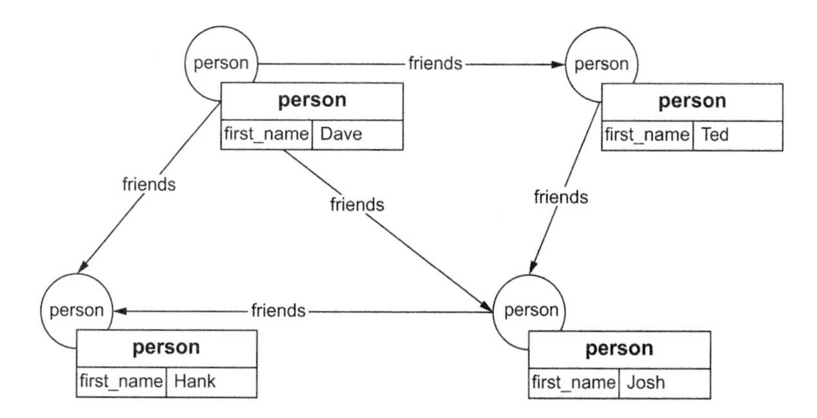

Figure 4.1 Our graph containing the small social network data set that we used in chapter 3

While this simple graph worked fine for the basic traversals in the last chapter, its data set is quite small. To demonstrate pathfinding algorithms, we need a graph with more data, so let's make it look more like figure 4.2.

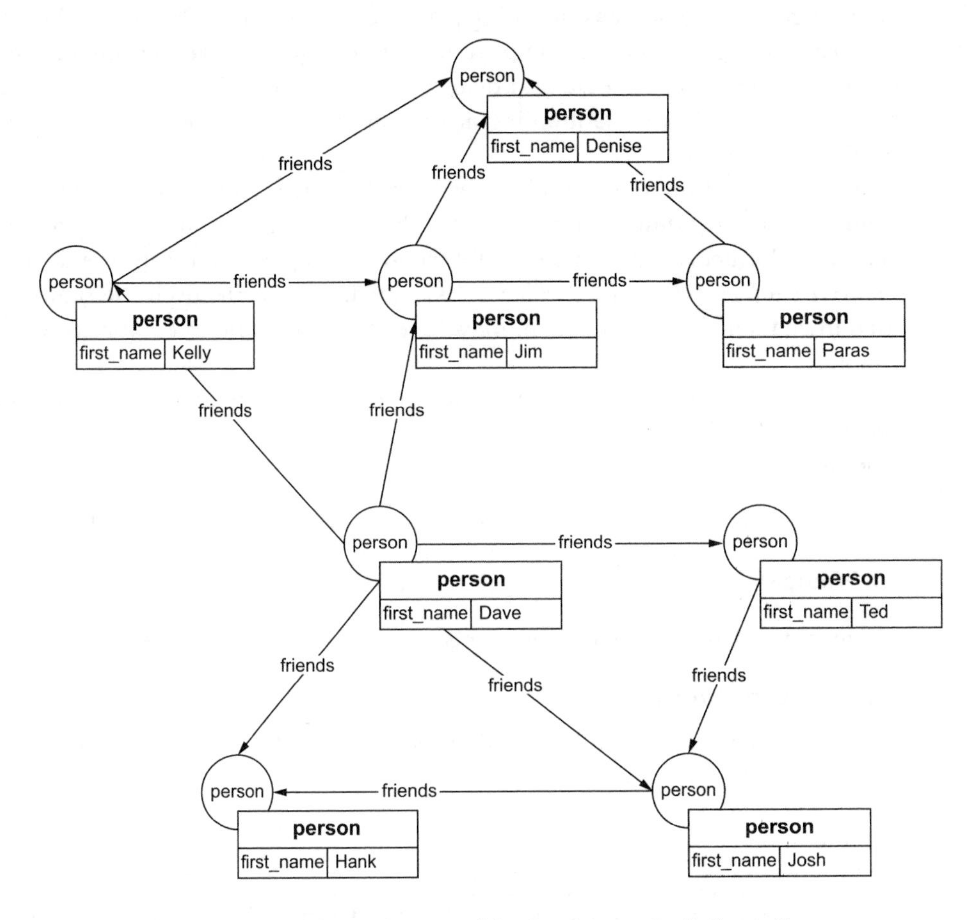

Figure 4.2 Our graph containing our larger social network to use for finding paths

Now that's really starting to look like a graph! But we have a new problem: How do we add this data to our graph?

4.1.1 Creating vertices and edges

The fundamental concept of creating entities in graph databases isn't all that different from the relational database world. Creating new vertex entities involves adding the appropriate elements and properties. However, creating new edges is a bit more complicated because we need to specify the vertex that belongs at each end of the edge.

ADDING VERTICES

Let's say we want to add a new person named Dave to our graph. Extending what we've learned in the last chapter about how graph traversals work, we can think of the process for adding a vertex as

- Given a traversal source `g`
- Add a new vertex of type `person`.
- Add a property to that vertex with the key `first_name` and the value `Dave`.

Well, that seems pretty straightforward, and it really is that easy. The query in terms of SQL is

```
INSERT INTO person (first_name) VALUES ('Dave');
```

The process in a graph database is nearly the same. If we look at figure 4.3, we see the corresponding process as a Gremlin traversal.

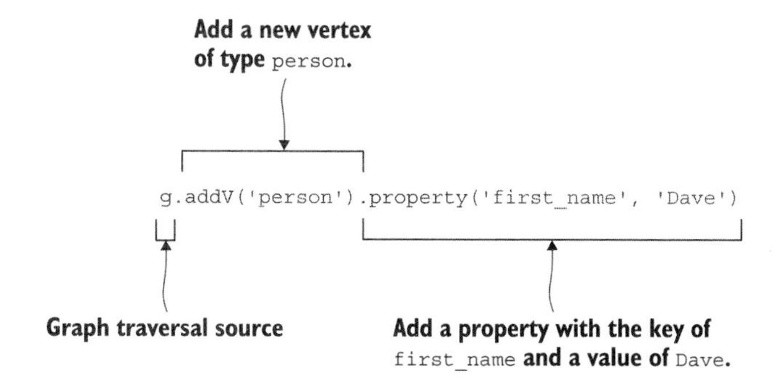

Figure 4.3 Mapping the plain text description for adding a vertex `Dave` to the corresponding Gremlin steps

From these steps, we can see the similarities in the process of adding a vertex and inserting a row in a relational database. This traversal also introduces you to the first two steps for mutating our graph:

- `addV(label)`—Adds a new vertex to the graph of the type `label` and returns a reference to the newly added vertex.
- `property(key, value)`—Adds a property to a vertex or an edge. The property includes the specified `key` and `value`. This step then returns a reference to the vertex or edge that entered the step, thus operating as a side effect.

If we take this traversal and run it in the Gremlin Console, we get

```
g.addV('person').property('first_name', 'Dave')
==>v[13]
```

Concerning mutations and graph steps

In the last chapter, we mentioned two graph steps, `V()` and `E()`, which we used in filtering traversals. We haven't utilized these when mutating traversals. Let's discuss why.

A mutating traversal, or mutation, is an operation that changes the graph's content or structure in some way. The first part of this chapter is focused on such operations, starting with the `addV()`, `addE()`, and `property()` steps.

The `V()` step not only represents the full set of vertices in the graph, but also returns the full set of vertices. The next step is then executed for each element output by the previous step. If we erroneously tried `g.V().addV('person')` in a graph, we'd get a `person` vertex added *for every existing vertex in the graph.*

This may be a desired outcome in some cases, but we suspect that those cases are rare. Most often, we just want to add a single vertex for each `addV()` step, so we don't include the `V()` graph step.

Yippee! We can see that we get back the unique identifier for the vertex. In our example, it returned `v[13]`.

NOTE If you ran this traversal and didn't get back the same vertex ID, don't worry. The ID values are internally generated based on the current state of the database. As long as you get back a value, it worked correctly.

Now that we have our `Dave` vertex, let's check to see that the vertex was added as expected. Let's verify this by running a traversal to find all the `person` vertices with the `first_name` of `Dave`:

```
g.V().has('person', 'first_name', 'Dave')
==>v[0]
==>v[13]
```

Wait, why did two vertices return? At the beginning of this chapter, we already had a vertex in our database with the `first_name` property of `Dave` (see figure 4.1). Running the `addV()` traversal added a second vertex with the `first_name` property of `Dave`. This is akin to adding the same row to a relational database with an auto-incrementing primary key column.

Vertex ID values: Generation and usage

In the previous example, we added a vertex to our graph, and it was automatically assigned the number 13: `g.V(13)`. This ID value is automatically generated by the database. Different databases handle this generation in different ways. Some databases, such as Gremlin Server, use a simple integer (32-bit) for vertices and a long integer (64-bit) for edges. Other graph database engines might use a UUID/GUID, an encoded string, or a hash of some sort. While it is possible to use these values in code, it is a best practice to not use these at all.

> The ID of a vertex should be considered "internal" to the graph database engine, and you should be extremely cautious when working with internals of a tool like a database engine. The internals are maintained by the engine in a manner seen as best by the engine's developers. Using these in application code for the application's own purposes is extremely dangerous!

Now that we know how to add vertices, let's see how to connect vertices with the edges.

ADDING EDGES

Connecting vertices to edges is venturing into new territory, as there isn't a corresponding concept in the relational world. In the relational world, the connections between entities are implicitly made when we populate a column containing a foreign key. In the graph world, these connections need to be explicitly added via edges. Let's say that we want to add a `friends` edge between the `Ted` and `Hank` vertices. The process for adding this edge is

- Given a traversal source `g`
- Add a new edge with a label `friends`.
- Assign the outbound vertex of the edge to the vertex with the key of `first_name` and the value of `Ted`.
- Assign the inbound vertex of the edge to the vertex with the key of `first_name` and the value of `Hank`.

As we can see, the main difference between adding a vertex and adding an edge is that when adding an edge, we need to specify the inbound and the outbound vertex for the edge. We'll leverage the approach we learned in the last chapter to filter and locate these vertices. Taking a look at figure 4.4, we see how this is coded in a Gremlin traversal.

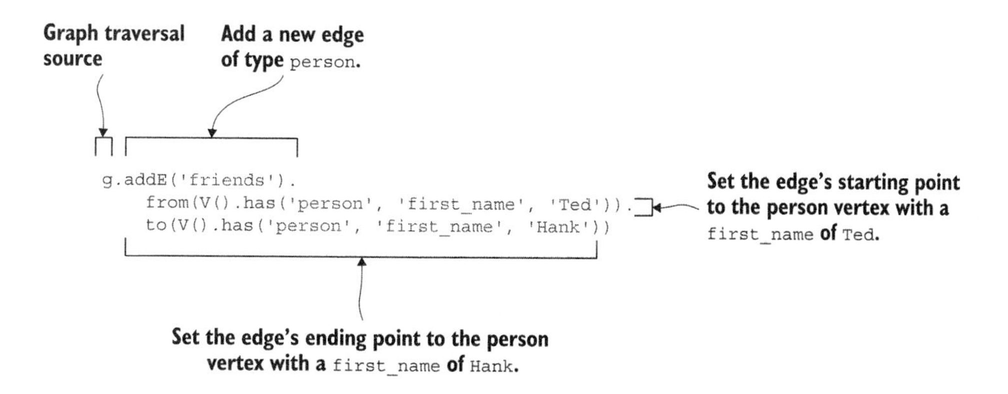

Figure 4.4 The Gremlin steps for adding an edge from the `Ted` vertex to the `Hank` vertex

As seen in figure 4.4, adding an edge requires our traversal to first insert the edge and then to specify the inbound and outbound vertices to connect with that edge. To accomplish this in Gremlin, we'll use a new step and a couple of step modulators:

- `addE(label)`—Adds a new edge with a label of `label`
- `from(vertex)`—Modulator that specifies the `vertex` where the edge will start; the source for the new edge
- `to(vertex)`—Modulator that specifies the `vertex` where the edge will end; the destination for the new edge

Step modulators like `from()` and `to()` cannot be used independently. We use these to provide some configuration to an associated step (in this case, the `addE()` step). While the term *modulator* may be specific to TinkerPop, the need to provide source and destination vertex details is not. Regardless of the engine or language used, creating edges requires knowing the starting and ending vertices.

Keen observers will also notice that we used a `V()` step in the middle of our traversal (figure 4.4). This ability to start another graph traversal from within a traversal is quite similar to performing a SELECT *inside* a SELECT in SQL like this:

```
SELECT * FROM table1 WHERE id = (SELECT id1 FROM table2);
```

Let's see what happens when we run the code in figure 4.4 in the Gremlin Console. In the following example, we added some arbitrary line breaks within the `from()` and `to()` modulators to highlight the sub-traversals:

```
g.addE('friends').
  from(
    V().has('person', 'first_name', 'Ted')
  ).
  to(
    V().has('person', 'first_name', 'Hank')
  )
==>e[15][4-friends->6]
```

Once again, it's time to do our happy dance because we added an edge to our graph. This is great! We now have the building blocks to add data. Now, let's see how to remove it.

4.1.2 Removing data from our graph

Just as with adding data, removing data from a graph isn't that different from deleting data in a relational database. Let's examine how to remove a vertex and an edge from a graph.

Removing a vertex

Removing a vertex is similar to deleting a row in a relational database. The process for removing a vertex is

- Given a traversal source `g`
- Find the vertex with an ID of `13`.
- Remove (or drop) that vertex.

In a relational database, we do this using SQL. For example

```
DELETE FROM person WHERE person_id = 13;
```

Looking at figure 4.5, we see how to perform the same process with Gremlin code.

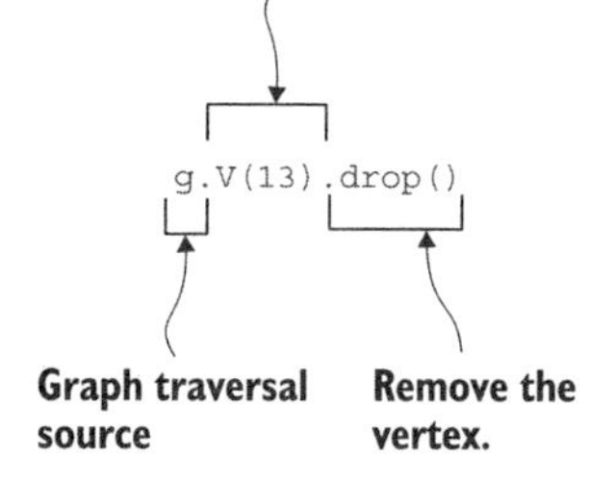

Figure 4.5 Mapping the plain text steps to the corresponding Gremlin steps for removing a vertex

Examining this traversal, we notice two new Gremlin steps:

- `V(id)`—Returns the vertex with the specified `id`. This `id` is an internal ID property assigned and maintained by the Gremlin Server (or the selected database).
- `drop()`—Deletes any vertex, edge, or property that's passed to it.

Let's run the code in figure 4.5 in the Gremlin Console to see what happens:

```
g.V(13).drop()
```

Nothing was returned. Are we sure it did anything? Let's check to see if the vertex still exists:

```
g.V().has('person', 'first_name', 'Dave')
==>v[0]
```

The drop command must have worked because we didn't get an error and we now have only a single `Dave` vertex (remember we had two previously), but why didn't the `drop()` step return anything? The `drop()` step is different from the other Gremlin steps we have learned so far. When it works, it doesn't return anything to the client. This is a bit unexpected as results go, but it did accomplish the removal of the vertex. In chapter 6, we build a traversal that also reports the number of vertices affected by the `drop()` step.

REMOVING AN EDGE

We can remove an edge from a graph in one of two ways. First, if we delete the starting or ending vertex, any edge associated with that vertex is also deleted; it's the graph database version of referential integrity. This is similar to a relational database because relationships are not explicitly created or destroyed; instead these are implicitly represented by the presence of foreign keys. The second way to remove edges from a graph is to explicitly remove or drop these, leaving the start and end vertices. To drop an edge via this method we would

- Given a traversal source `g`
- Find the edge with an ID of `15`
- Remove (or drop) that edge

In a graph, we map this process to the corresponding Gremlin traversal as shown in figure 4.6.

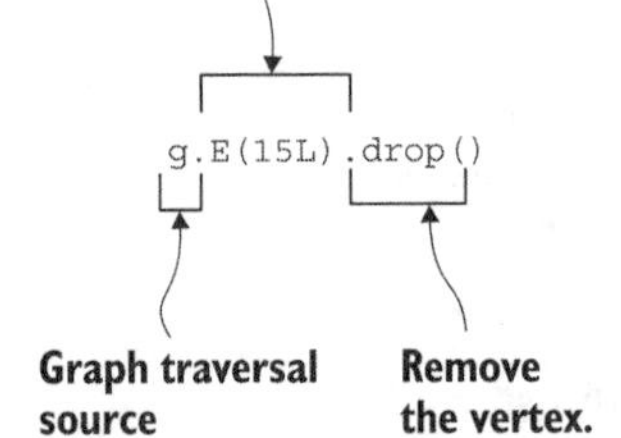

Figure 4.6 Mapping the plain text steps to their corresponding Gremlin steps for removing an edge

> **NOTE** The default implementation of `g.E()` in TinkerPop requires a `Long`, not an `int`, and Java's funny about these things.

Upon examining this traversal, we notice that it looks almost exactly the same as the one we used to drop a vertex. The similar syntax shows that vertices and edges are of equal importance inside a graph database.

There are other approaches to removing vertices and edges that don't require knowing their internal ID values, such as filtering vertices and dropping them. We'll cover those when we get to the implementation details in chapter 6. For now, we wanted to illustrate the basic use of the `drop()` step.

4.1.3 Updating a graph

Up until now, we've worked on how to add and remove vertices and edges to our graph. This leaves us with one major mutation operation left: updating properties in our graph. If you followed along and typed everything in perfectly, you're doing well. However, what if you accidently misspelled "Dave" as "Dav" when adding a vertex? How do we fix this mistake? If we think about how to perform this sort of update, we would

- Given a traversal source g
- Find the vertex with the key of first_name and the value of Dav.
- Update the property to that vertex with the key of first_name and the value of Dave.

In SQL, this would be similar to

```
UPDATE person SET first_name = 'Dave' WHERE first_name = 'Dav';
```

Fortune is with us because we already learned the steps to set a property value. We know how to find a vertex with a specific property using has(), and we know how to set the value of a property using property().

EXERCISE Take a minute and see if you can write the traversal to change a property value.

Hopefully, you were able to write this traversal on your own. It should look like this:

```
g.V().has('person', 'first_name', 'Dav').
      property('first_name', 'Dave')
==>v[18]
```

If this traversal makes sense, you are starting to see how we combine the basic operations of graph query languages into more complex operations. At the beginning of this chapter, we ran a script that added data to a graph. Let's examine each of the operations in that script. We reproduce it here with just the basic Gremlin operations, without the connection logic or code comments:

```
g.V().drop().iterate()

dave = g.addV('person').property('first_name', 'Dave').next()
josh = g.addV('person').property('first_name', 'Josh').next()
ted = g.addV('person').property('first_name', 'Ted').next()
hank = g.addV('person').property('first_name', 'Hank').next()

g.addE('friends').from(dave).to(ted).next()
g.addE('friends').from(dave).to(josh).next()
g.addE('friends').from(dave).to(hank).next()
g.addE('friends').from(josh).to(hank).next()
g.addE('friends').from(ted).to(josh).next()
```

SCRIPTING MUTATIONS

It's good to know how to add vertices and edges one by one, but what if we wanted to be a bit more efficient and add these together as part of a script or as a bulk action? Well, that's actually what we did at the beginning this chapter when we ran the script 4.1-simple-social-network.groovy as we started Gremlin Console. This script ran several mutation operations together to load data into our graph. Let's take a look at that script and see how to chain together multiple mutations. We will skip over the first five lines with the `:remote` statements that configure the Gremlin Console connection to Gremlin Server, the details of which are covered in appendix A.

The first line of Gremlin code is `g.V().drop().iterate()`, which is used to clear out all the data in our graph. This statement allows the script to be rerunnable because it always starts by removing all existing data. This should look familiar, except for the `iterate()` step. The `iterate()` step and the similar `next()` step both cause the traversal to execute. The key difference between these is that the `iterate()` step does not return a result, while the `next()` step returns the result of the traversal. We can think of this line as, "For each vertex in my graph, drop it and don't return anything."

In a few pages, we'll look in more detail at the `next()` step and these types of terminal steps in general. Because the `drop()` step doesn't return a value, the `iterate()` step is a better fit than the `next()` step. Now we can get to the meat of the script where we add data. Let's look at the line

```
dave = g.addV('person').property('first_name', 'Dave').next()
```

This is different from the mutation traversals we just wrote; specifically, what's this `dave` value and why include the `next()` step? Well, this is a script that creates some elements, some of which we want to reuse. Namely, we want to later use the vertices that were added at the start when adding edges between these. We need the `dave` value and the `next()` step for this reuse requirement. We explain this shortly, but first, if we were to write this traversal in English, we would write something like the following (figure 4.7 demonstrates how these steps are coded as Gremlin steps):

- Declare a variable `dave` that holds the result of the traversal.
- Given a traversal source `g`
- Add a new vertex with the label `person`.
- Add a property to that vertex with the key `name` and the value `Dave`.
- Execute the steps and return the first (next) item in the iterable as the result.

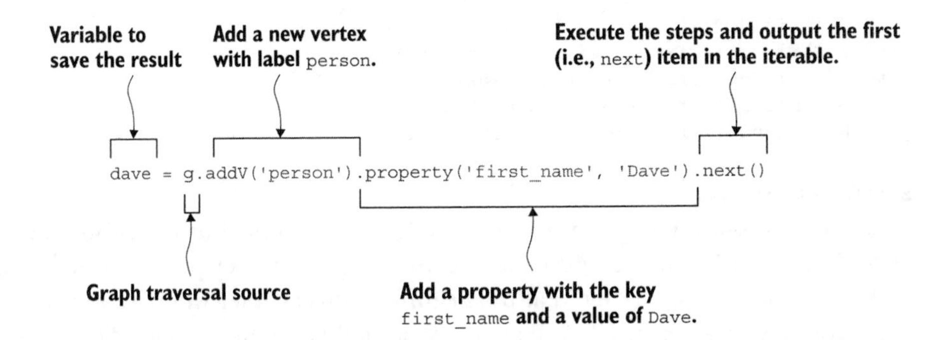

Figure 4.7 Mapping the plain text steps to their corresponding Gremlin steps to add a `person` vertex with a `first_name` of `Dave`

First, `dave` is a variable to which we're assigning the output of the traversal. In this case, it's a reference to the vertex that can be used later in the script.

> **Gremlin Console and Groovy**
>
> One of the features of the Gremlin Console is that it can work with Groovy constructs. Groovy (http://www.groovy-lang.org/) is a superset of the Java programming language. Technically, the Gremlin Console is a Groovy REPL (read–eval–print loop). As a Groovy REPL, it has the ability to assign the output of statements to variables without having to declare the type of those variables. Isn't that *groovy*?

Variables are another source of variance within the graph world. Other query languages, such as Cypher, don't support these across requests, and even TinkerPop-enabled graph databases have varying levels of support. For example, neither Azure's CosmosDB nor Amazon Neptune has this functionality, while JanusGraph and DataStax Enterprise Graph fully support it. If the query language and database support variables, then we recommend using these because variables can simplify some operations quite a bit, such as chaining together units of work like adding vertices and edges, as we do in this chapter.

The second difference with our script in figure 4.7 is the `next()` step. This is a terminal step like `iterate()`. We can think of it as a step that forces evaluation of the traversal, so another step we can add to our tool belt is

- `next()`—A terminal step that takes the iterable traversal source composed from the previous steps, iterates it once, and returns the first or next item in the iterable.

Because Gremlin is lazily evaluated, we need to iterate our traversal in order to get a result. Otherwise, all we have is an iterable that contains the desired result but isn't of any use until it iterates.

> **The Gremlin Console and terminal steps**
>
> The Gremlin Console, provided by the Apache TinkerPop project, is a nifty tool for using Gremlin and for interacting with an in-memory graph (e.g., TinkerGraph) or with a server (e.g., Gremlin Server). In fact, it's so helpful that it *automatically* iterates the results.
>
> Each step in Gremlin takes a traversal source and returns a traversal source, which is a type of iterable. Think of an iterable as a package that contains results. What we want are just the results, but what we get is a package containing the results. Gremlin Console is like an elf that cheerfully unwraps that package by giving just the results. In other words, Gremlin Console automatically iterates the results for us. This is so critical that we're going to repeat it again with emphasis: *the Gremlin Console automatically iterates the results!*

> ***(continued)***
>
> This is all well and good until we work with Gremlin without the use of a Gremlin Console (such as in chapter 6, when we write our application). In that case, we need to unwrap the package ourselves in order to get the results. We use terminal steps such as `next()` for that unwrapping.
>
> Throughout this and the next couple of chapters, we'll omit the terminal steps to enhance readability, and because the Gremlin Console automatically iterates the results for us. However, when we assign the results to a variable, we have to include a terminal step. Otherwise, the iterable gets assigned to the variable, not the result.

The script runs this traversal after starting the Gremlin Console, and there are two results. First, the graph has a newly added vertex. Second, the Gremlin Console now has the variable dave assigned to the output of the traversal; in this case, a reference to the added vertex. The full traversal can be rerun, but it creates another vertex and assigns it to the existing dave variable (this is not an idempotent operation):

```
dave = g.addV('person').property('first_name', 'Dave').next()
==>v[16]
```

We can check to see what the dave variable contains. Simply enter it at the prompt:

```
dave
==>v[16]
```

We now have a reference to the newly added vertex stored in a variable. Looking at the next three lines of the script, we see the same pattern for additional vertices:

```
josh = g.addV('person').property('first_name', 'Josh').next()
ted = g.addV('person').property('first_name', 'Ted').next()
hank = g.addV('person').property('first_name', 'Hank').next()
```

Running the script up to this point results in a graph with four vertices and no edges, as shown in figure 4.8.

Let's take a look at the next line to see how to use the variables when adding edges:

```
g.addE('friends').from(dave).to(ted)
```

Written in English, the traversal is

- Given a traversal source `g`
- Add a new edge with the label `friends`
- From the vertex referenced by the dave variable
- To the vertex referenced by the `ted` variable

As figure 4.9 depicts, we can see how the plain text steps map to the code in Gremlin.

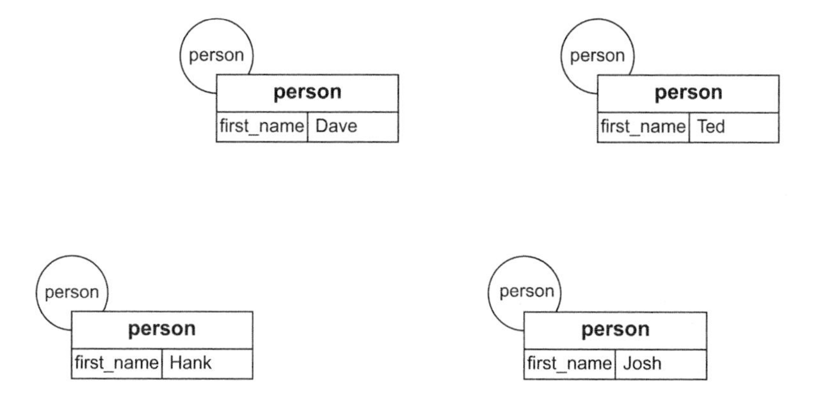

Figure 4.8 The graph our script creates if we stop it after adding the four vertices

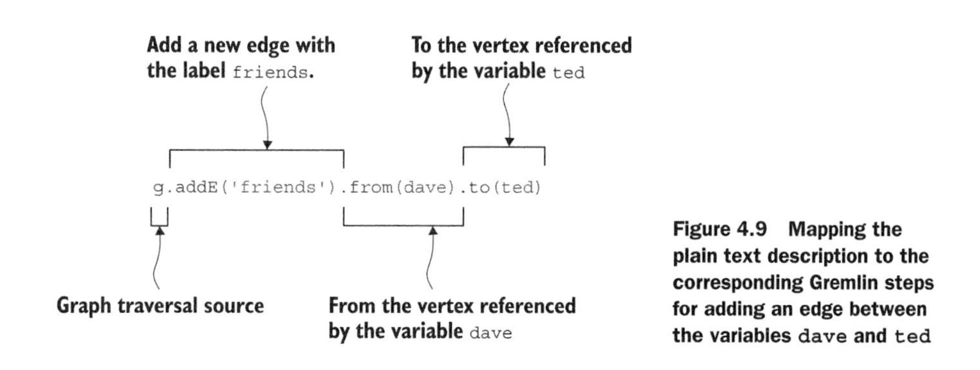

Figure 4.9 Mapping the plain text description to the corresponding Gremlin steps for adding an edge between the variables `dave` and `ted`

This looks similar to the earlier `addE()` traversal we wrote, except that instead of having to find the `Ted` and `Hank` vertices as we did previously, we reference the variables we assigned earlier in the script. This ability to reference vertices more clearly later in the script is why we create the variables in the first place. Also, with larger graphs, it can be more performant to have the variable in memory than to do repeated lookups or searches. Finally, the remainder of the script adds the last of the edges required to get the graph in figure 4.10.

We are able to write scripts that add larger amounts of data, all at the same time. However, this only works if our database allows for variables. How do we accomplish the same thing if it does not support variables?

CHAINING MUTATIONS

In graph databases, mutations can be chained together to perform multiple changes simultaneously. In the last section, we saw how to script multiple individual mutation traversals to create a graph. We chose this approach to simplify the script for teaching

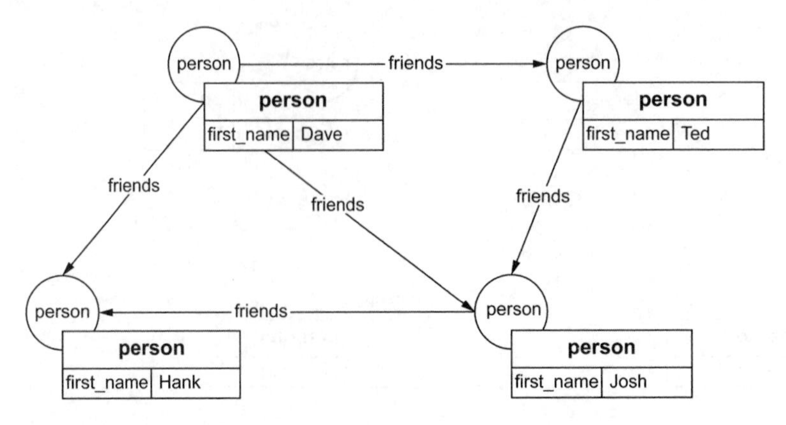

Figure 4.10 The graph generated by the `4.1-simple-social-network.groovy` script

purposes. While sufficient, it is also possible to add all of the edges with a single traversal by chaining together multiple mutations.

Whether across databases, and even within TinkerPop-enabled databases, the way a mutation traversal performs depends on the vendor's implementation. In some databases, chaining multiple mutations allows these to act atomically; that is, as a single logical operation. This is similar to a transaction in a relational database, where either all or none of the mutations can be made. In other databases, each mutation is handled independently even though chained together in a single traversal. Check with the database vendor to understand how this works with their specific implementation. Let's see how the mutations to add the remaining edges look if, instead, we chained these together in one traversal:

```
g.addE('friends').from(dave).to(josh).
  addE('friends').from(dave).to(hank).
  addE('friends').from(josh).to(hank).
  addE('friends').from(ted).to(josh).iterate()
```

Written in English, the traversal is

- Given a traversal source `g`
- Add a new edge with the label `friends` from `dave` to `josh`.
- Add a new edge with the label `friends` from `dave` to `hank`.
- Add a new edge with the label `friends` from `josh` to `hank`.
- Add a new edge with the label `friends` from `ted` to `josh`.
- Now apply all of these changes and don't return anything.

Take a look at figure 4.11. There we see code in a single traversal with the steps annotated.

```
g.addE('friends').from(dave).to(josh).                  <-- Adds an edge from dave to ted
  addE('friends').from(dave).to(hank).                  <-- Adds an edge from dave to hank
  addE('friends').from(josh).to(hank).                  <-- Adds an edge from josh to hank
  addE('friends').from(ted).to(josh).iterate()          <-- Adds an edge from ted to josh and
                                                            iterate the full traversal
```

Figure 4.11 Mapping the chaining together of multiple mutations with Gremlin

In figure 4.11, notice how each line (except the last line) ends with a period. Each of these `addE()` steps was chained together. All of this was executed within a single traversal starting from one traversal source.

> **IMPORTANT** For composing complex operations into a single statement, chaining steps is a fundamental strategy in Gremlin, as well as with other query languages. The concept is that each step takes in data passed to it from the previous step, performs work on the data, and passes it on to the next step. Gremlin is able to do this because every step takes as input an iterable `GraphTraversal`, and nearly every step emits as its output a `GraphTraversal`. For those with functional programming experience, this should all be quite familiar.

This ability to chain multiple statements together into a single query isn't something that's possible, as far as these authors are aware, in SQL or other graph query languages such as Cypher. In a relational database, this would be equivalent to running multiple INSERT statements in a single statement or using multiple common table expressions (CTEs) to combine several complex operations together. In SQL when you submit multiple statements at the same time, separated by a semicolon, this is executed as multiple independent queries that run in sequence. On the other hand, the Gremlin shown in figure 4.11 is a single traversal: it contains multiple mutations, all of which are executed at the same time.

4.1.4 *Extending our graph*

Remember that back at the beginning of this chapter, we mentioned that we needed to extend our graph to add additional data before we begin our work with paths. So far, we've learned to create vertices, edges, and properties; however, we only added the same four vertices and five edges that we used in chapter 3. Because we now have all the tools to add data to a graph, let's put those to use. In figure 4.12, we highlighted the data that still needs to be added before we can move on to understanding paths.

> **EXERCISE** Write the necessary Gremlin traversals to add the highlighted data in figure 4.12 to the graph.

We hope you took the time to try and work this out for yourself before taking a look at our answer. Either way, the following shows the set of traversals we wrote to add the

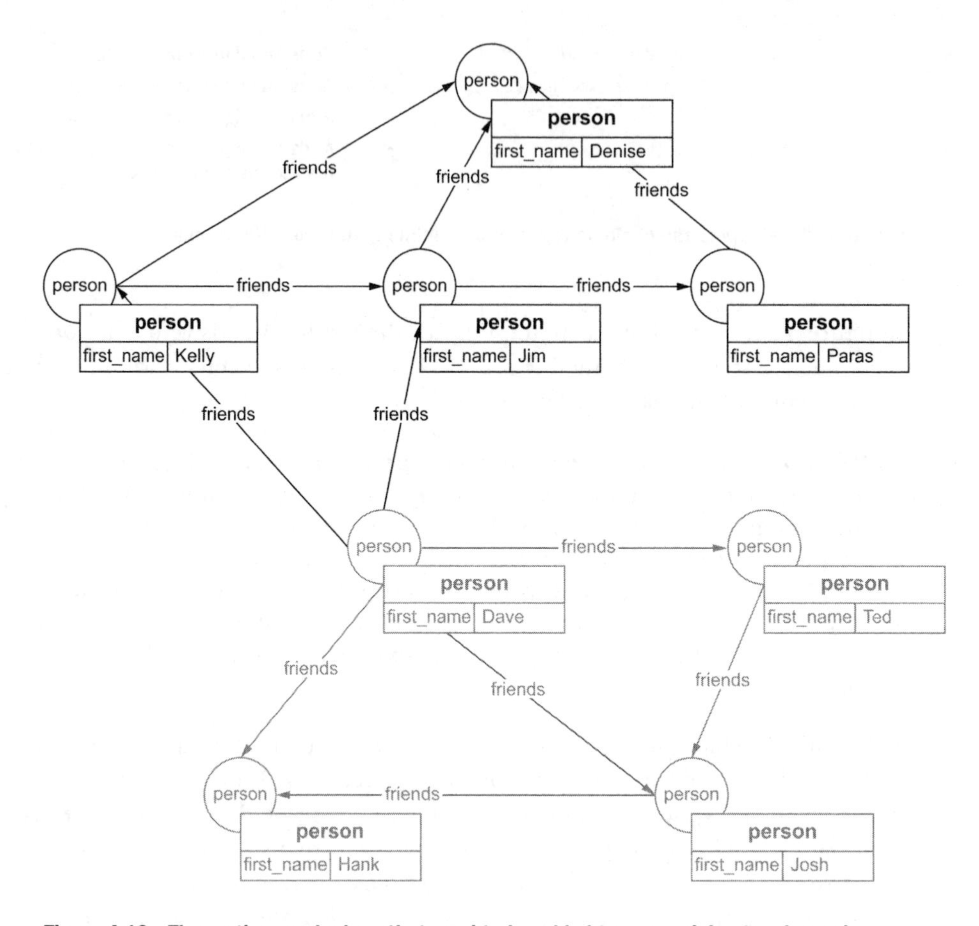

Figure 4.12 The vertices and edges that need to be added to our social network graph

essential elements to our graph. We can add this data to our graph by restarting the Gremlin Console with the script

```
chapter04/scripts/4.2-complex-social-network.groovy
```

or by using the following commands:

```
//Adds a person vertex with a name of Kelly and saves it to a variable
kelly = g.addV('person').property('first_name', 'Kelly').next()

//Adds a person vertex with a name of Jim and saves it to a variable
jim = g.addV('person').property('first_name', 'Jim').next()

//Adds a person vertex with a name of Paras and saves it to a variable
paras = g.addV('person').property('first_name', 'Paras').next()

//Adds a person vertex with a name of Denise and saves it to a variable
denise = g.addV('person').property('first_name', 'Denise').next()
```

```
//Adds additional friends edges
g.addE('friends').from(dave).to(jim).
  addE('friends').from(dave).to(kelly).
  addE('friends').from(kelly).to(jim).
  addE('friends').from(kelly).to(denise).
  addE('friends').from(jim).to(denise).
  addE('friends').from(jim).to(paras).
  addE('friends').from(paras).to(denise).iterate()
```

Figure 4.13 shows the graph after running these traversals.

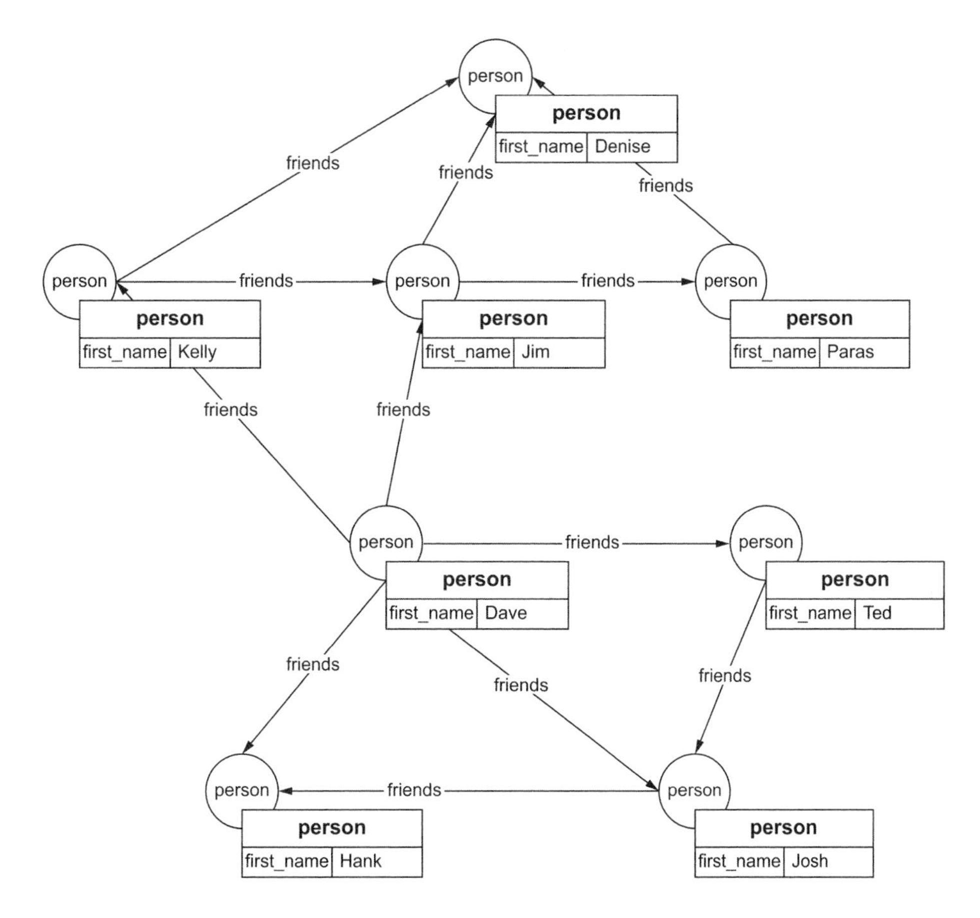

Figure 4.13 Our social network graph with the additional information added (a total of 8 vertices and 12 edges)

Whew, it took a while to get all the necessary data added to our graph! Now that we have enough data in our graph, let's extend what we learned in the last chapter about navigating around a graph to work with paths.

In the next section, we'll use the graph in figure 4.13, so we need to make sure it is loaded into database. If you have followed along and typed in everything correctly, then you're already there. If not, or if you want to ensure that your graph is correct, quit the Gremlin Console with a `:q` and run this command:

```
bin/gremlin.sh -i $BASE_DIR/chapter04/scripts/4.2-complex-social-network.groovy
```

4.2 Paths

In this section, we cover paths in depth. Paths offer a description of the series of steps that a traverser takes to get from the start vertex to the end vertex. This means that not only can we find out which two vertices are connected, as shown in chapter 3, but we can also determine exactly how to get from the start to the end. It's reasonable to think of a path in a graph in much the same way as we consider GPS driving directions in a mapping application. We enter a starting location and an ending location, and we get back the series of turns required to move from start to finish.

When working with path algorithms, we begin by specifying a start vertex, an end vertex, and which edges to traverse between the two. The traversal returns all possible sets of directions that go from the start vertex to the end vertex. Figure 4.14 is an example of paths inside a simple graph.

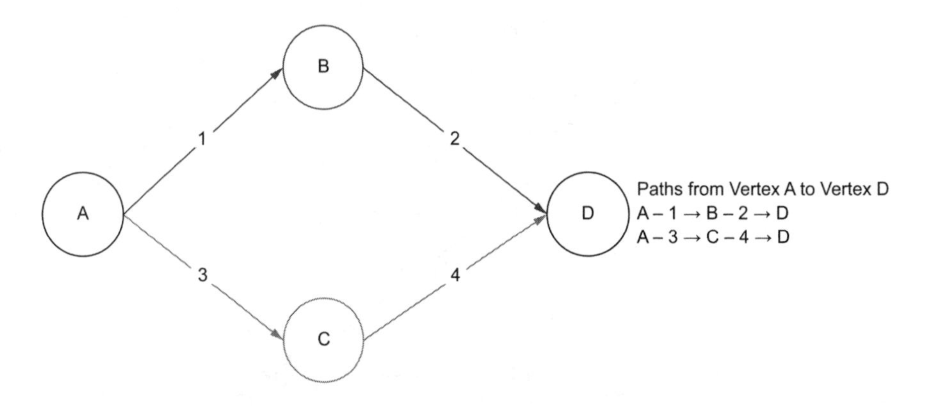

Figure 4.14 A simple example of all the paths from our start vertex `A` to the end vertex `D` in a graph

Let's say that we wanted to find which set of friends Ted needs to go through to get introduced to Denise. To accomplish this task, we need to

1. Find the `Ted` vertex.
2. Traverse across each incoming and outgoing `friends` edge.
3. Check to see if the vertex we're on is the `Denise` vertex.

4 Repeat steps 2 and 3 until we reach the `Denise` vertex.
5 Return the path—the series of vertices and edges we traversed to get from Ted to Denise.

This looks much like the recursive looping traversals we wrote in the last chapter. Using that knowledge, let's write a recursive looping traversal to move from Ted to Denise in our graph, which appears to solve steps 1–4, but which introduces some other problems.

This is where most of us go with our first recursive loop through highly connected data, and we often discover that the data has more connections than we anticipated. This next traversal generates an error. We'll explain the error after addressing another one of its deficiencies:

```
g.V().has('person', 'first_name', 'Ted').
  until(has('person', 'first_name', 'Denise')).
  repeat(
    both('friends')
  )
```

What this traversal doesn't provide is the list of vertices and edges, or *path*, from Ted to Denise. To retrieve this information, we need to introduce another step:

- `path()`—Returns the history of the vertices (and optionally the edges) a specific traverser visits as the traversal runs

NOTE Using the `path()` step in Gremlin requires additional resource overhead on the server because each traverser needs to maintain the entire history of the steps it visits. For performance reasons, only use `path()` when you desire the full path data.

Adding the `path()` step to the end of our traversal makes it look like this:

```
g.V().has('person', 'first_name', 'Ted').
  until(has('person', 'first_name', 'Denise')).
  repeat(
    both('friends')
  ).path()
```

And running this traversal in the Gremlin Console returns this:

```
Script evaluation exceeded the configured 'scriptEvaluationTimeout' threshold
     of 30000 ms or evaluation was otherwise cancelled directly for request
     [g.V().has('person', 'first_name', 'Ted').
  until(has('person', 'first_name', 'Denise')).
  repeat(
    both('friends')
  ).path()]
Type ':help' or ':h' for help.
```

Well, that's not good. We see that our traversal timed out, but why? What we just accidently tripped over is a cycle in our graph. Not only that, there's a fair chance we got the fan spinning in our laptop, and it's even possible that Gremlin Console died (or was killed? Wow, this got morbid really quickly).

> **Stuck in Gremlin Console**
>
> If we forget a parenthesis or somehow get stuck in a Gremlin Console at a prompt, try using the `:clear` command to clear the buffer and start over on the traversal.
>
> If you lost your Gremlin Console, we're sorry. But we'd like to point out that because your Gremlin Server is still running, you haven't lost any data. Just restart your Gremlin Console and then reconnect to the Gremlin Server. You can use these two `:remote` commands in a Gremlin Console. To connect to the running server, type
>
> ```
> :remote connect tinkerpop.server conf/remote.yaml session
> ```
>
> To connect to the Gremlin Server and send your commands to the server, type
>
> ```
> :remote console
> ```
>
> If you look at this chapter's scripts, you will notice that these always start with those commands. Now let's get back to the material at hand—paths.

4.2.1 Cycles in graphs

The root cause of all of this death and destruction (or minor inconvenience of electrons and excessive use of fans) is a concept in graph theory known as a *cycle*. A cycle is a path of vertices and edges in a graph that contains one or more vertices that are reachable from themselves, as figure 4.15 illustrates.

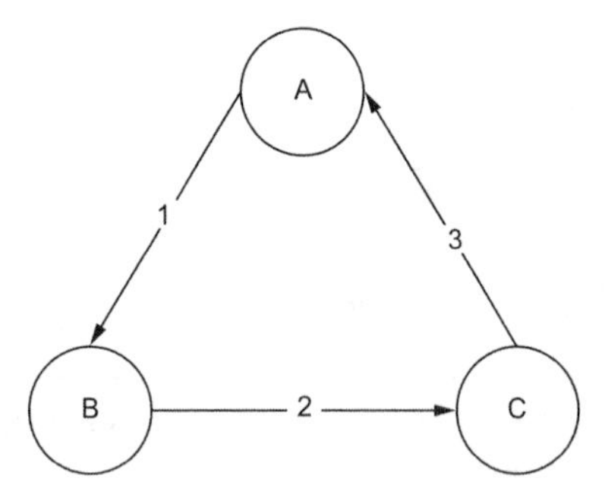

Figure 4.15 A simple graph that has a cycle from A back to itself

In the example graph shown in figure 4.15, we see that by traversing the edges, each vertex (`A`, `B`, `C`) can be reached from itself. Vertex `A` could be reached by traversing [`A → 1 → B → 2 → C → 3 → A`] or by [`A → 3 → C → 2 → B → 1 → A`]. Vertex `B` and vertex `C` can also be accessed from themselves using similar traversal patterns. By applying this knowledge of cycles to our previous graph, we see that there are multiple cycles within our graph, one of which is highlighted in figure 4.16.

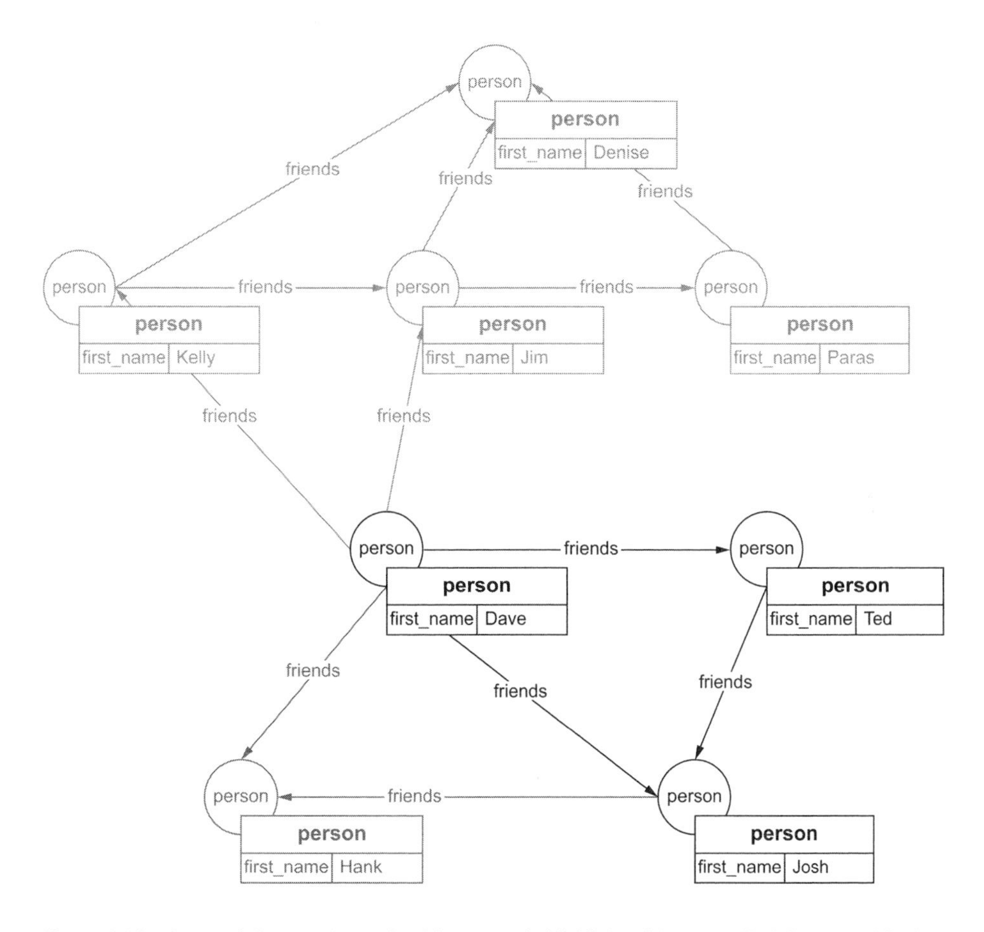

Figure 4.16 Our social network graph with one cycle highlighted between Ted, Dave, and Josh

From this image, we see that there's a cycle running between Ted, Dave, and Josh. This isn't unusual; cycles in a graph are common. However, as demonstrated, cycles cause problems when running traversals if we don't account for these. Some traversers can get caught up in an endless loop, which lead to a timeout. So how do we write a traversal without getting stuck in an endless loop?

4.2.2 Finding the simple path

In graph theory, there's a concept known as the *simple path*. A simple path is a path that doesn't repeat any vertices, meaning that we only get results that are not cyclical. When looking for the simple path between two vertices, each traverser maintains a history of all the items it visits. If it comes across an item it has already visited, it knows it's in a cycle and removes itself. Only traversers pursuing paths devoid of cycles continue to completion.

This sounds exactly like what we need in order to find our path from Ted to Denise without blowing up our CPUs. To update our traversal to find the simple path, we introduce another Gremlin step:

- `simplePath()`—Filters out traversers that visit the same vertex more than once

Using this step, we can update our traversal to find the simple path between Ted and Denise. We do this by adding the `simplePath()` step within our `repeat()` step and running it in Gremlin Console:

```
g.V().has('person', 'first_name', 'Ted').
  until(has('person', 'first_name', 'Denise')).
  repeat(
    both('friends').simplePath()
  ).path()

==>path[v[4], v[0], v[15], v[19]]
==>path[v[4], v[0], v[13], v[19]]
==>path[v[4], v[2], v[0], v[15], v[19]]
==>path[v[4], v[2], v[0], v[13], v[19]]
==>path[v[4], v[0], v[15], v[17], v[19]]
==>path[v[4], v[0], v[15], v[13], v[19]]
==>path[v[4], v[0], v[13], v[15], v[19]]
==>path[v[4], v[2], v[6], v[0], v[15], v[19]]
==>path[v[4], v[2], v[6], v[0], v[13], v[19]]
==>path[v[4], v[2], v[0], v[15], v[17], v[19]]
==>path[v[4], v[2], v[0], v[15], v[13], v[19]]
==>path[v[4], v[2], v[0], v[13], v[15], v[19]]
==>path[v[4], v[0], v[13], v[15], v[17], v[19]]
==>path[v[4], v[2], v[6], v[0], v[15], v[17], v[19]]
==>path[v[4], v[2], v[6], v[0], v[15], v[13], v[19]]
==>path[v[4], v[2], v[6], v[0], v[13], v[15], v[19]]
==>path[v[4], v[2], v[0], v[13], v[15], v[17], v[19]]
==>path[v[4], v[2], v[6], v[0], v[13], v[15], v[17], v[19]]
```

Why do we add the `simplePath()` within the `repeat()` step instead of at the end? To see if we're in a cycle, we evaluate both our current position in the graph as well as our historical path through the graph at the end of each loop's iteration. If we put `simplePath()` at the end of the traversal, which is outside of the looping logic, then we have traversers that are stuck in cycles with no way to break out. This is analogous to creating a `for` loop in Java that iterates the counter variable outside of the `for` loop. With this addition, we now see all the different simple paths from Ted to Denise.

Note that we only see the vertices that connect the start and end vertex. We stated earlier that paths return the *vertices and edges* between the start and end vertex. In the next section, we show how to include edges in the results, but first we must introduce some additional capabilities for traversing edges, capabilities that also open up new ways to filter edges. For those of you interested in the details (e.g., the names) of the returned vertices in the paths, we'll cover that and other formatting functionality in chapter 5.

4.3 Traversing and filtering edges

To get the edge information as part of the path, we go from vertex to edge to vertex. We must be explicit about stepping onto the edge and then stepping off of the edge. These operations are rolled up in the traversing steps we introduced in the last chapter: `in()`, `out()`, and `both()`. Now we introduce additional steps to break these out.

Our simple social graph has only modeled friendships, but our social circle often includes our professional connections as well. In this section, we take a small detour from the DiningByFriends model to discuss how to traverse and filter edges. We temporarily extend our graph with some additional edges to demonstrate these concepts. At end of this section, we include the edge information in the returned paths.

For this section only, we'll use a specific graph, as illustrated in figure 4.17. To set your local graph up with the correct data, exit the Gremlin Console with `:q` and restart it with this command:

```
bin/gremlin.sh -i $BASE_DIR/chapter04/scripts/4.3.1-complex-social-network-
     with-works-with-edges.groovy
```

This loads the graph with the same eight vertices and the `friends` edges, but also with additional `works_with` edges.

4.3.1 Introducing the E and V steps for traversing edges

Hypothetically, let us say that many of the people in our graph also worked together at one time or another. One way to model this is to add a new relationship between people called `works_with` and give it properties to track the start and end year of those relationships. By including these edges in our graph, we get a graph like figure 4.17.

Let's say that we want to answer a question like, "Who did Dave work with before the job he started in 2018?" Based on the new version of the graph, how would we go about finding this information?

- Given a traversal source `g`
- Find the vertex with the key of `first_name` and the value of `Dave`.
- Traverse the `works_with` edges that have a `start_year` that's less than or equal to `2018`.
- Traverse to the adjacent vertex.
- Return the `first_name`.

We already know how to do some of these steps, specifically steps 1, 2, and 5. What we're missing is how to traverse from a vertex stop on the edge, filter based on a

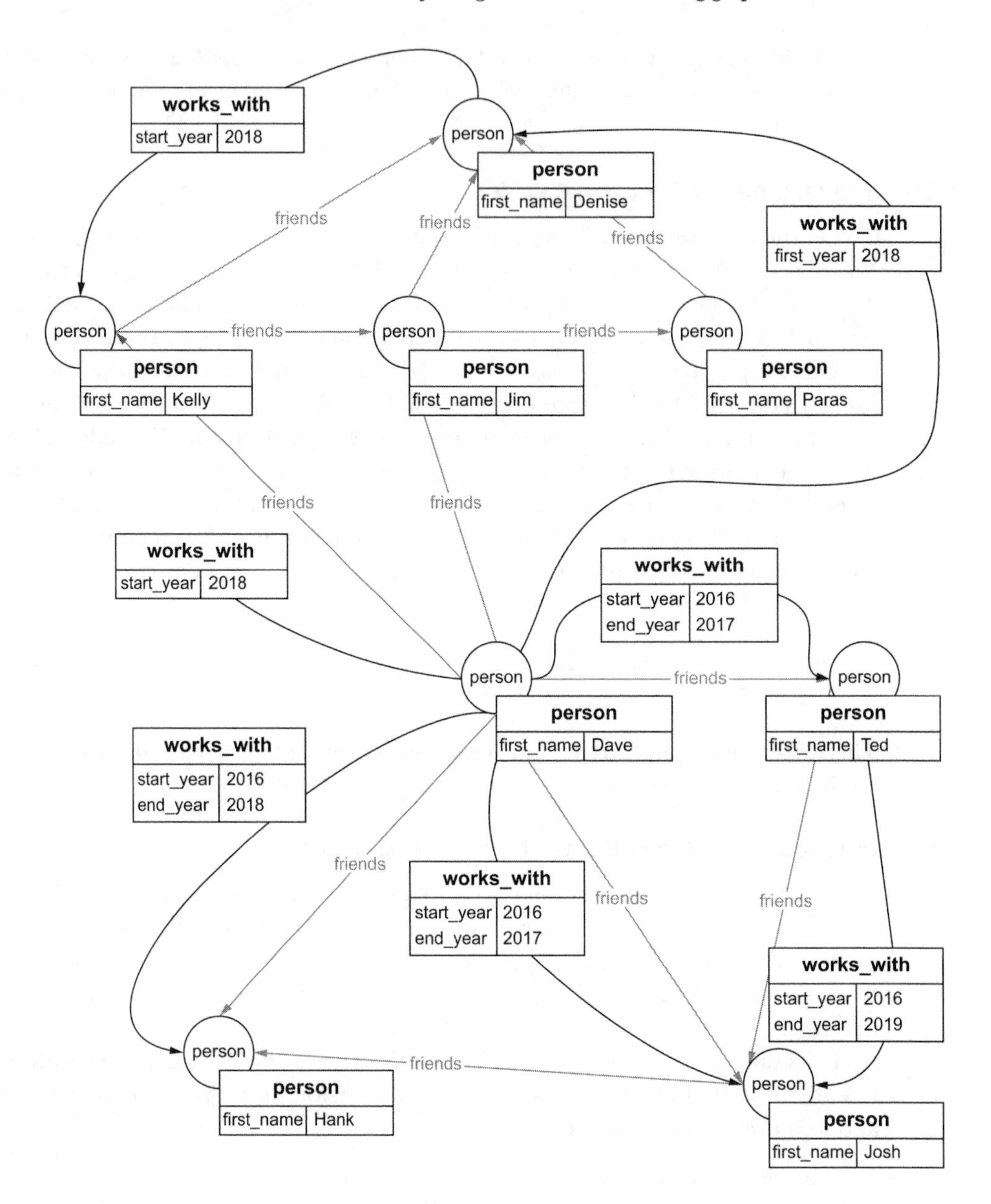

Figure 4.17 An extended version of our social network graph with the `works_with` edges highlighted

property on that edge, and then go to the adjacent vertex. The key is to traverse not from a vertex to the incident vertex, but to stop on the edge itself, look around a bit, then traverse to the next vertex. Figure 4.18 shows how to map these steps to the corresponding steps in Gremlin.

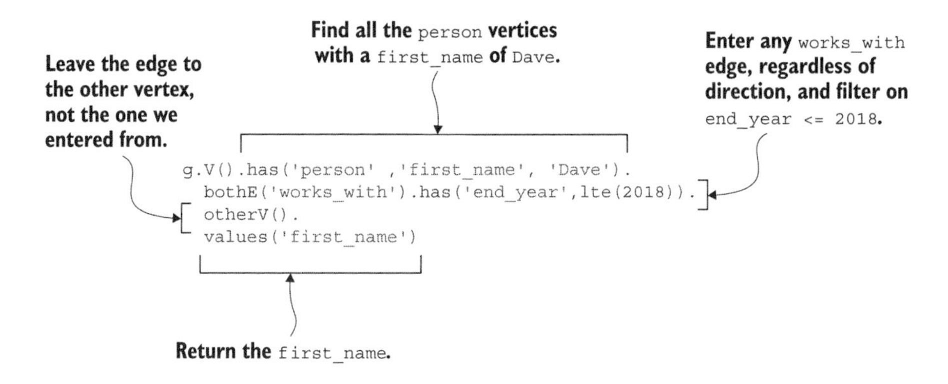

Figure 4.18 Mapping the plain text description to the coded Gremlin steps for filtering and traversing our edges

In the traversal in figure 4.18, we see one of several Gremlin steps specifically created for working with edges, and these steps all end with an `E`. The three `E` steps are

- `inE(label)`—Traverses from the current vertex onto the *incoming* incident edges. If a `label` is specified, then filters to only traverse to edges of that type.
- `outE(label)`—Traverses from the current vertex onto the *outgoing* incident edges. If a `label` is specified, then filters to only traverse to edges of that type.
- `bothE(label)`—Traverses from the current vertex onto the incident edges, *regardless of direction.* If a `label` is specified, then filters to only traverse to edges of that type.

These steps each start on a vertex, traverse to an edge, and stop on the edge. This is a bit different than the `in()`, `out()`, and `both()` traversal steps we learned in chapter 3 because these steps end with us located on the edge instead of on the adjacent vertex as demonstrated in figure 4.19.

The location at the end of the step is the crucial difference between `out()` and `outE()`. This leaves us with another question: "How do we get back to the vertex?" To do that, Gremlin provides companion `V` steps to accompany the `E` steps:

- `inV()`—Traverses from the current edge to the incoming vertex. It's commonly paired with the `outE()` step.
- `outV()`—Traverses from the current edge to the outgoing vertex. It's commonly paired with the `inE()` step.
- `otherV()`—Traverses to the vertex that isn't the vertex that's used to traverse onto the edge (e.g., the other vertex). It's commonly paired with the `bothE()` step.
- `bothV()`—Traverses from the current edge to both of the incident vertices. Rarely used.

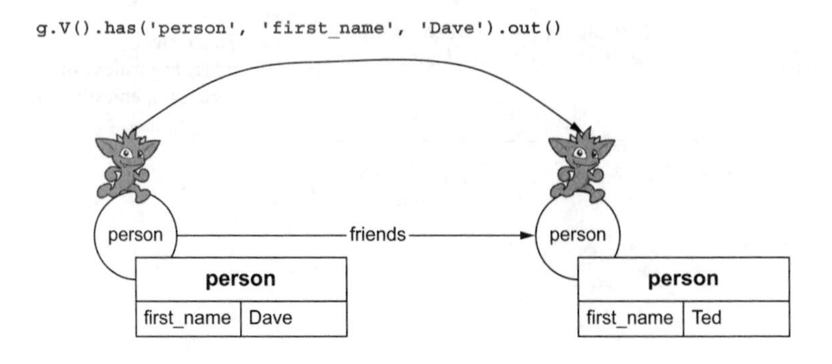

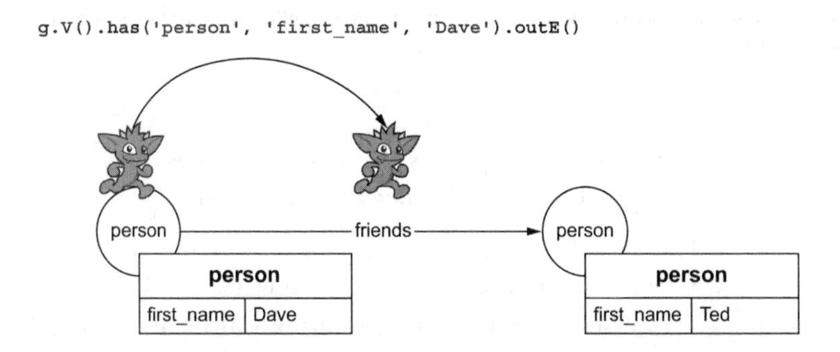

Figure 4.19 Comparing the behavior of the `out()` step, which ends on a vertex, to the `outE()` step, which ends on the edge

In each of these V steps, we notice that there's no input or modification. That's because these are designed to be paired with an E step, usually the opposite one. For example, `inE()` combines with an `outV()` or `outE()` combines with `inV()` to complete the traversal to the adjacent vertex.

In the case of `bothE()`, we might be tempted to use `bothV()`, but that would be a mistake. If we used a `bothV()`, we would end up with two traversers: one on the start vertex and one on the end vertex, as shown in figure 4.20.

With `bothE()`, the best choice is `otherV()`, which simply takes us to "the other vertex," which isn't the one we came from when traversing the edge, as shown in figure 4.21.

NOTE The use of `otherV()`, while common, does incur some performance overhead as each traverser needs to retain state that contains the originating vertex. If performance is critical for a specific traversal, then it's best to avoid the use of `otherV()`, assuming that the traversal can be written another way.

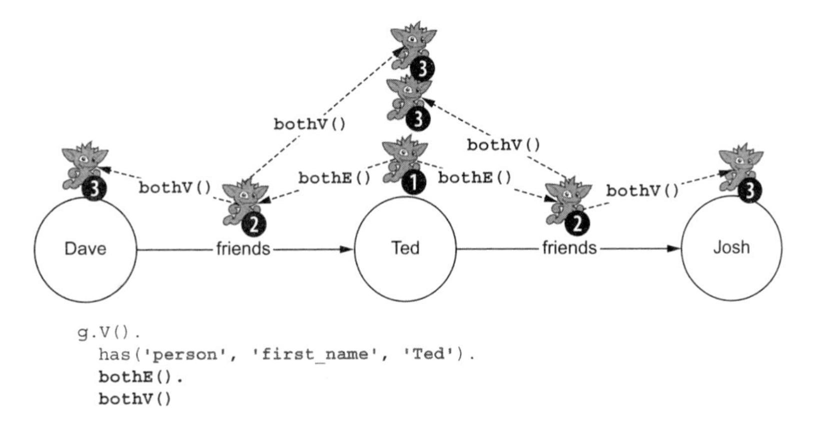

Figure 4.20 Demonstrating that the `bothE()` and `bothV()` combination ends on both the start and end vertex of the edge

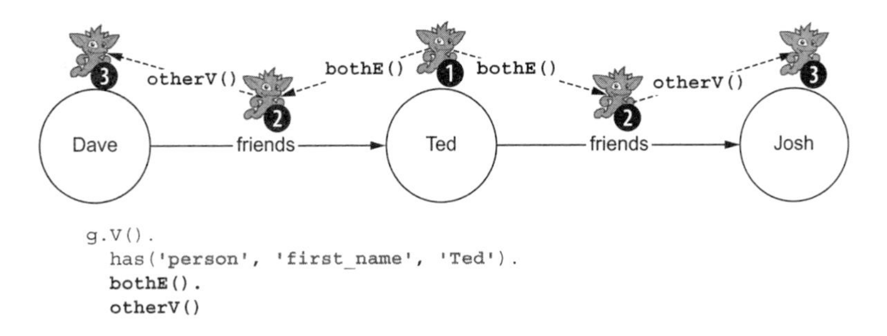

Figure 4.21 Demonstrating that the `bothE()` and `otherV()` combination ends on the opposite vertex of the edge

Now let's put this into practice and obtain the names of all of Dave's co-workers in the graph. To do that, we use the following traversal:

```
g.V().has('person','first_name','Dave').
  bothE('works_with').otherV().values('first_name')
==>Ted
==>Josh
==>Hank
==>Kelly
==>Denise
```

Note that because we didn't perform any actions on the edge, it's more succinct to use the following instead:

```
g.V().has('person', 'first_name','Dave').
  both('works_with').values('first_name')
==>Ted
==>Josh
==>Hank
==>Kelly
==>Denise
```

Why would we ever use the `E` and `V` steps (that Apache TinkerPop calls *vertex steps*) if can we do everything with the `in()`, `out()`, and `both()` steps? We find that there are three common use cases for the `E` and `V` steps, and we'll discuss each of these cases in the following sections:

- Filtering on edge properties using the `has()` filtering method from chapter 3
- Including edges in `path()` results
- Performant edge counts and denormalization

4.3.2 Filtering with edge properties

Filtering with edge properties usually comes in two flavors: time-based filters or weight-based filters. With the addition of the `works_with` edges, we create a simple time-versioned graph that allows us to traverse edges based on a provided time input. Weight-based filtering is another common pattern used in performing analytics and algorithms for full-graph processing.

For our purposes, let's used time-based filtering to find out who Dave worked with before the job he started in 2019. Looking at our graph in figure 4.17, it looks like Dave changed jobs in 2018, but not after that, so we look for cases where the `end_year` is less than or equal to `2018`. Let's break this traversal down into steps:

1. Find the `Dave` vertex.
2. Traverse onto the `works_with` edge ignoring direction.
3. Filter on where the `end_year` property is less than or equal to `2018`.
4. Complete the traversal of the edge with the `otherV()` step to the adjacent vertex.
5. On the adjacent vertex, return the value of the `first_name` property.

Let's code these in a traversal. Figure 4.22 shows how to transform these steps into a language that Gremlin understands.

When we run this traversal, we get the following results:

```
g.V().has('person','first_name','Dave').
  bothE('works_with').has('end_year',lte(2018)).
  otherV().
  values('first_name')
==>Josh
==>Ted
==>Hank
```

Not only have we filtered based on a property on the edge, but we also slipped in a sweet little predicate step: `lte()`, which stands for "less than or equal to." Predicate

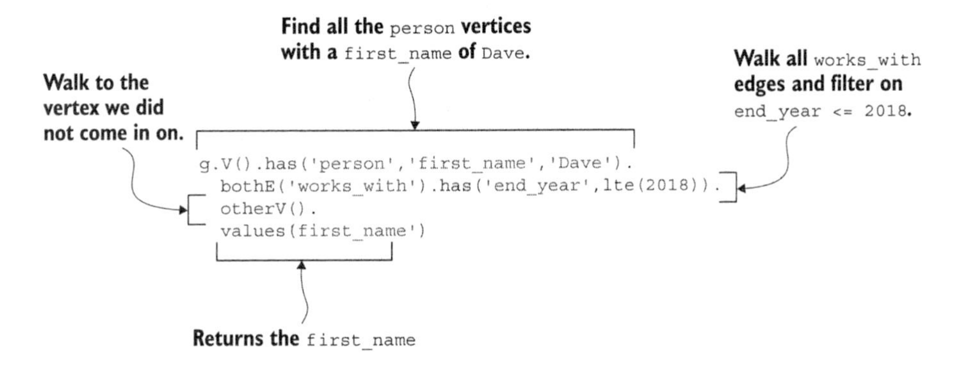

Figure 4.22 Adding the edge filtering steps to the code of the Gremlin traversal

steps are handy tools for managing flow control with more sophistication than the simple value matching we've used up to this point. You can find a complete list of predicate steps on the Apache TinkerPop reference site (http://mng.bz/mNgn).

4.3.3 Include edges in path results

Returning to our unanswered question from section 4.2.2, "When using the `path()` step, how do we also include the edges traversed?" We need to use the `bothE().otherV()` traversal pattern to explicitly traverse on to the edge. Let's find the paths from Ted to Denise, but limit the path to only use the `works_with` edge:

```
g.V().has('person', 'first_name', 'Ted').
  until(has('person', 'first_name', 'Denise')).
  repeat(
    bothE('works_with').otherV().simplePath()
  ).path()

==>path[v[4], e[29][0-works_with->4], v[0], e[33][0-works_with->19], v[19]]
==>path[v[4], e[29][0-works_with->4], v[0], e[32][0-works_with->13], v[13],
     e[34][19-works_with->13], v[19]]
==>path[v[4], e[30][2-works_with->4], v[2], e[28][0-works_with->2], v[0],
     e[33][0-works_with->19], v[19]]
==>path[v[4], e[30][2-works_with->4], v[2], e[28][0-works_with->2], v[0],
     e[32][0-works_with->13], v[13], e[34][19-works_with->13], v[19]]
```

As expected, the previous code and results show four paths to get from Ted to Denise, and each of the results includes the edges traversed as well as the vertices. This use of the `path()` step is common when the edges contain important domain details. A good example for this is air traffic routes, where the vertices signify airports and the edges represent the flights between them. In such cases, it's vital to return flight details such as airline name, flight number, departure time, and arrival time.

4.3.4 Performant edge counts and denormalization

We think that performance optimizations should only be applied after core functionality is established. We consider *established* to mean that the functionality is working, with good test coverage, and deployed with a production-similar data set. For that reason, we'll only discuss these points briefly because we'll cover performance optimization in chapter 10.

Recall how in the previous chapter we described the conceptual perspective of a Gremlin sitting on a vertex as being in an escape room with

- *A chest of drawers with labels on them*—These are the vertex properties.
- *A series of doors also with labels*—These are the incident edges.
- *Each door has sets of drawers with labels*—These are the edge properties.

We use this analogy to emphasize that vertex properties, edges, and edge properties are essentially local to a vertex, so the cost to use these is basically free. But everything outside of this room can only be accessed by traversing an edge (walking through a door in our mental model) to get to another vertex. Depending on the implementation, accessing anything outside the room could involve a cache hit, a disk operation, or possibly a network call.

Due to this additional cost, when possible, don't traverse an edge to the other vertex. This means is that if we stay in the current room (e.g., remain on the current vertex), we avoid those additional cache hits, disk operations, and network calls. The `E` steps allow us to do this. Using an `E` step is akin to looking at the doors without actually opening them. That's the approach of the following traversal:

```
g.V().bothE().count()
```

But counting with the `both()` step is usually a more expensive operation:

```
g.V().both().count()
```

In the first case, we count the edge doors that we see from our vertex room. In the second case, we go through each door and count the rooms (vertices) on the other side. Because there's a one-to-one correspondence between the doors (edges) and the other rooms (vertices), counting the doors returns the same value as counting the rooms.

We'll cover denormalization more thoroughly in chapter 7, but for the purpose of this discussion, *denormalization* in the graph is a matter of copying an often-accessed vertex property onto an adjacent edge. Denormalization avoids taking the cost of a full traversal when reading that property; it can be helpful for certain types of read-intensive activity. Yes, there's overhead in maintaining two copies of a property value. But maintaining multiple copies of data is denormalization, which always comes with additional maintenance overhead regardless of whether we use a relational database or a graph database. That's a quick look at a couple of performance optimizations that we employ using the `V` and `E` steps. And remember, we'll cover performance optimizations in more detail in chapter 10.

In this chapter, we first discussed how to return paths from a graph, as well as how to avoid infinite queries caused by cycles in our graph. Finally, we reviewed the `E` and `V` steps and how to do filtering of edge properties. Now that we know how to perform necessary traversals on our graph, the next chapter will explore how to manipulate the results and types of data that these traversals return.

Summary

- Adding vertices to a graph is similar to adding entities to a relational database.
- Adding edges to a graph requires that we not only add the edge but also add or identify the vertex on each end.
- Mutation operations in graph traversals allow for chaining together multiple mutation operations into a single operation, unlike SQL.
- Paths in a graph represent the series of vertices and edges that connect two elements.
- Cycles in a graph refer to a path that has repeated vertices and are a common cause of long-running recursive and pathfinding queries in graph traversals.
- A simple path is a path in a graph that does not repeat any vertices.
- Edges can be traversed to and filtered on directly, without having to traverse to the adjacent vertex.

5 Formatting results

This chapter covers

- Retrieving values from our vertices and edges
- Aliasing vertices and edges for later use in the traversal
- Crafting custom result objects by combining static and computed values
- Sorting, grouping, and limiting our results

Finding data within the graph is one skill, but returning it efficiently presents a whole new set of challenges. While it's entirely possible to send raw, unorganized data to the client, in most cases, it's best to do as much data processing at the database layer as possible. Client applications are quite busy handling user interactions.

In this chapter, we'll focus on the different methods of collecting, formatting, and outputting traversal results at the database level. We'll review the value steps introduced in chapter 3 and illustrate why these are required. Then we'll discuss how to return values from elements that are located in the middle of a traversal, as well as crafting custom objects. Finally, we'll wrap up this chapter by demonstrating how to sort, group, and limit results for efficient communication with client applications.

If you haven't done so already, download the corresponding source code for this chapter: https://github.com/bechbd/graph-databases-in-action. The code relevant to this chapter is located in the chapter05 folder. All examples begin with the assumption that our social network data set is loaded. To accomplish this, run this command:

```
bin/gremlin.sh -i $BASE_DIR/chapter05/scripts/5.1-complex-social-network.groovy
```

5.1 Review of values steps

We start with the most common of formatting tools: the `values()` and `valueMap()` steps, which we introduced in chapter 3. But before we get to these, we look at the default behavior in TinkerPop. Most of the traversal examples return the ID of the elements. Looking at the following traversal, the ID value of `4` is returned as part of a `toString()` construct:

```
g.V().has('person', 'first_name', 'Ted')
==>v[4]
```

In relational database terms, this is the equivalent of running this SQL query:

```
SELECT ROWID FROM person WHERE first_name = 'Ted';
```

We rarely want to return just the ID. In most cases, we aim to return all or a subset of the properties. So, if the normal use case is to retrieve the attribute values, then why not return all the attributes by default?

The reason is quite practical. If a database returns all the attributes by default, we transmit a lot of unrequired data. Generally, we want to transmit only the data we need, so most databases require that we specify the particular attributes to return. We know it is an (unfortunately) common practice in SQL to use a wildcard (`*`) in the `SELECT` clause as seen here:

```
SELECT * FROM person WHERE first_name = 'Ted';
```

But the preferred method is to specify the column names like this:

```
SELECT first_name FROM person WHERE first_name = 'Ted';
```

So, what's the graph equivalent of this SQL? Let's say that we want to return all the properties for the `Hank` vertex in our graph. We already know the basic steps for this:

1. Given a traversal source `g`
2. Find vertex of type `person` with a `first_name` of `Hank`.
3. Return the properties of that vertex (in this case, there is only one, `first_name`).

At this stage, we are already old pros at handling the first two steps. Back in chapter 3, we also used the `values()` step and discussed the `valueMap()` step, often at the end of the traversal, to retrieve the properties of a vertex. Both the `valueMap()` and `values()` steps return the property values of a vertex or an edge. As you'll remember, the `values()` step returns only the *values* of the properties. The `valueMap()` step returns a map, which is a collection of key-value pairs of the specified properties. (In some programming languages, a map is known as a dictionary. This is the same concept here.)

> **Why is values() plural?**
> Although plural, `values()` is most often used to return scalars; specifically, the value of a single property. Because it returns only the value portion of the property without a key or label, the requesting code must know which property is called for. It's plural because `values()` is designed to work on one or more properties and to distinguish it from the rarely used `value()` step.

Let's take a look at the difference between the two values steps by returning all the attributes for Hank in our graph. First, using `values()`

```
g.V().has('person', 'first_name', 'Hank').values()
==> Hank
```

and then employing `valueMap()`:

```
g.V().has('person', 'first_name', 'Hank').valueMap()
==>{first_name=[Hank]}
```

As shown, while we receive the same basic data back for each of these, it's returned with slightly different formats. The `valueMap()` step differs from the `values()` step because it returns the data as a key-value pair (or map) instead of just the value. Generally, we find that having the keys for the property values makes it much easier to work with the results. Additionally, the `valueMap()` step returns one row per traverser while the `values()` step returns one row per property per traverser. In our work, we usually prefer the `valueMap()` step.

> **Empty values() steps**
> In our sample traversals, note that we don't specify the properties to return in our `values()` step. For example
>
> ```
> g.V().has('person', 'first_name', 'Hank').values()
> ```
>
> Having an empty `values()` step is generally a bad idea. It's the equivalent of running a SQL SELECT query with a wildcard like this:
>
> ```
> SELECT * FROM person WHERE first_name = 'Hank';
> ```
>
> As with SQL, although this is allowed in graph traversals, it has potentially significant drawbacks:

- The values we receive can change over time. When a new property is added to these vertices, it automatically gets included in the results.
- There's no guarantee of ordering the properties with graph databases.
- We can end up getting significantly more data than required, slowing down the application.

For these reasons, as in SQL, it's a best practice to always specify the properties to return with a comma-separated list of property keys as illustrated here:

```
g.V().has('person', 'first_name', 'Hank').values('first_name')
```

Now that we've reviewed the `values()` and `valueMap()` steps, why are these needed at all? In SQL we don't do anything extra to get the values, so why do we need to do that with graph databases? It's necessary because of a crucial difference between how SQL engines process queries and how graph database engines process traversals:

- In a graph database, only the values of the current vertices or edges are retrieved.
- In a relational database, all the values from all the joined tables can be included in the results.

This difference arises from the disparity in how the engines process queries. Understanding this difference is critical for creating effective and efficient graph queries. To demonstrate, let's look at how a relational database handles a query and compare that to how a graph traversal works.

Let's use an example of an order-processing system, consisting of orders and products. It's likely that we're familiar with this sort of simple hierarchical relationship in a relational model. If we model a simplified version of this system for a relational database, we'd design two tables: `Orders`, containing the orders, and `Products`, containing the products. These two tables are connected with a linking table, `ProductsInOrder`. Figure 5.1 shows this relationship.

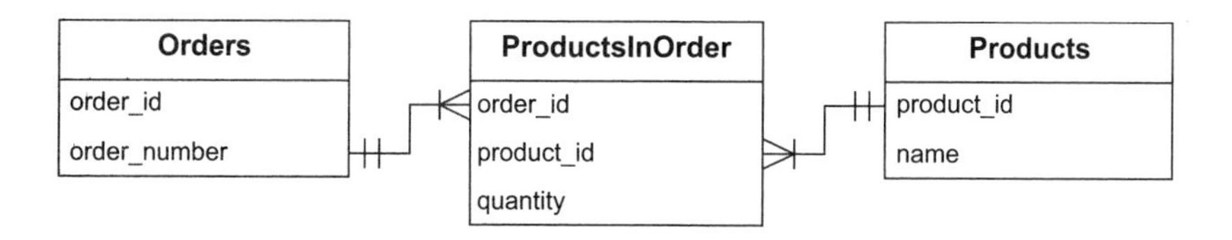

Figure 5.1 A sample Entity Relationship Diagram (ERD) for our order-processing system, which contains two tables, `Orders` and `Products`, linked together with a foreign key.

Populating these tables with some sample data, we get something like figure 5.2.

Orders	
order_id	order_number
1	ABC123
2	DEF234

ProductsInOrder		
order_id	product_id	qty
1	1	5
1	2	10
2	2	4
2	3	6

Products	
product_id	name
1	widget 1
2	widget 2
3	widget 3

Figure 5.2 Our example order-processing system, which contains sample data for a relational database model

A common question to answer with our relational database system is, "What are all the orders and the products that were ordered?" To answer this, we join the `Orders` table and the `Products` table with the `ProductsInOrder` table as in this SQL query:

```
SELECT *
FROM Orders
  JOIN ProductsInOrder ON ProductsInOrder.order_id = Orders.order_id
  JOIN Products ON Products.product_id = ProductsInOrder.product_id;
```

When we run this query, the SQL engine generates tabular output by combining the rows from the `Orders`, `ProductsInOrder`, and `Products` tables, where both the `order_id` values and the `product_id` values match. The following table provides the output.

order_id	order_number	order_id	product_id	qty	product_id	name
1	ABC123	1	1	5	1	widget 1
1	ABC123	1	2	10	2	widget 2
2	DEF234	2	2	4	2	widget 2
2	DEF234	2	3	6	3	widget 3

The critical point to notice is that our result set contains the data from both tables involved in the join operation. If our SQL query contained additional join clauses, then the columns from the additional tables are also included in the result set by default because of the use of the wildcard.

Now let's look at how we represent the same order-processing system in a graph. Taking what we learned in chapter 2, we create a schema as represented in figure 5.3. It consists of two vertices, `order` and `product`, and one edge, `contains`.

If we then populate our graph with the data used in our SQL tables, we get the graph presented in figure 5.4.

Applying what we learned about how graph traversals work, we know that our first step is to find all `order` vertices in the graph. Figure 5.5 illustrates this step.

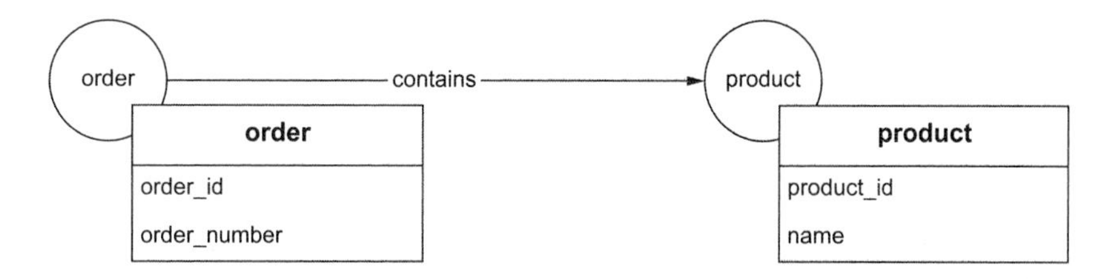

Figure 5.3 A graph schema for our order-processing system with two vertices, `order` and `product`, and one edge, `contains`

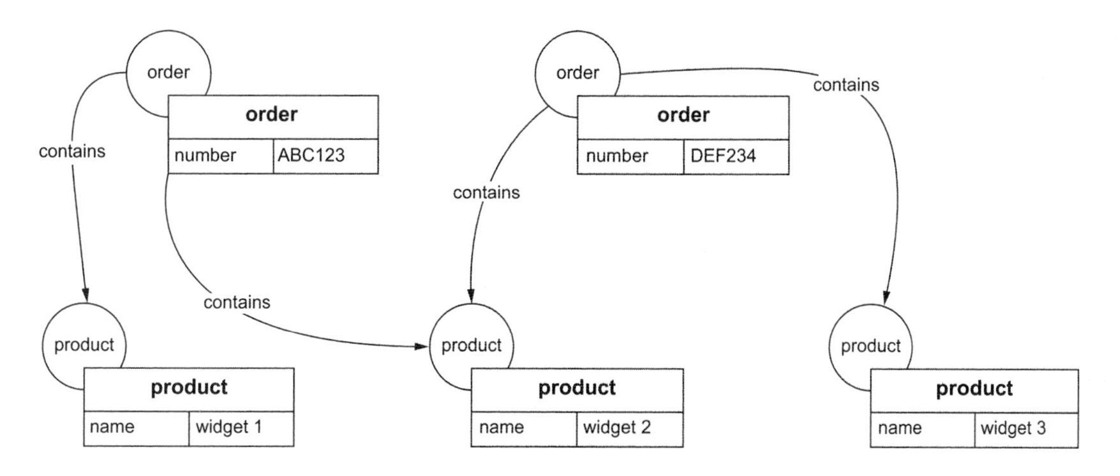

Figure 5.4 Our order-processing graph populated with the same data used in our SQL example.

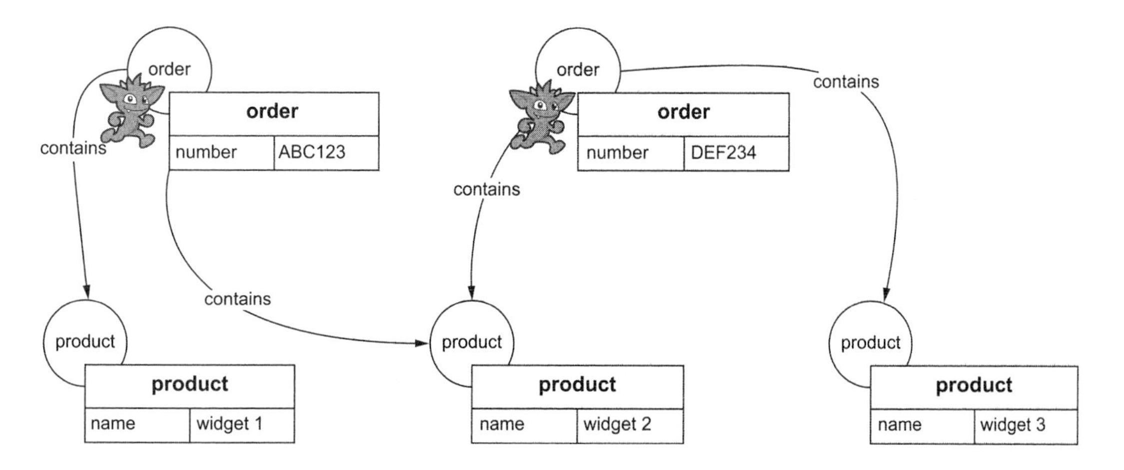

Figure 5.5 Our traversal finds all the `order` vertices in our order-processing graph.

Our next step is to traverse out all the `contains` edges to the adjacent `product` vertices. Figure 5.6 shows this step.

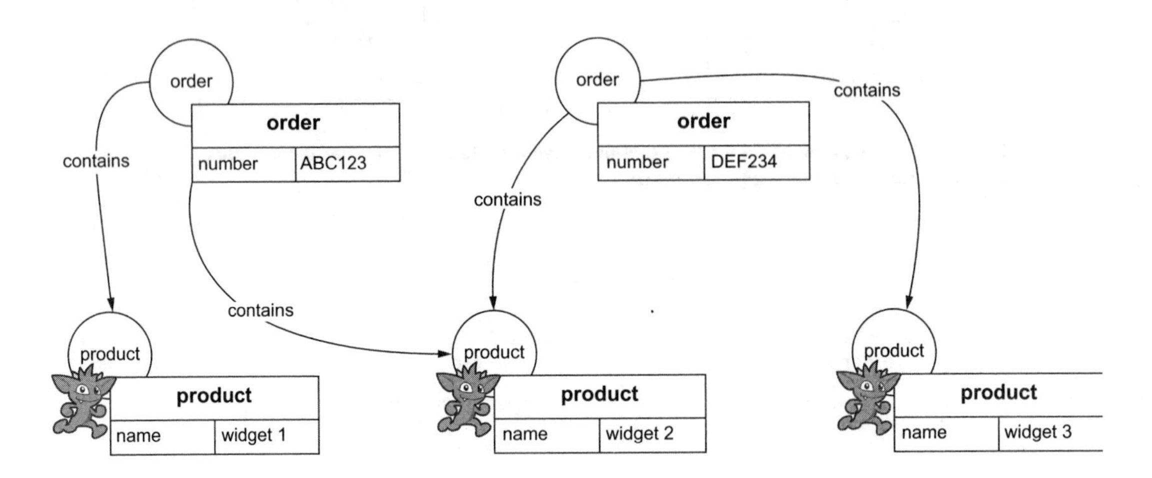

Figure 5.6 Traversing out the `contains` edges from all `order` vertices to the `product` vertices

In Gremlin, we'd write a traversal like the following:

```
g.V().hasLabel('order').out('contains')
```

If we compare the final location of our graph traversal to the final result set of our SQL query, we'll notice that although the SQL results have information about both the orders and the products, the graph results only have the properties of the `products` vertices. This represents a fundamental difference between querying a relational database and traversing a graph. Further, in a relational database, the output of a join operation is the combination of all of the joined tables. In a graph database, the output of any step of a traversal is the current set of vertices or edges. How do we return both the order and product information for a graph?

5.2 Constructing our result payload

To return both the `order` and `product` vertices, we use an alias on the `order` vertices. An *alias* in a graph database is a labeled reference to a specific output of a step, either a vertex or an edge, that can be referenced by later steps. In our order-processing graph, the steps to get a combined order/product result are as follows:

1. Find all the `order` vertices in the graph.
2. Give these an alias labeled `o`.
3. Traverse out the `contains` edge to the `product` vertices.

4 Give these an alias labeled P.
5 Return all the properties from the elements labeled O as well all the properties from the elements labeled P.

It might appear strange to also alias the `product` vertices, not just the `order` vertices. When returning the aliased elements, all these elements must have an alias, not just the mid-traversal elements. If we look at our order-processing graph after we complete steps 1 and 2 (shown in figure 5.7), we have a graph with all our `order` vertices labeled as O.

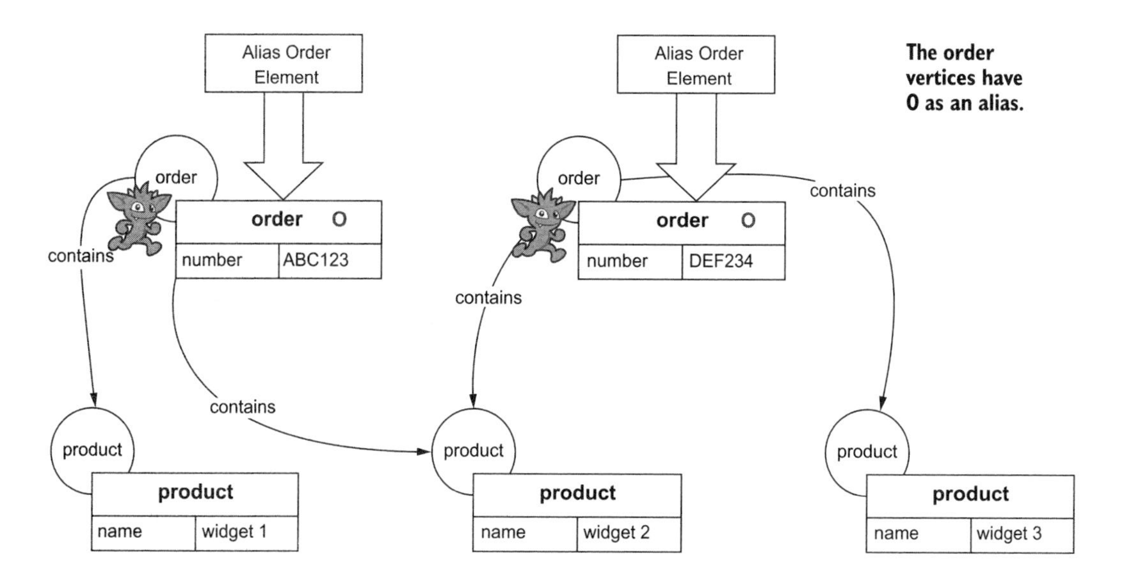

Figure 5.7 Our order-processing graph with all the `order` elements labeled O

Now we can traverse the `contains` edge and alias the adjacent `product` vertices (steps 3 and 4). This provides a graph where all our `order` vertices aliased as O and all our `product` vertices aliased as P. Figure 5.8 exhibits this part of the traversal.

So far, it looks like we're on the right track. We have references to both the `order` and `product` vertices (step 5). Next, we'll select our O and P vertices and return their properties. To retrieve these values, we'll refer to the alias and the properties we want to return for each desired attribute. While this makes sense conceptually, let's dive into a concrete example from our social network and see how this all works using our social network graph.

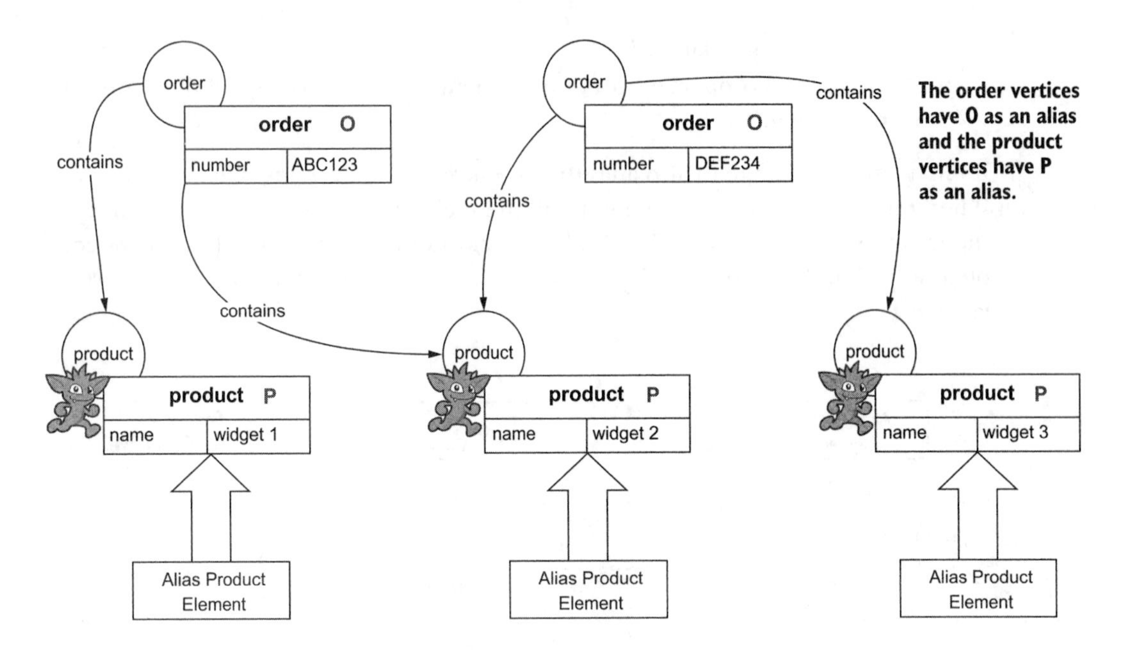

Figure 5.8 Our order-processing graph with the `order` vertices labeled `O` and the `product` vertices labeled `P`

5.2.1 *Applying aliases in Gremlin*

Having covered the concept of aliases, let's apply this to our friends-of-friends traversal we created in chapter 3. In section 3.3, we crafted the traversal:

```
g.V().has('person', 'first_name', 'Ted').
   repeat(
     out('friends')
   ).times(2).
  values('first_name')
```

The `Ted` vertex has just a couple of connections in our sample graph, so it does not have many results as is. Let's move to the middle of the graph and search for Dave's friends-of-friends instead of Ted's. Also, instead of just returning the friends-of-friends name, let's also return the friend's name. Our results should be a list of objects, each with a friend's name and with a friends-of-friends name. Let's start this exercise by first reminding ourselves what our social network graph looks like, illustrated in figure 5.9.

NOTE In figure 5.9, the vertices are labeled using only the `first_name` property to simplify the visual presentation.

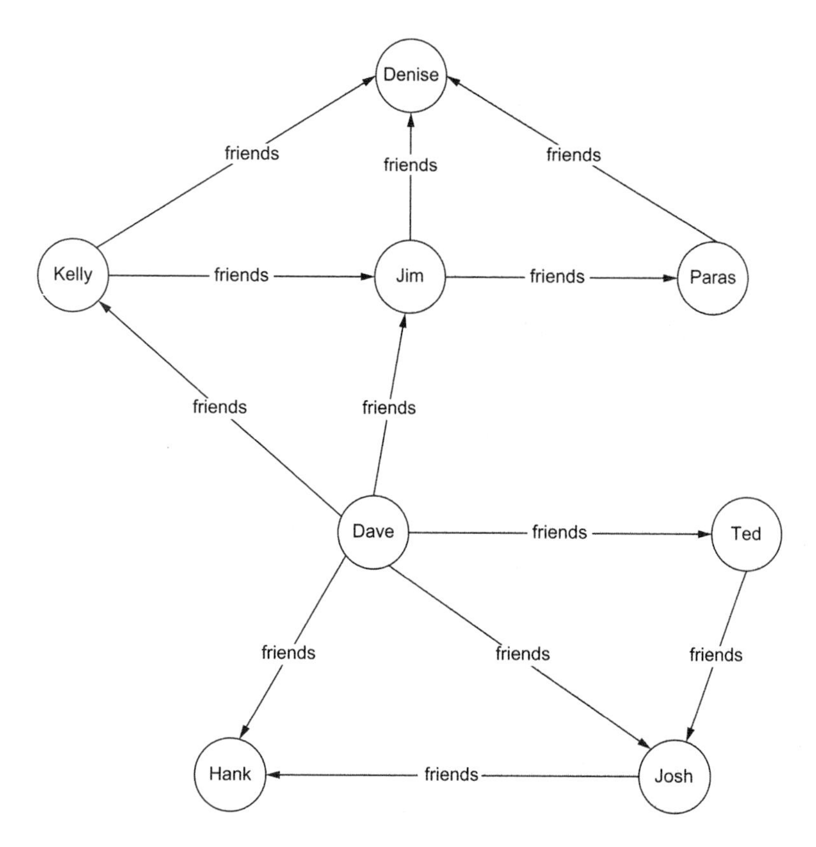

Figure 5.9 Our simplified social network with `person` vertices labeled with the person's name

Applying what we learned, we come up with the following steps for answering this friends-of-friends question for Dave:

1. Find the `Dave` vertex.
2. Traverse out the `friends` edges.
3. These are Dave's friends, so alias them with the label `'f'`.
4. Traverse out the `friends` edge again.
5. These are Dave's friends-of-friends, so alias them with the label `'foff'` (for friends-of-friends).
6. For each result, return the `first_name` property of the element labeled `'f'` and the `first_name` property of the element labeled `'foff'`.

Figure 5.10 shows what this looks like in our social network graph. In this figure, we recognize Dave's friends shown by the end vertices with the solid lines (representing steps 1–3), and his friends-of-friends denoted by the end vertices with the dashed lines (representing steps 4 and 5).

Figure 5.10 Finding Dave's friends, shown as the endpoints with the solid lines, and their friends, depicted as the endpoints with the dashed lines, within our social network graph

Traversing our graph in this manner returns one result for each of the six solid arrows in figure 5.10 and includes a value for each result based on the solid arrow traversed. Because there are five solid arrows, some of the `friends` vertices, namely `Jim` and `Kelly`, end up being duplicated in our results. Note also that there's an solid arrow to `Hank` but no solid arrow from `Hank`. We shouldn't expect to find Hank in the friends category, only in the friends-of-friends, owing to the solid arrow from Josh.

Because there isn't an outgoing `friends` edge from the `Hank` vertex, we don't get a friends-of-friends result for Dave's friend Hank. Put another way, there is no vertex that satisfies the pattern: `Dave -> Hank -> ???`. This is a good example of how the edge directions can influence results in sometimes unexpected ways. Let's write our traversal starting with the friends-of-friends traversal from section 3.3 but replacing `Ted` with `Dave`.

TIP It's best to write code in small chunks and test early and often.

```
g.V().has('person', 'first_name', 'Dave').
  out().
  out().
  values('first_name')
==>Denise
==>Denise
==>Paras
==>Jim
==>Josh
==>Hank
```

Comparing the outcome of this traversal with what we expected from our graph, we confirm that the results match. However, this graph just returns the name of the friends-of-friends. We want the combination of the friend and friends-of-friends vertices. To get the missing pieces, two new pieces are required: first, aliasing an element in the middle of a traversal, and second, using aliased elements later in the traversal to choose properties.

ALIASING ELEMENTS MID-TRAVERSAL USING AS()

The first concept, aliasing elements mid-traversal, is what enables us to retrieve the friend's name. We use a new Gremlin step, the `as()` modulator:

- `as()`—Assigns a label (or labels) to the output of the previous step, which can be accessed later in the same traversal.

Think of `as()` in Gremlin as similar to assigning an alias to a table in SQL. For example, with SQL

```
SELECT alias_name.* FROM table AS alias_name;
```

would be represented in Gremlin as

```
g.V().hasLabel('table').as('alias_name')
```

Both approaches use the keyword as and alias a specific portion of the data (in SQL, it's a table; in a graph, it's a reference to an element) to use a simple reference later. Let's add the `as()` step to alias our vertices after each of the `out()` steps. Then, in figure 5.11, we isolate and describe the portions of the traversal where we alias the vertices:

```
g.V().has('person', 'first_name', 'Dave').
  out().as('f').
  out().as('foff')
```

As figure 5.12 illustrates, each vertex is aliased with the assigned name after we traverse each `friends` edge.

For each vertex we traverse, label the current vertex as `f`.

```
out( ).as('f').
out( ).as('foff')
```

For each vertex we traverse, label the current vertex as `foff`.

Figure 5.11 Portions of the traversal where we alias the vertices at each step away from the starting vertex

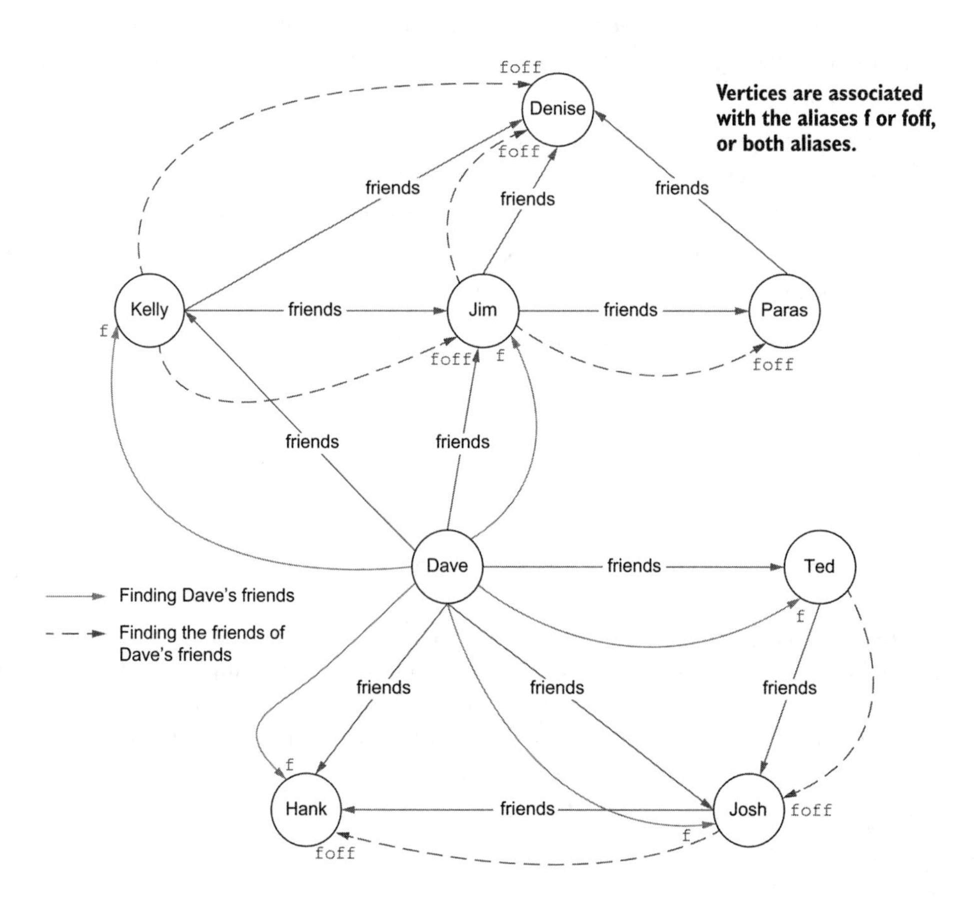

Figure 5.12 Our social network graph depicting each vertex being aliased as `f` or `foff` for each step in our traversal

While it might be tempting to assign an `as()` at each step in a traversal, its use comes at a cost. As each traverser moves through the graph, it carries a reference to each element that's aliased. The more aliases we create, the more the traverser has to keep track of with each additional step. Therefore, it's best practice to only alias steps that we plan to retrieve later in the traversal. Now that we've wrapped our heads around how to alias elements mid-traversal, let's examine the second new concept: retrieving the aliased elements.

RETURNING ALIASED ELEMENTS

Retrieving the aliased elements in our traversal requires two different steps. First, we need to specify what aliased elements to retrieve, and second, what properties of each to return. In the case of our friends-of-friends traversal, we

1. Return all elements labeled `'f'`.
2. Return all elements labeled `'foff'`.
3. For each of the returned elements, return the `first_name` property.

To return aliased elements, let's turn to a new Gremlin step:

- `select(string[])`—Selects aliased elements from earlier in the traversal. This step always looks back to previous steps in the traversal to find the aliases.

The `select()` step takes an array of strings, which are the aliases to retrieve. In our example, we specify `select('f', 'foff')` to use both sets of vertices in our results. To specify what properties to return, we introduce another new Gremlin step, or more accurately, another modulator, `by()`. Like the `from()`, `to()`, and `as()` modulators, the `by()` modulator only works in the context of another step; in this case it will be working with the `select()` step (although it can work with others, as we will see later):

- `by(key)`—Specifies the `key` of the property to return the value from the corresponding aliased element
- `by(traversal)`—Specifies the `traversal` to perform on the corresponding aliased element

There are two forms of `by()`. The first form takes the property key and returns the corresponding property value from the labeled element. This is a bit of Gremlin syntax sugar because `by(key)` is equivalent to `by(values(key))`. The second form takes a traversal that allows us to perform additional steps on the labeled element, such as a `valueMap()` or `out().valueMap('key')`. It's also possible to use complex traversals within a `by()` modulator to format results. We demonstrate more complex uses in later chapters.

The `by()` modulator specifies what to do with the corresponding aliased elements from a step like the `select()` step. In our case, we apply the first form to specify that we want the `first_name` property from each of the aliases referenced in our `select()` step. Putting the `select()` and `by()` steps on our previous friends-of-friends traversal, we get

```
g.V().has('person', 'first_name', 'Dave').
  out().as('f').
  out().as('foff').
```

```
  select('f', 'foff').
    by('first_name').
    by('first_name')
==>{f=Jim, foff=Denise}
==>{f=Jim, foff=Paras}
==>{f=Kelly, foff=Jim}
==>{f=Kelly, foff=Denise}
==>{f=Ted, foff=Josh}
==>{f=Josh, foff=Hank}
```

With these two concepts, we can create complex results by combining elements from different points in our traversal. In this scenario, it means including not only the name of the friends-of-friends of Dave, but also which of Dave's friends they're connected to as well. Our traversal yields the six results we expected with the `Jim` and `Kelly` vertices referenced twice as friends. Also, the `Hank` vertex was not included as a one of the friends because there was no corresponding friends-of-friends returned. This is awesome, but why are there two `by()` steps?

One confusing aspect of using `by()` statements is that each aliased element we specify in a `select()` statement should have a corresponding `by()` statement to indicate the operations to perform on it. Additionally, the order of the `by()` step corresponds to the order of the aliases specified.

In our example, `select('f', 'foff')`, our traversal must have two `by()` statements. The first `by()` performs actions on the elements labeled as `'f'`; the second `by()` performs actions on the elements labeled as `'foff'`. Figure 5.13 demonstrates how the `by()` steps correlate in our example.

> **NOTE** Strictly speaking, it is possible to have greater or fewer `by()` statements than referenced elements. In these scenarios, the `by()` statements are used in a round-robin fashion. This can lead to confusion, so we always match the number of `by()` statements to the number of aliased elements to be clear about what should happen for each alias.

Upon examination, in figure 5.13, we notice that the first `by()` statement returns the `first_name` property from our vertices labeled as `'f'`, and the second `by()` statement yields the `first_name` property from our vertices labeled as `'foff'`.

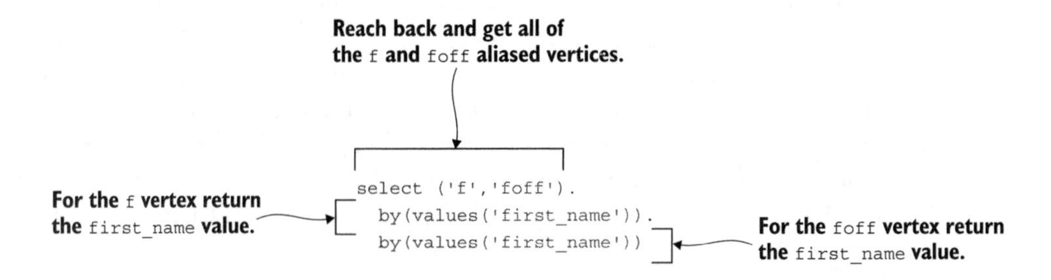

Figure 5.13 Diagram showing the portions of the traversal where we select the labeled vertices

5.2.2 Projecting results instead of aliasing

Sometimes, instead of looking back in the traversal for earlier results, it is preferable to project results forward from the current elements. Projecting results differs from retrieving the previous results in a simple, but somewhat subtle way. When we retrieve (or select) data, we can only get information that we already traversed and aliased. When we project results, we create new results, possibly branching to items not yet traversed. Let's start by looking at projection in contrast to selection:

- *Selection is the process of working with vertices, properties, or additional traversal expressions to return results from previously labeled steps.* Selection always looks back to earlier parts of the traversal.
- *Projection is the process of working with vertices, properties, or additional traversal expressions to create results from the input to the current step.* Projection always moves forward, taking the incoming data as the starting point.

Understanding the difference between these two items is crucial. Selection is generally used to combine results from elements traversed earlier in the traversal. Projection is generally used to group or aggregate data starting from the current location in the graph (for example, finding the `degree` property of each member of a set of vertices, which we do later in this section).

Let's return to our order-processing graph for an illustration. For this example, let's answer the question, "For each of the products in order *ABC123*, how many times has that product been ordered?" Using what we already know from the previous section about selecting results, let's use the following process:

1. Find the order vertex `ABC123`.
2. Traverse the `contains` edge to each of the `product` vertices, aliased as `p`.
3. Traverse out all the contains edges, aliased as `c`.
4. Return a selection of the name of `p` with the count of `c`.

Completing step 1 has our traverser located on the `order` vertex. Figure 5.14 illustrates this step.

Moving along, completing step 2 of the process has two traversers, one located on each adjacent `product` vertex. Figure 5.15 shows this step.

With our next to last step (step 3), the traversers are located on the `contains` edges. Figure 5.16 shows this step.

With everything we've learned about using aliases and returning results, this is the correct way to think about the problem. However, if you count the gremlins in figure 5.16, you see there are three. Each of these returns its own results, based on what it knows. What they each know is the widget they came from and how many edges these occupy. This approach yields the following:

```
{name: widget 1, count: 1}
{name: widget 2, count: 1}
{name: widget 2, count: 1}
```

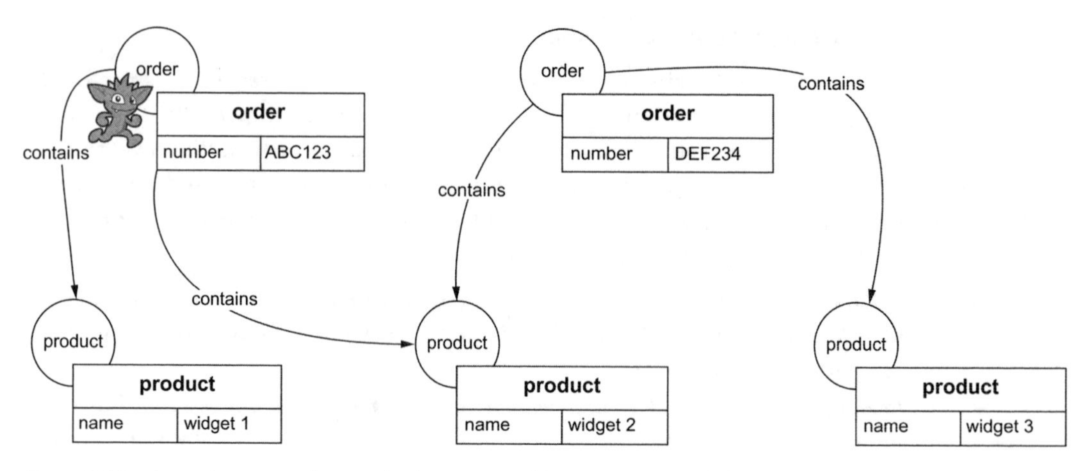

Figure 5.14 Our order-processing graph after step 1 with our traversers sitting on the `order` vertices

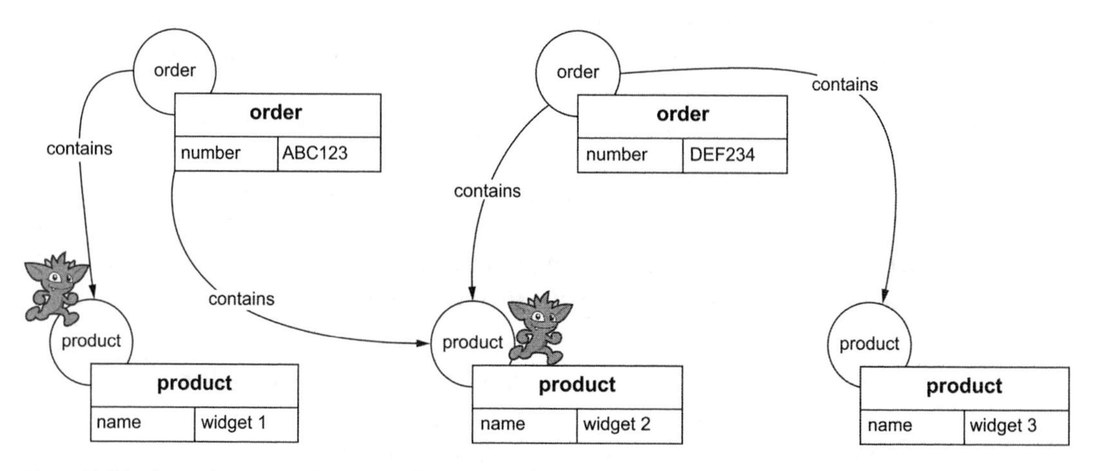

Figure 5.15 Our order-processing graph after step 2 with our traversers sitting on the `product` vertices

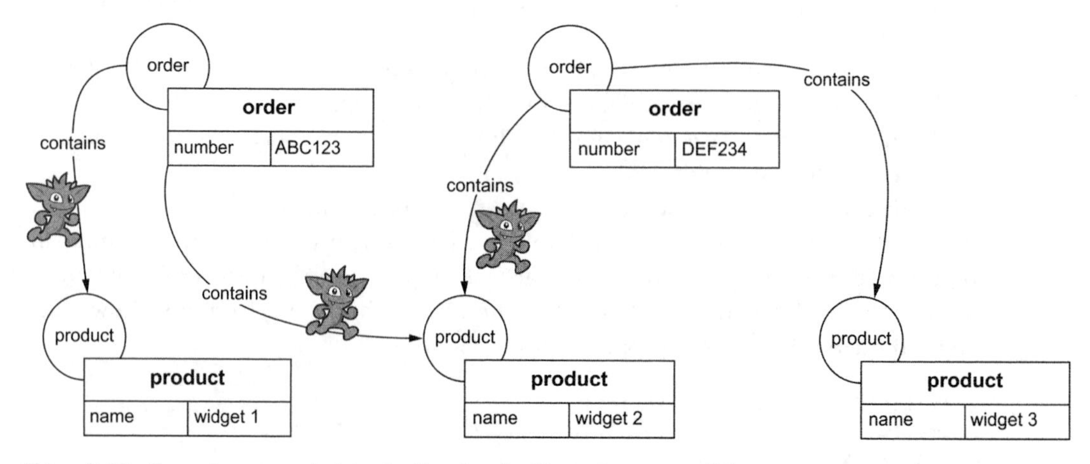

Figure 5.16 Our order-processing graph after step 3 with our traversers sitting on our `contains` edges

That is one result for each of the three gremlins. What we want instead is a result like this:

```
{name: widget 1, count: 1}
{name: widget 2, count: 2}
```

Why didn't the traversal return the expected results? Why did we end up with three gremlins and three results each with a count of 1? Previously, the return values from our traversals were generated by selecting the values of previously traversed and labeled elements. In the case of the previous order-processing traversal, we end up with three traversers. So, what do we do instead? The steps shown are mostly valid, but require a few tweaks:

1. Find the order vertex `ABC123`.
2. Traverse the `contains` edge to each of the `product` vertices.
3. Traverse out all the `contains` edges.
4. Return a projection of the product name with a count of the incident `contains` edges.

Studying the difference between the two processes, we'll notice a few specifics:

- We're no longer aliasing elements as we traverse these.
- We don't have to traverse back out to the `contains` edge a second time.

In the example traversal, we use projection to end up with the count of `contains` edges for a specific product because we want to branch our logic at the `product` vertex.

Whew, we know this is a lot to take in, so let's see what this looks like in a practical example from our social network graph. Let's say we applied the same concepts to this question: "Find the `degree` property for every `person` vertex in my graph." To answer this question, we need to do the following:

1. Find all the `person` vertices in the graph.
2. Create a new result object with the `name` and `degree` keys.
3. For the `name` key, return the `first_name` of the person.
4. For the `degree` key, count all the edges for the person.

We also need a new step, the `project()` step:

- `project(string[])`—Projects the current object into a new object or objects as specified by the criteria in the `by()` modulators

Let's apply what we learned about the `by()` modulators and projection. Figure 5.17 shows the traversal we should get from this.

As we did with `select()`, we'll use the `by()` modulators to instruct the `project()` step on how to return its results.

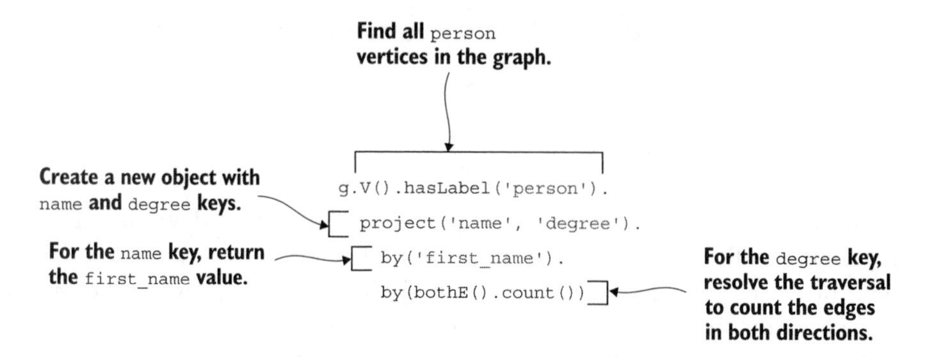

Figure 5.17 Diagram showing the portions of the traversal describing the `project()` step

> **IMPORTANT** In this traversal, instead of specifying a property name in the second `by()` statement, we specify additional traversal steps. This takes the incoming element (in this case, the `product` vertex) and then performs additional traversal steps from that point in the graph. This ability to specify additional traversal steps within a `by()` step isn't unique to `project()`. We can specify these sub-traversals with either a `select()` or other Gremlin steps. This is quite powerful—the ability to do complex operations within a traversal or steps within steps.

When we run this traversal on the complex social network graph, we get the following results:

```
g.V().hasLabel('person').
  project('name', 'degree').
    by('first_name').
    by(bothE().count())
==>{name=Dave, degree=5}
==>{name=Paras, degree=2}
==>{name=Josh, degree=3}
==>{name=Denise, degree=3}
==>{name=Ted, degree=2}
==>{name=Hank, degree=2}
==>{name=Kelly, degree=3}
==>{name=Jim, degree=4}
```

To review, let's summarize the differences between the two methods: selection versus projection. Then we'll compare these side by side in figure 5.18 before we move on:

- *Selection uses the* `select()` *step to create a result set based on previously traversed elements of a graph.* To use the `select()` step, we alias each of the elements with the `as()` step for later use.
- *Projection uses the* `project()` *step to branch from the current location within the graph and creates new objects.* In our present example, we had one element remain

static, the person's name, but we needed the other elements to be calculated through further traversing of the graph to return the number of friends.

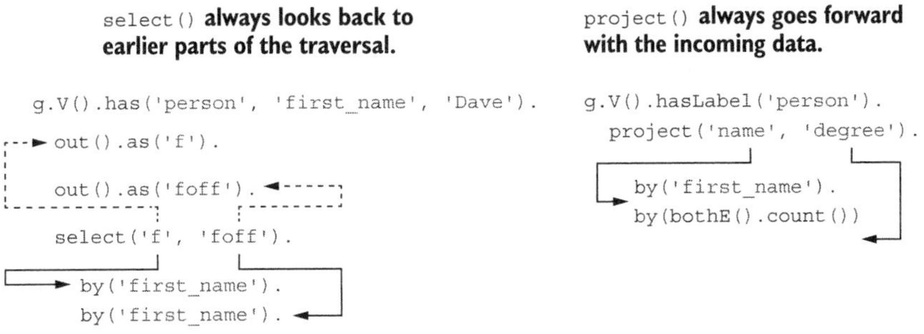

`select()` **always looks back to earlier parts of the traversal.**

```
g.V().has('person', 'first_name', 'Dave').
  out().as('f').
  out().as('foff').
  select('f', 'foff').
    by('first_name').
    by('first_name').
```

`project()` **always goes forward with the incoming data.**

```
g.V().hasLabel('person').
  project('name', 'degree').
    by('first_name').
    by(bothE().count())
```

Figure 5.18 The `select()` step looks back to previously aliased steps, and the `project()` step takes the incoming data and moves forward with it.

Don't worry if you don't feel comfortable with how to manipulate results right away. Our experience tells us that, as with many powerful software concepts, it takes some practice to know which technique produces the desired result. It does become more natural with regular practice, and by *regular practice*, we mean trial and error until you get the desired results. Now that we know how to construct complex result structures, how do we go about returning these in a predictable way?

5.3 Organizing our results

In this section we investigate two other mechanisms to manipulate results: ordering and grouping. Most clients want nice, clean, ordered data. However, as in most relational databases, graph databases don't guarantee the order of the results by default. This leads us to the three most common requirements for organizing our result data:

- Ordering the results
- Grouping the results
- Limiting the size of the results

5.3.1 Ordering results returned from a graph traversal

Clients often expect the returned data to be sorted by one or more properties. For example, when displaying everyone in the graph by name, people usually expect to see the results in alphabetical order. This leaves us with a couple of options.

The first option is to return all the names and sort these client-side, in memory, within the application. While this works, it's undesirable. For example, let's say our application only shows the first 10 names of a possible 100. Sorting all the data client-side means that we return all 100 values, sort these, and then choose the top 10. This is inefficient and adds load not only on the client, but also on the database and the network. While there are scenarios where this might make sense, such as if we were caching all the names in the application for repeated reuse, we normally want to reduce unneeded work.

This leaves us with the second option: sorting the names first on the server-side. This is the method frequently taken in a relational database and is also common in a graph database. In SQL, we use the `ORDER BY` clause like this:

```
SELECT *
FROM person
ORDER BY first_name;
```

The syntax in a graph database is similar. In fact, to order results in Gremlin, we use the following step:

- `order()`—Collects all objects up to this point of the traversal into a list, which is ordered according to the accompanying `by()` modulator

This new step, used in conjunction with the `by()` modulator, specifies how to arrange our data and which property to use to sort the results. For example, to order the names of every `person` vertex in the graph by `first_name`, we use this traversal:

```
g.V().hasLabel('person').values('first_name').
  order().
    by()
==>Dave
==>Denise
==>Hank
==>Jim
==>Josh
==>Kelly
==>Paras
==>Ted
```

The `order()` step defaults to sorting in ascending order. To sort by descending order, specify the `decr` parameter in the `by()` step as shown:

```
g.V().hasLabel('person').values('first_name').
  order().
    by(decr)

==>Ted
==>Paras
==>Kelly
==>Josh
==>Jim
==>Hank
==>Denise
==>Dave
```

While sorting in either ascending or descending order is common, there are times when we want to order data randomly, such as when sampling. For this, we use the `shuffle` parameter in the `by()` step:

```
g.V().hasLabel('person').values('first_name').
  order().
    by(shuffle)
==>Dave
==>Jim
==>Ted
==>Paras
==>Kelly
==>Hank
==>Denise
==>Josh
```

Ordering is probably the most frequent requirement for formatting data. Another typical need is to group or count the number of items in a group.

5.3.2 Grouping results returned from a graph traversal

If we return to our previous friends-of-friends traversal, the client might want to return that list grouped by which of Ted's friends they're friends with. In this scenario, we're left with the same choices as with ordering data: either perform the work on the client-side in the application or on the server-side in the database.

There's a natural desire, and one which we encourage, to push as much of this work as close to the data as possible. In SQL, we accomplish this by using the GROUP BY clause:

```
SELECT f.person_id, count(foff.*)
FROM person
 INNER JOIN friends AS f ON f.person_id = person. id
 INNER JOIN friends AS foff ON foff.person_id = f.friend_id
WHERE person.first_name = 'Ted'
GROUP BY f.person_id;
```

Similar to ordering data, the syntax in a graph database for grouping is comparable when using Gremlin. To perform these grouping operations, we can use either of the following steps:

- `group()`—Groups the results based on the specified `by()` modulator. Data is grouped by using either one or two `by()` modulators. The first one specifies the keys for the grouping. The second one, if present, specifies the values. If not present, the incoming data is collected as a list of the values associated with the grouping key.
- `groupCount()`—Groups and counts the results based on the specified `by()` modulator. It takes one `by()` modulator to specify the keys. The values are always aggregated by the `count()` step.

In the following, we apply these steps to group all of the friends-of-friends of Dave by his friends:

```
g.V().has('person', 'first_name', 'Dave').
  both().
  both().
  group().
    by('first_name')

==>{Denise=[v[19], v[19]], Ted=[v[4]], Hank=[v[6]], Paras=[v[17]], Josh=[v[2],
     v[2]], Dave=[v[0], v[0], v[0], v[0], v[0]], Kelly=[v[13]], Jim=[v[15]]}
```

We can see that our traversal returns a `Map` containing arrays of vertices for each name. Because we didn't specify a second `by()` modulator, it simply collected the references to the vertexes into a list. To make this a bit easier to read, let's use the `unfold()` step:

- `unfold()`—Unrolls an interable or map into its individual components

Applying the `unfold()` step to our results unwinds those into individual records for each name. For example

```
g.V().has('person', 'first_name', 'Dave').
  both().
  both().
  group().
    by('first_name').
  unfold()
==>Denise=[v[19], v[19]]
==>Ted=[v[4]]
==>Hank=[v[6]]
==>Paras=[v[17]]
==>Josh=[v[2], v[2]]
==>Dave=[v[0], v[0], v[0], v[0], v[0]]
==>Kelly=[v[13]]
==>Jim=[v[15]]
```

Instead of returning the actual vertices for each name, what if we were more interested in discovering which of Dave's friends is the most popular? To do aggregated grouping, we need the count for a group by name, so we use the `groupCount()` step:

```
g.V().has('person', 'first_name', 'Dave').
  both().
  both().
  groupCount().
    by('first_name').
  unfold()

==>Denise=2
==>Ted=1
==>Hank=1
==>Paras=1
==>Josh=2
==>Dave=5
==>Kelly=1
==>Jim=1
```

The groupCount() step is just a little syntax sugar for the most common use of the group() step—aggregating counts of things. As a quick point of comparison, and a nice demonstration of how we can use a traversal in a by() modulator, the group() version of the groupCount() step is

```
g.V().has('person', 'first_name', 'Dave').
  both().
  both().
  group().
    by('first_name').
    by(count()).
  unfold()
==>Denise=2
==>Ted=1
==>Hank=1
==>Paras=1
==>Josh=2
==>Dave=5
==>Kelly=1
==>Jim=1
```

Note how we were able to use a traversal of one step (the count() step) in our second by() modulator. The group() step applied by() to all of the incoming vertices that shared the same first_name value. As these examples demonstrate, grouping and ordering results in a graph database is similar to the process used in a relational database.

5.3.3 Limiting results

The final topic in organizing results is returning a subset of the data. This is commonly used to minimize the result size or for pagination functionality. For example, let's say that we want to return all the names for people in our graph, but our graph contains one million people. Can any application display all at the same time? Usually we want to limit the initial results and then allow the user to move through the data set in groups of records. This approach is standard in a number of types of applications.

As with grouping or ordering, the question remains: Do it on the client-side or do it on the server-side? In this case, it's almost always better to limit data on the server before returning it to the client. That provides a drastic reduction in resources across the whole stack, from the database to the network to the application. In SQL, we use the LIMIT clause for this:

```
SELECT *
FROM person
LIMIT 10;
```

As before, the approach in a graph database is similar. Gremlin, however, has several steps depending on the desired outcome: first *X* results, last *X* results, or *X* results from within the data set:

- `limit(number)`—Returns the first `number` of results
- `tail(number)`—Returns the last `number` of results
- `range(startNumber, endNumber)`—Returns the results from `startNumber` (inclusive, zero-based) to `endNumber` (not inclusive)

These three steps are usually paired with an ordering step because graph traversals don't guarantee the order of the data returned. Let's say that we want to return only the top three names from our graph, ordered by `first_name`. If we extend the traversal we built in the ordering section to add the `limit()` step, we get a traversal that looks like this:

```
g.V().hasLabel('person').values('first_name').
  order().
    by().
  limit(3)
==>Dave
==>Denise
==>Hank
```

What if we want to do the reverse and return the last three names? How would we accomplish this task?

EXERCISE Take a minute and think about what you've learned about manipulating our results. What would you do to answer the question?

We see two ways to do this. The first is to use the previous traversal, but arrange names in descending order instead of ascending order; then limit our results to the top three. The second way is to use the `tail()` step instead of the `limit()` step like this:

```
g.V().hasLabel('person').values('first_name').
  order().
    by().
  tail(3)
==>Kelly
==>Paras
==>Ted
```

Both methods accomplish the same goal. It's therefore up to your discretion to decide which one you want to choose.

The last requirement to discuss is pagination of results. Let's say we want everyone ordered by first name, but only three at a time. We use the `range()` step and specify the first and last result number to return:

```
g.V().hasLabel('person').values('first_name').
  order().
    by().
  range(0, 3)
==>Dave
==>Denise
==>Hank
```

By manipulating the `startNumber` and `endNumber` values, we can page through our results. For example, if we wanted to move to a second page of results, we could accomplish this by incrementing the values in our range step to `range(3, 6)`.

5.4 Combining steps into complex traversals

Given all these different ways of manipulating our data, we'll share one last example that combines these concepts together to answer the question, "What three friends-of-friends of Dave have the most connections?" We'll start with the answer and then break it down into its component parts. First, the traversal to answer this question follows:

```
g.V().has('person', 'first_name', 'Dave').
  both().
  both().
  groupCount().
    by('first_name').
  unfold().
  order().
    by(values, desc).
    by(keys).
  project('name', 'count').
    by(keys).
    by(values).
  limit(3)

==>{name=Dave, count=5}
==>{name=Denise, count=2}
==>{name=Josh, count=2}
```

That's a lot to take in when you are used to traversals with at most five steps. This one has nine steps plus five modulators. In chapter 8, we introduce a methodology for developing more involved traversals such as this one. In this final example for this chapter, we walk through our approach to understand the traversal that is already written.

The first step is to ascertain the traversal writer's intent. Here, we know that the traversal is supposed to answer a specific question: "What three friends-of-friends of Dave have the most connections?" We may not always know the intent when looking at someone else's traversals. Out of consideration for the developers who might someday need to support the traversal you write, we recommend either having descriptive method names (e.g., `getTop3FriendOfFriendsByEdgeCount`) or including a helpful comment in the code. We discuss adding comments a bit later in this section.

Next, determine the traversal's starting point. It is a single vertex or a type of vertices. We do this by looking at *all* of the filtering steps at the start. Remember that filtering steps such as `has()` can be chained together efficiently. In this example, the traversal starts at a single vertex, the `Dave` vertex, before it traverses to other parts of the graph:

```
g.V().has('person', 'first_name', 'Dave')
```

Next, we want to take each step, or collection of steps, in their order. While reading through the steps, we want to develop our mental view of our position within the graph at each point. Sometimes, we find it helpful to add comments to the code as a way of taking notes. For example, with the first few steps, we might add comments like the following:

```
g.V().has('person', 'first_name', 'Dave').  // single person: Dave
  both().                                   // friends
  both()                                    // friends of friends
```

This helps us to keep track of the output at each step. The only problem with this is that not all traversal processors support inline comments like this. For example, while our IDE recognizes this as valid Groovy code, the Gremlin Console gives us an error if we attempt to run it. To be able to run the code in Gremlin Console, we need to change the comment to something like this:

```
// single person: Dave
g.V().has('person', 'first_name', 'Dave').
  // friends
  both().
  // friends of friends
  both()
```

But that makes the traversal verbose in ways that are almost unhelpful. In these situations, we are tempted to keep two versions of the traversal in our IDE or editor, one with comments and the other for testing. Either way, we can see that the first three lines get us to the friends of Dave's friends. Testing in the Gremlin Console gives us

```
g.V().has('person', 'first_name', 'Dave').
  both().
  both()
==>v[19]
==>v[17]
==>v[0]
...
```

Let's look at the next set of steps, the `groupCount()` step, it's `by()` modulator, and the `unfold()` step. We can run just those steps through the Gremlin Console and review the results:

```
g.V().has('person', 'first_name', 'Dave').
  both().
  both().
  groupCount().
    by('first_name').
  unfold()
==>Denise=2
==>Ted=1
==>Hank=1
==>Paras=1
```

```
==>Josh=2
==>Dave=5
==>Kelly=1
==>Jim=1
```

From this, we can see that we get a series of key-value pairs, where the key is the `first_name` of the friends-of-friends vertex, and the value is the number of times that it appears in the results. Referring back to our starting question, this covers the friends-of-friends of Dave and their number of connections. The `groupCount()` step handled both the needed grouping (by friends-of-friends' first names) and the count aggregation. An `unfold()` step is tacked on to simplify the ordering, which is next:

```
g.V().has('person', 'first_name', 'Dave').
  both().
  both().
  groupCount().
    by('first_name').
  unfold().
  order().
    by(values, desc).
    by(keys)
==>Dave=5
==>Denise=2
==>Josh=2
==>Hank=1
==>Jim=1
==>Kelly=1
==>Paras=1
==>Ted=1
```

The ordering statement gets interesting. First, it orders by the values in descending order and then by the keys. The ordering by the keys is a nice little tie breaker to ensure somewhat deterministic results. We're not sure that an alphabetical bias by first name is the best approach, but it certainly works for ensuring consistency. Now let's reformat these results so that it's easier to parse with a client program:

```
g.V().has('person', 'first_name', 'Dave').
  both().
  both().
  groupCount().
    by('first_name').
  unfold().
  order().
    by(values, desc).
    by(keys).
  project('name', 'count').
    by(keys).
    by(values)
==>{name=Dave, count=5}
==>{name=Denise, count=2}
==>{name=Josh, count=2}
```

```
==>{name=Hank, count=1}
==>{name=Jim, count=1}
==>{name=Kelly, count=1}
==>{name=Paras, count=1}
==>{name=Ted, count=1}
```

Here, we use a `project()` step so that we have objects with clear labels on the properties. This probably isn't necessary if working with data within a single traversal, but it is a great practice when returning results to another program. Now, the only operation that remains is to limit the results:

```
g.V().has('person', 'first_name', 'Dave').
  both().
  both().
  groupCount().
    by('first_name').
  unfold().
  order().
    by(values, desc).
    by(keys).
  project('name', 'count').
    by(keys).
    by(values).
  limit(3)

==>{name=Dave, count=5}
==>{name=Denise, count=2}
==>{name=Josh, count=2}
```

This example demonstrates how we brought together the concepts and constructs in this chapter to format results for client software. And, as a bonus, we walked through how to examine a traversal piece by piece to understand its operations. In the next chapter, we'll bring the skills from the last few chapters together to build a working application.

Summary

- By default, properties aren't returned from graph elements, so we must explicitly ask for those. In Gremlin, we use steps such as `values()` and `valueMap()` to retrieve the values in the desired form.
- Aliases in the traversal allow for referencing results from earlier steps in later steps, supporting composition of powerful traversals.
- Selecting and projecting steps create complex results from multiple vertices or edges, allowing for the composition of intricate result structures.
- Selection creates a result set based on previously traversed elements of a graph. To use the `select()` step, we alias with the `as()` step elements for use in later steps.
- Projection operates from the current location within the graph and creates new objects with either static or calculated properties.

- Ordering, grouping, or counting by group are common ways to transform results using the `order()`, `group()`, and `groupCount()` steps.
- The `limit()` step limits the number of results, the `tail()` step returns the last *X* records, and the `range()` step allows for result pagination.
- Combining different steps performs complex manipulation and transformation of traversal results in the database prior to returning the results to a client. Used appropriately, this improves performance in the database, across the network, and in the application itself.

6 Developing an application

This chapter covers

- Setting up a project
- Choosing the database driver and connecting to the database
- Translating recursive and pathfinding traversals into Java methods
- Processing traversal results within an application

In chapters 3, 4, and 5, we covered the process of writing traversals. But writing traversals is only part of what is required to create an application. Applications also require handling tasks, such as connecting to a database, managing user input, and processing the traversal results into a usable form. While the process for doing these is similar to how we work with relational databases, there are a few critical differences, so let's find out how to approach these tasks when working with graph databases.

In this chapter, we'll use the traversals we built in the earlier chapters to demonstrate the process for translating these into a console application written in Java. We'll start by setting up our project, including selecting the proper graph database driver. Next, we'll walk through how to connect to our graph database. Finally, we'll show how to translate our Gremlin traversals into the equivalent Java code and process the results. By the end of this chapter, we'll have a fully functioning application based on our DiningByFriends social network.

NOTE If you haven't done so already, download the corresponding source code for this chapter available in the repository https://github.com/bechbd/graph-databases-in-action. The code relevant to this chapter is located in the chapter06 folder.

All examples begin with the assumption that our complex social network data set is loaded. To accomplish this on MacOS/Linux, run this command:

```
bin/gremlin.sh -i $BASE_DIR/chapter06/scripts/6.1-complex-social-network.groovy
```

Or on Windows, run

```
bin\gremlin.bat -i $BASE_DIR\chapter06\scripts\6.1-complex-social-network.groovy
```

The chapter06/java directory contains three different versions of the application. Each is contained in a separate folder:

- *skeleton*—A skeleton of the application without any code, only stubs for methods. This option is the right choice for anyone who wants to write the code themselves.
- *commented*—All necessary application code for this project is included but commented out. This option is the right choice for one who follows along in the code but doesn't want to do all the typing.
- *completed*—A complete, functional version of the application. This option is the right choice for those readers who just want to examine the finished product.

We recommend reading the README.md file located within each version's folder because it contains the details on any prerequisites, as well as the specific steps to build and run the application. To follow along throughout this chapter, the best approach is to have two projects set up in your preferred IDE, likely each in its own instance or window. For example, you can display your completed project on the left side of the screen as a reference, and then your own code repository, perhaps started as a copy of the skeleton version, on the right side of the screen where you do your own typing.

6.1 Starting the project

When we start a data-backed software development project, we must address a near-universal set of fundamental needs. No matter if we are building for a relational database or a graph database, each project needs to work through the following concerns:

- Select our tools: development language and database.
- Set up the software project.

- Obtain the appropriate driver for the database.
- Prepare the database server instance.

Each of these is required to get a framework in place to write and test code. Because our end goal is to build a functioning application for our DiningByFriends social network, this section discusses the decisions we made for our graph database project. While this process is similar to that of relational databases, there are a few key differences we highlight along the way.

If you have a lot of experience with setting up this type of work, then feel free to skip to section 6.2 and look at the skeleton version of the code. An experienced Java developer should be able to skim through this version and complete their own setup within a few minutes. Developers familiar with other languages might have to read this section a little more closely to see how the Java ecosystem compares to other software language environments.

6.1.1 Selecting our tools

When selecting the tools for any data-backed application, there are two major decisions to make: what language to use and what database to build on. This section looks at each.

USING JAVA FOR OUR DEVELOPMENT LANGUAGE

Due to its popularity, we chose the pairing of Java (version 8), a perennial top programming language, and Maven (version 3.5), a commonly used build tool. This is also the combination we encounter most often when working with graph databases.

> **NOTE** If you are not a Java developer, don't despair. Most graph databases support a variety of popular development languages, and the concepts and constructs discussed in this chapter transfer to different languages, albeit with some language-specific tweaks.

USING THE GREMLIN SERVER FOR OUR DATABASE IMPLEMENTATION

There are a number of commercial graph databases on the market, and many provide an Apache TinkerPop-compatible interface. In the interest of simplicity and ease of local development, we decided to use the Apache TinkerPop reference implementation, called Gremlin Server. You can find the installation and configuration instructions in appendix A. Although this may not be your chosen database, we are confident that the code and examples port easily to any Apache TinkerPop-compatible graph database.

If you chose to use a graph database that isn't directly compatible with Apache TinkerPop, then it might be a bit more difficult to make use of the provided code samples. While the concepts discussed in this chapter are nearly universal, other databases have their own methods for accomplishing tasks, such as connecting to a database, running traversals, and processing results.

6.1.2 Setting up the project

The particulars of setting up a project for development are specific to the programing language. In our case, we chose Java and Maven, so we expect to find a folder for the project containing a pom.xml file for dependency and build instructions and an App.java file for the source code. As a matter of good development hygiene, we like to include both a README.md with the specific build instructions and a .gitignore file to keep the source control system clean.

6.1.3 Obtaining a driver

One of the first items we need for our application is an appropriate driver to connect to our database. This isn't a unique requirement; a driver is needed for any application that uses a database, whether relational or graph. In our case, we use the Apache TinkerPop Gremlin Driver to connect to the Gremlin Server. As we've opted to build a Java application with Maven, we need to add a reference to the Apache TinkerPop Gremlin Driver in our pom.xml file, including these lines:

```
<dependency>
    <groupId>org.apache.tinkerpop</groupId>
    <artifactId>gremlin-driver</artifactId>
    <version>3.4.6</version>
</dependency>
```

NOTE You can find the Gremlin Driver at this site: http://mng.bz/Nne2.

If this were a C# project, we'd use a tool like NuGet to add a dependency for the Apache TinkerPop Gremlin.Net driver. This process depends on the "norms" of your selected language and build tools. Let's ensure that a basic build of our project works. We accomplish this by using the following command in a terminal window in the directory of your application:

```
mvn clean compile
```

The build results should show a `BUILD SUCCESS` as its output:

```
[INFO] Scanning for projects...
[INFO]
[INFO] ----------------< com.diningbyfriends:diningbyfriends >-------------
[INFO] Building diningbyfriends 1.0
[INFO] --------------------------------[ jar ]----------------------------
[INFO]
[INFO] --- maven-clean-plugin:2.5:clean (default-clean) @ diningbyfriends -
[INFO]
[INFO] --- maven-resources-plugin:2.6:resources (default-resources) @ ga --
[INFO] Using 'UTF-8' encoding to copy filtered resources.
[INFO]
[INFO] --- maven-compiler-plugin:3.8.1:compile (default-compile) @ ga ---
[INFO] Changes detected - recompiling the module!
```

```
[INFO] Compiling 1 source file to chapter06/skeleton/target/classes
[INFO] ------------------------------------------------------------------------
[INFO] BUILD SUCCESS
[INFO] ------------------------------------------------------------------------
[INFO] Total time:  2.962 s
[INFO] Finished at: 2019-12-07T15:18:39-06:00
[INFO] ------------------------------------------------------------------------
```

A successful build's output text

If you don't get a BUILD SUCCESS result, review your setup or compare it with the skeleton code version.

6.1.4 Preparing the database server Instance

The final step before we're ready to start our application is to configure the database that our application uses. With an RDBMS, we create a new database and explicitly define the schema, or data model elements (e.g., tables, columns, views, keys, indexes and so forth) prior to writing code. The Apache TinkerPop Gremlin Server doesn't require a predefined schema to be applied to the graph, although the implementations of other graph database vendors might.

Because we use Gremlin Server, we can start writing code without having to first apply a schema or data model to our graph; that is, of course, if you loaded the database as mentioned earlier in this chapter. If you want to check to see if your server is up and running correctly, you can do this using the command g within the Gremlin Console as shown here:

```
         \,,,/
         (o o)
-----oOOo-(3)-oOOo-----
plugin activated: tinkerpop.server
plugin activated: tinkerpop.utilities
plugin activated: tinkerpop.tinkergraph
gremlin> g
==>graphtraversalsource[tinkergraph[vertices:8 edges:12], standard]
```

Recall that g is our GraphTraversalSource. The data loading script uses the g that is preconfigured on the Gremlin Server. The response shows us that the server's graph has 8 vertices and 12 edges.

For the rest of this chapter, all of our interactions with the server use the application code. You can leave the Gremlin Console up and running if you like; we often do this while writing code, as it's helpful for investigating the data in a more interactive manner. This is similar to running Toad®, Microsoft's SQL Server Management Studio, or Oracle's Developer Tools along with your IDE to aid in investigating the state of the data or in debugging an application. With all this in place, we're ready to start building our application.

6.2 Connecting to our database

Now that our project is set up with the appropriate driver and a server is waiting for us to connect to it, our next step is to write the code to connect to our graph database. For this application, as with most, we connect to our graph database through a network-accessible endpoint, similar to how a JDBC driver connects to a relational database. Connecting to our graph database requires two steps:

- Setting up the database configuration object, called the `Cluster`
- Setting up the `GraphTraversalSource`

6.2.1 Building the cluster configuration

The first step in connecting to our graph database is to build the equivalent of a JDBC connection string for a relational database. In a TinkerPop-enabled graph database, this is known as the cluster configuration. This configuration specifies the server and port that our application uses to communicate with our graph database server.

To create the cluster configuration, let's look at it line by line. For those following along in the code, this section demonstrates the construction of the `connectToDatabase()` method. Let's begin by examining the required import to create a cluster configuration:

```
import org.apache.tinkerpop.gremlin.driver.Cluster;
```

This line should be quite familiar to any Java developer as it imports the required `Cluster` class from the Gremlin driver library. Similar to the JDBC `Connection` object, we use the `Cluster` object to send our traversals to the database. Examining the first line of the `connectToDatabase()` method, we see that it creates a `Cluster.Builder` instance:

```
Cluster.Builder builder = Cluster.build();
```

A `Cluster.Builder` instance is where we set the configuration parameters to connect to our database. The `Cluster` object uses the builder design pattern to allow for step-by-step construction of complex configurations. The next two lines show how to use the builder pattern to add the server address and port:

```
builder.addContactPoint("localhost");
builder.port(8182);
```

The first line specifies that we are going to connect to a server listening on everyone's favorite host, localhost. The second line defines the connection port as 8182, which is the default port for the Gremlin Server. As with any database, there are numerous other configuration options available for the cluster builder, such as specifying a username and password for authentication. However, to keep things simple, we only defined the two essential ones required for a connection.

Specifying a graph to connect with?

It may surprise you that we don't need to specify the name of a database (or a graph) in the connection configuration. It is common in relational databases to specify the name of the database because relational servers often host more than one database. However, most graph database servers only support a single graph per instance. With only a single graph, we don't need to specify which one to use.

There are several graph databases that do provide for more than one graph per server. In these databases, you must use the vendor-provided drivers. These include additional configuration parameters, like the graph name.

Our code now has a properly configured `Cluster` object, but we haven't yet connected to our server. Connecting to our server is performed by calling the `create()` method on our `builder` instance as shown here:

```
builder.create();
```

Putting this code together gives us this method:

```
import org.apache.tinkerpop.gremlin.driver.Cluster;        // Imports the required classes

public static Cluster connectToDatabase() {
    Cluster.Builder builder = Cluster.build();             // Creates our Cluster builder instance
    builder.addContactPoint("localhost");                  // Sets the server location
    builder.port(8182);                                    // Specifies the port to connect to on the database

    return builder.create();                               // Connects to our database
}
```

Woo-hoo, this is great! We're now able to connect to our server, which is a massive step in the right direction.

6.2.2 Setting up the GraphTraversalSource

With our configuration complete and our server connection created, the last step before we start running traversals is to create the traversal source we'll use to run those traversals. As discussed in chapter 3, a traversal source is an object upon which all our traversals operate, and a `GraphTraversalSource` object is how we represent the traversal source in Java. In other words, the `GraphTraversalSource` is the `g` in our `g.V()` and `g.E()`.

In previous chapters, we used a traversal source that came preconfigured on the Gremlin Console as part of our startup script. To achieve this same functionality in Java, we create a `GraphTraversalSource` object in our application. This section demonstrates the construction of that object using the `getGraphTraversalSource` method.

The first step to instantiating our GraphTraversalSource is to create a Traversal object. In Java, this uses a static function of the traversal.AnonymousTraversalSource class, traversal(). When called, traversal() returns a GraphTraversalSource object.

Because we connect across a network, we need to use the withRemote() method of our Traversal object to specify that we are making a remote connection. To accomplish this we pass in the cluster object from the previous section to the static Driver-Remote-Connection.using() method as shown here:

```
traversal().withRemote(DriverRemoteConnection.using(cluster));
```

Putting this all together, we get the following:

```
public static GraphTraversalSource getGraphTraversalSource(Cluster cluster) {
  return traversal().
    withRemote(
      DriverRemoteConnection.using(
          cluster)
    );
}
```

- return traversal(): Creates an instance of the static Traversal object
- withRemote(: Specifies that we want to create a remote connection
- DriverRemoteConnection.using(: Specifies that we want to use the hostname and port number
- cluster): Passes in the previously created cluster object

Excellent! This method returns a GraphTraversalSource object that we can use as the starting point for all of our traversals. We know that this seems a bit archaic. There are actually a few different ways to build and configure a remote connection to a traversal source, and sometimes a vendor provides specifics for use with their implementation.

The GraphTraversalSource modes: Remote/detached or embedded/API

The GraphTraversalSource's toString output is a little perplexing. What we need to keep in mind is that we configured this GraphTraversalSource for a remote connection. In the TinkerPop world, this is sometimes referred to as "being detached." What is meant by "detached" is that the graph database isn't located in the same processing/memory space as the driver and application. With that in mind, TinkerPop can run in two different modes:

- *Embedded mode*—The graph operates in the application's memory and process space instead of being accessed as an external resource. This configuration requires that we add some additional libraries to our project. We don't take this approach in this book.
- *Network mode*—The graph database operates as an external resource communicating via a network. This mode is how we most often think of working with other databases. It's the method we use when working with all popular graph database offerings on the market.

We use the network mode throughout this book because it is portable across most graph database vendors, as long as these support the Apache TinkerPop APIs.

Putting these two methods (`connectToDatabase` and `getGraphTraversalSource`) together, we can make the following simple program that configures and connects to our network-enabled graph database and that prints some basic information:

```
public static void main( String[] args ) {
    Cluster cluster = connectToDatabase();
    System.out.println("Using cluster connection: " + cluster.toString());
    GraphTraversalSource g = getGraphTraversalSource(cluster);
    System.out.println("Using traversal source: " + g.toString());
    cluster.close();
}
```

Remember that, by convention, we use the variable `g` to represent our `GraphTraversalSource`, but that's an arbitrary designation. The terms *gts, traversal,* or *graphSource* are just as suitable. We're a couple of conventional guys, so we stick with `g`. Running our program yields the following:

```
$ mvn -q clean compile exec:java -Dexec.mainClass="com.diningbyfriends.App"
Using cluster connection: localhost/127.0.0.1:8182
Using traversal source: graphtraversalsource[emptygraph[empty], standard]
```

Yay! Based on these results we see that we can connect to our database.

More details about GraphTraversalSource and TinkerPop's strategies

One aspect you might have noticed when looking at the previous results is that our traversal source is empty and that it had a second parameter `standard`. What's that about?

We mentioned that a `GraphTraversalSource` is a process that knows how to navigate the data store (e.g., the graph). There are several ways to find data within our graph, with two of the most common approaches of traversing a graph being depth-first and breadth-first. Additionally, many other optimizations can be applied, depending on the specifics of the graph database implementation.

Apache TinkerPop calls these different approaches *strategies* and has developed several of these as part of TinkerPop. When a `GraphTraversalSource` is created, two things happen:

- It's associated with a specific instance of a graph database.
- It sets up a collection of strategies optimized for finding data within the associated graph.

For the second part, by default, the `GraphTraversalSource` applies a standard set of strategies. There's a lot more that could be said about strategies, but that material is beyond the scope of this book (see http://mng.bz/DzN9 for details). All that's important to understand is that each vendor provides their own strategies, which allow them to optimize how you traverse through the graphs in their implementations.

Now that we can communicate with our database, let's start running some traversals on our DiningByFriends social network graph. Before we do that, however, here's one last tip about the `GraphTraversalSource`: setting it up is an expensive process in much the same way as creating an ODBC/JDBC connection to a database. Therefore, the best practice is to create one `GraphTraversalSource` object and reuse it for each traversal for the lifespan of the application.

6.3 Retrieving data

Now that we're set up, it's time to write Java code to traverse the graph. In many development scenarios, developing traversals and adding these to an application would be done simultaneously with the traversal building. However, to ease the learning process, we decided to separate these into three steps to break down the process into its component parts: data modeling (done in chapter 2), traversal writing (done in chapters 3–5), and now application development.

General implementation process

Although the focus in this chapter is implementing the Gremlin traversals in Java code, this requires some basic plumbing for each traversal. Here, we briefly outline the changes needed in the implementation of each feature of the application; then we cover these in detail with the first example. After that, we only focus on the Gremlin implementations for the remaining cases in the chapter because these details are basic application development concerns and not specific to working with graph databases. Each feature implementation requires the following set of changes:

- Add a menu entry in `showMenu()`.
- Insert a switch case in `displayMenu()`.
- Add a method to make the graph traversal and return the results.

If you're using the skeleton version of the code, then you need to develop the relevant portions of the code as we move through this chapter. If you're using the commented version of the code for this chapter, you need to uncomment the relevant portions as we work through this chapter. And finally, if you're using the completed version of the code, then everything exists, so you only need to follow along. Now that we know our general approach, let's work through our examples.

6.3.1 Retrieving a vertex

As we did in chapter 3, let's start with the least complicated traversal, which is retrieving a single vertex. Because you're already old pros at building this type of traversal, using this as a starting place allows us to focus on demonstrating the aspects unique to doing this in code; namely, how to traverse our graph from code and how to process the data we receive from our graph. To begin, in order to find Ted in our application, we need to take the following steps:

1. Connect to our database.
2. Create our `GraphTraversalSource`.

3 Run our traversal to find Ted.
4 Process the results.

Luckily for us, we accomplished the first two steps in the last section. We already have code that connects to our database and creates our `GraphTraversalSource`, so the next question is how we run our traversal to find Ted. Fortunately, we already know how to write the traversal:

```
g.V().has('person', 'first_name', 'Ted').valueMap()
```

But how do we go about actually running this from our Java code? If we were to write this for a relational database application, our code would look like this:

```
Statement stmt = conn.createStatement();
ResultSet rs = stmt.executeQuery("SELECT * FROM person WHERE first_name='Ted'");
while(rs.next()) {
    //Process results
}
```

How does this look like for a graph-backed application? Actually, running this traversal in code closely resembles the Gremlin statement. The Java equivalent of the traversal is

```
//This returns a list of the properties
List properties = g.V().
        has("person", "first_name", name).
        valueMap().toList();
```

That's right; just a single line of Java replaces the entire code block for a relational database application. Our Java equivalent not only performs the traversal, but it also returns our results. In this case, the result is a list of objects, each containing all of the key-value pairs of each property. That's surprising, isn't it? It's probably the first time in this entire book that doing something with a graph database is less work and more straightforward than its relational counterpart. Comparing the two code examples, we notice two major differences:

- The traversal is built with Java functions instead of strings.
- The traversal ends with a `toList()` step instead of a `ResultSet`.

That's right! We built our traversal using Java functions from the TinkerPop driver. This is instead of manipulating strings and submitting the strings as is common with an SQL query in JDBC or ODBC. This use of functions instead of strings is a feature unique to Gremlin and is known as the Gremlin Language Variant (GLV). For more details, see http://mng.bz/nz2v.

NOTE The next section discusses the GLVs that are a Gremlin-specific construct. This construct does not have a comparable alternative in other query languages such as Cypher. If you are not using a TinkerPop-enabled database, feel free to skip over this section if this concept does not apply to your database.

6.3.2 Using Gremlin language variants (GLVs)

A GLV is a TinkerPop-only feature that provides language-specific implementations of the TinkerPop interfaces. Gremlin was built to be embedded into multiple programming languages and uses the constructs of that language to represent the traversal. This allows the user to compose a traversal by using functions rather than string manipulations as is standard in other graph and relational databases. The fact that GLVs are language-specific means that we can leverage our IDE's code-completion functionality, as well as gain the ability to have typed objects both as part of our traversals and in the results.

If you're familiar with .NET, GLV seems similar to the LINQ (Language Integrated Query) component that's prevalent in that ecosystem. At the time of this writing, there are GLVs for Java, Groovy, JavaScript, Python, and .NET. In practical terms, this means that our traversals in our graph application are a series of functions strung together instead of the all-too-familiar paradigm of concatenating strings to build an SQL query.

Note that it is possible to take the string concatenation approach with Gremlin. Instead of establishing our own `GraphTraversalSource`, we simply create a string of the traversal and use a client object to submit it to the cluster, as in the following example:

```
Cluster cluster = Cluster.open();
Client client = cluster.connect();
ResultSet results = client.submit("g.V().hasLabel('person')");
```

This approach goes by several names: string submission, Groovy script, or script submission. It can be parameterized, and the `submit()` method is overloaded to handle various request options. However, this approach is discouraged for several reasons:

- Serializing and deserializing of strings incurs additional overhead, which can be significant, which is why prepared statements are recommended for use in SQL.
- The use of string concatenation opens your code up to SQL injection ("Gremlin injection?") types of attacks, unless there's a consistent use of parameterization. We all know what havoc little Bobby Tables or his cousin Jimmy G dot V Drop Iterate can wreak with his last name's legal spelling of `; g.V().drop() .iterate()`. (This is a little joke based on the ideas of the XKCD comic about Bobby Tables: https://xkcd.com/327/.) This code deletes all the data in the graph!
- There's no ability to use the code validation and completion capabilities of modern IDEs.
- `ResultSet` results must still be coerced into a type of object, but with GLVs, the response is automatically strongly typed.

We find that using a GLV provides significant advantages that greatly simplify the development process. It does require that we translate each of the traversals we write to be handled in the programming language of choice, such as Gremlin Console, for example. This translation is the most simple with Java because it usually means adding a terminal step like `next()` or `toList()` and then handling the strongly-typed response appropriately.

We believe that the advantages of using GLV outweigh the additional work required to translate the strings to a native language. Additionally, the future of support for string-based traversals in TinkerPop is unclear at this time. Because of this, we strongly encourage you to use GLV-style traversals whenever possible. However, at the time of writing this book, not all TinkerPop-enabled databases support GLV-based traversals. Some only support string-based ones, which is why we provide an example for that approach as well.

6.3.3 Adding terminal steps

Returning back to the other difference from our example in section 6.3.1, in the application implementation, we added the `toList()` step. This step is another of the terminal steps we discussed in chapter 4.

> **NOTE** When writing traversals in an application, we must end our traversal with a terminal step (http://mng.bz/v9ex). We can't emphasize this requirement enough. Gremlin is a lazily-evaluated language, so if you don't end your traversal with a terminal step, the traversal won't return a result.

When we use Gremlin Console, it automatically adds terminal steps to force the evaluation. When constructing an application, however, the impetus is on us to add terminal steps. If we don't, we get a `GraphTraversal` object, which isn't helpful when we want our data.

> **IMPORTANT** Forgetting to force the evaluation of a traversal is one of the most common problems we encounter when debugging applications. This oversight bites even the most experienced of us from time to time.

This "always take a `GraphTraversal` as input, always return a `GraphTraversal` as output" feature of Gremlin steps allows for great flexibility with the composition of complex statements. However, it does require that we take the extra step to tell it we're done and want our results. For example, the following two code snippets perform the same steps and produce the same outcome:

```
return g.V().count().next();
```

and

```
GraphTraversal t = g.V();
t = t.count();
return (Long)t.next();
```

In the second snippet, the first line creates a `GraphTraversal`, called `t`, and defines it with the vertex step, `V()`. Then `t` is reassigned to itself, followed by a `count()` step.

Finally, the code returns the result, iterated with the `next()` step, which is coerced to a `Long` value.

This ability to chain together various `GraphTraversal` objects with multiple statements can be useful. We could utilize this approach to include a filter step under certain conditions in the middle of a Gremlin traversal. But we find it's more natural to handle certain types of branching and flow control as part of Java logic to compose a simple traversal for a complex use case than it is to write the corresponding Gremlin to handle each possible permutation required by our application. We make this judgement call regularly with relational databases as well, choosing to write some business logic in the application and maybe other logic with SQL, depending on the specific use-case requirements.

There are several other terminal steps to be aware of as these provide convenient mechanisms for returning data from your traversal in easily consumable ways. Some of the other terminal steps include

- `hasNext()`—Returns a boolean value: `true` if there are available results, `false` if there are no results.
- `tryNext()`—A convenience method that is a combination of the `hasNext()` and `next()` steps to execute the traversal if there are available results. It returns a Java `Optional` and is only available in JVM languages like Java and Groovy.
- `toList()`—Returns the results of the traversal as a Java `List`.
- `toSet()`—Returns the results of the traversal as a Java `Set`.

NOTE Not all of these terminal steps are available in the non-Java GLV.

6.3.4 Creating the Java method in our application

Now that we have looked at the pieces required to create a Java method for our traversals, let's see what it takes to create a method in our application to find Ted. To demonstrate this method creation process, we create a method named `getPerson()`, which finds a vertex by a person's first name (e.g., Ted). This method needs to take the following steps:

1. Pass in our `GraphTraversalSource`
2. Get the name of the person we want to find (Ted)
3. Run our traversal to find Ted
4. Process the results

As we can see from these steps, to create this Java method, we want to pass in the `GraphTraversalSource` because we want to reuse it for the lifetime of the application. We then require some necessary boilerplate code to read the name of the person we want to find from the command line, because it won't always be Ted. We then take this input and run our traversal to give us the list of properties. Finally, we return the properties to the calling method (in your code sample, this is the `getPerson()` method):

```
public static String getPerson(GraphTraversalSource g) {
    Scanner keyboard = new Scanner(System.in);
    System.out.println("Enter the first name for the person to find:");
    String name = keyboard.nextLine();

    //This returns a list of the properties
    List properties = g.V().
            has("person", "first_name", name).
            valueMap("first_name").toList();

    return properties.toString();
}
```

Passes in our GraphTraversalSource

Boilerplate code to read the person's name

Runs our traversal and returns a list

Returns the list of first_name properties

NOTE While we went into detail on how we implemented this method for this example, we won't do so for each of the additional methods we add. For each method, we point out the name and call attention to any critical portions that are relevant, but we leave it as an exercise for the reader to take a look at any details of the boilerplate code one might be curious about.

Now that our application can retrieve data from our graph database, let's revisit our traversals from the last chapter and see how we add, modify, and delete vertices and edges from the graph.

6.4 Adding, modifying, and deleting data

Among the most common tasks applications perform are adding, updating, and deleting data. In chapter 4, we went through the process of creating these traversals, and in this section, we walk through each option and show some of the unique aspects of how to process these within a graph application.

6.4.1 Adding vertices

In chapter 4, we went through how to mutate our graph to add data. Let's look at adding a new vertex to our graph. Remembering back, we wrote a traversal to add a person to our graph:

```
g.addV('person').property('first_name', 'Dave')
```

Now, how do we go about taking that traversal and turning it into code? In the last section, we showed how simple it was to use the GLV in our code compared to the JDBC approach or to a Groovy script. What we didn't discuss was precisely how we translated the Gremlin script we used in the Gremlin Console to the Java code, so we're going to rectify that now.

TRANSLATING YOUR GREMLIN TRAVERSAL INTO THE GLV

To translate the string-based traversal to a GLV-type traversal, we take each step in the string traversal and replace it with an identically named Java method. Yes, seriously, that's it! It's a bit anticlimactic, isn't it? The good news is we really aren't kidding.

Because Gremlin is built on the JVM, the syntax is the same, at least in Java. The other piece that we do need to add is the common imports (http://mng.bz/4BMB). While we haven't needed to include any of these so far, the imports are required to write more complex traversals.

We know we said that it's as easy as copying and pasting our script to translate from the Gremlin Console to an application, right? Well, in Java, that's also the case, but the same isn't always true for other languages.

Each GLV is built to provide a native experience in the target language. This native experience also means that we inherit the native casing, reserved words, and other features of each language. So, for example, if we build our application in .NET, we need to capitalize the first letter of each method instead of using camel case (e.g., `HasNext()`, not `hasNext()`). Or, if we use Python, we must postfix an underscore (_) to functions such as `and()`, `as()`, `from()`, and so forth because these are reserved words in Python (e.g., `from_()`).

Moving from GLV to Java

In our opinion, the ability to quickly build and maintain our traversals consistently with native language standards far outweighs the minor inconsistencies across target languages. Using this methodology, let's take the Gremlin traversal we used in the console:

```
g.addV('person').property('first_name', 'Dave')
```

Translating that into a Java statement, this becomes

```
g.addV("person").property("first_name", name).next();
```

A keen observer will notice that we changed from single quotes to double quotes. Gremlin, owing to its Groovy roots, accepts either double or single quotes for strings. But we're using the Java GLV, and Java has stronger opinions, leading to the double quotes for strings.

Well, that was pretty easy, but what does our statement return? Remembering back to when we ran this in the Gremlin Console, we get a reference to a vertex:

```
g.addV('person').property('first_name', 'Dave')
==>v[13]
```

How's this vertex reference returned in Java? It's returned as a `Vertex` object. Inside the TinkerPop framework, there are objects to represent many of the graph-specific structures (e.g., `Vertex`, `Edge`, `Path`, etc.) that we've discussed so far. This inheritance is one of the main benefits of using GLVs: results aren't generic objects that require a lot of additional coding to handle correctly. As a `Vertex` object, our vertex comes with several built-in properties and methods that simplify common operations. Putting this all together, we get the following method (for those following along in the code, it's called `addPerson`):

```
public static String addPerson(GraphTraversalSource g){
    Scanner keyboard = new Scanner(System.in);
    System.out.println("Enter the name for the person to add:");
    String name = keyboard.nextLine();

    //returns a Vertex type
    Vertex newVertex = g.addV("person").
           property("first_name", name).next();

    return newVertex.toString();
}
```

Adds a vertex and returns a strongly-typed object

In our case, we decided to use a `toString()` method to return the value. If this were a real application, we'd likely want to massage our `Vertex` object result in an appropriate form to return our customers. But because we have a strongly-typed object to work with, we're in familiar territory for Java developers.

6.4.2 Adding edges

Now that we know how to add vertices to our graph application, let's look at how to add edges. Looking back to section 4.1, we wrote this traversal to add edges:

```
g.addE('friends').
  from(
    V().has('person', 'first_name', 'Dave')
  ).
  to(
    V().has('person', 'first_name', 'Josh')
  )
```

Let's take this traversal and translate it into Java code using the process discussed in the last section. Doing that, we get this Java code (assuming that from and to names are read into the variables; for example, `fromName` and `toName`):

```
Edge newEdge = g.addE("friends").
        from(V().has("person", "first_name", fromName)).
        to(V().has("person", "first_name", toName)).
        next();
```

This code is great, except that our editor is full of red squiggly lines, telling us something is wrong. When we hover over the lines, we see an error message that says something like "Cannot resolve method V()." Wait. That doesn't make sense. We used `V()` earlier in this traversal. So how is it *not* a valid method?

OK, you caught us; we didn't exactly lie when we said that translating from Gremlin Console to Java code was as easy as just replacing the string value with the identical Java method, but we did omit a few details. One of those details is that in some cases, such as this one, we need to add some code in as well. In this case, we must add a double underscore in front of the mid-traversal `V()`'s so that our Java code now looks like this:

```
Edge newEdge = g.addE("friends").
        from(__.V().has("person", "first_name", fromName)).
        to(__.V().has("person", "first_name", toName)).
        next();
```

Whew, that removes those red squiggles! But the more obvious question is, why are the squiggles gone?

The anonymous traversal

The double underscore element is called an *anonymous traversal*. It's a feature of the Gremlin language and, as far as we know, doesn't have a direct counterpart in other graph languages. It's like an anonymous function in languages such as Java or JavaScript. We use an anonymous traversal when we need to start another traversal within an existing one.

There are two common places where we use an anonymous traversal. The first is in the example in this section. The `to()` step is a modulator that can take either a string (usually referring to a previous `as()` step) or a traversal itself. In our case, we start a new traversal with the `to()` step and cannot reuse the `g`, so we instead start with the anonymous traversal.

The other case where we use the anonymous traversal is when a traverser must start with a step that is also a keyword in Groovy, such as the `as()`, `in()`, and `not()` steps. We don't have any examples like this at present, but we'll note the need for the anonymous traversal when we encounter these.

Now let's return to our construction of `addFriendsEdge()` method. To do that, we put this all together and take a look at how we add a `friends` edge in our application:

```
public static String addFriendsEdge(GraphTraversalSource g){
    Scanner keyboard = new Scanner(System.in);
    System.out.println("Enter the name for the person at the edge start:");
    String fromName = keyboard.nextLine();
    System.out.println("Enter the name for the person at the edge end:");
    String toName = keyboard.nextLine();

    //This returns an Edge type
    Edge newEdge = g.addE("friends").
      from(
        __.V().has("person", "first_name", fromName)      <-- Uses an anonymous traversal for fromName
      ).
      to(
        __.V().has("person", "first_name", toName)        <-- Uses an anonymous traversal for toName
      ).
      next();

    return newEdge.toString();
}
```

In the code, we introduce a new object, the `Edge` object. As we've said throughout, edges are first-class citizens in graph databases, so our `addE()` step returns an `Edge` object, just as in the last section, the `addV()` step returned a `Vertex` object.

6.4.3 Updating properties

Our next task, updating a property on a vertex, is a good opportunity for you to try to translate one of these traversals on your own. In this case, let's take the traversal we created to update a person's name and translate it into Java.

> **EXERCISE** Translate the following Gremlin traversal into the appropriate Java code:
>
> ```
> g.V().has('person', 'first_name', 'Dav').property('first_name', 'Dave')
> ```

Trust us. We promise you there are no hidden surprises. Go ahead, create an `updatePerson` method and debug it like you would any other method. Take your time and work through it. We'll wait . . . .

Great, you're back! Now that you've successfully written your Java function, compare it to our version:

```
public static String updatePerson(GraphTraversalSource g){
    Scanner keyboard = new Scanner(System.in);
    System.out.println("Enter the name for the person to update:");
    String name = keyboard.nextLine();
    System.out.println("Enter the new name for the person:");
    String newName = keyboard.nextLine();

    //This returns a Vertex type
    Vertex vertex = g.V().
      has("person", "first_name", name).        <- Finds a person by first_name
      property("first_name", newName).          <- Sets their newName
      next();                                   <- Remembers the terminal step

    return vertex.toString();
}
```

How did your solution compare to ours? Hopefully, it's similar. But just like everything else with coding, there's more than one correct way to solve a problem. In the end, as long as we both get it working correctly, we're both right!

6.4.4 Deleting elements

So far, we've dealt with how to perform the addition of vertices and edges within our application, and you were adventurous enough to create a method to update properties by yourself! We worked through three of the four everyday mutation actions. Now it's time to introduce the last one—deleting elements from our graph.

Let's say that in addition to finding, adding, and updating people, our application also needs to remove people. This means that we want to delete a `person` vertex from our graph by `first_name`. Applying what we learned in chapter 4, we come up with a Gremlin traversal like this:

```
g.V().has('person', 'first_name', 'Dave').drop().iterate()
```

Translating that traversal gives us the following Java code:

```
g.V().has("person","first_name", "Dave").drop().iterate()
```

Running this in our application, we get the result `null`. As we discussed in chapter 4, this is because `drop()` steps in Gremlin don't return a value. However, a user will likely want to receive some sort of feedback from the application—probably something that tells them the delete operation completed and made the change to the graph. While this seems like a typical pattern for an application, it requires a little extra work to accomplish.

So how do we go about returning the count from a step that does not provide it to us? This is an excellent question, and the answer ends up being more complicated than you might expect. To return the count of deleted vertices, we need to discuss a new concept: *side effects*. For that, we need a few additional steps.

TinkerPop's official documentation and picking Gremlin steps

We approach this side-effect example as we commonly see others do who are new to Gremlin. This approach may even bear a striking resemblance to the path we took the first time we had to build this type of functionality.

The starting place for figuring out all of this is, of course, this book. But because, sadly, not everyone has a copy of this book (we certainly didn't have a copy when we started), new Gremlinistas turn to the Apache TinkerPop reference web site, http://tinkerpop.apache.org/docs/current/reference/.

TinkerPop's reference site is excellent because it's the authoritative source of documentation and is replete with examples for each Gremlin step. However, we find that the site is sometimes confusing, as it starts with a fanciful summary of TinkerPop's history. This can leave you wondering if you're in the right place for technical documentation. It also organizes the steps in alphabetical order. This order is perfect for a reference, but it's all too easy to find a step that "appears" to do something like what's needed and to miss a step that's more appropriate for our needs.

Side-effect steps are different. These take in a `GraphTraversal`, perform some set of operation based on that input, and then return the input as the output. The practical result is that no matter what occurs inside the side effect, the original input data is output precisely as it was input. We can take an action on data without altering the original inputs. We're sure this is a bit confusing, so let's look at a specific example using side effects to get a count of deleted vertices. We'll look at a few steps that appear to be good candidates for counting the number of vertices dropped:

- `sideEffect(traversal)`—Processes the provided traversal as an additional process without effecting the results passed to the next step.
- `store(alias)`—Stores the results of the traversal collection specified by the alias.
- `cap(alias)`—Emits the results collection specified by the alias.

Our first attempt to get the count of dropped vertices likely involves using the `store()` step to obtain the count of vertices removed:

```
Object vertexCount = g.V().
    has("person","first_name", name).       <-- Finds a person by first_name
    sideEffect(__.count().store("x")).       <-- Stores the number of people with this name as "x"
    drop().iterate().                        <-- Deletes everyone with this name
    cap("x").next();                         <-- Returns the count stored in "x"
```

When we run this code we get an error! The traversal strategies are complete, and the traversal can no longer be modulated. Why the error? What this error is stating is that we try to call a step (in this case, `cap()`) after the traversal has already hit a terminal step, `iterate()`. There are two key aspects to terminal steps we haven't discussed yet:

- A traversal can only have one terminal step.
- No further steps can be called after a terminal step.

Looking at our traversal, we see that our traversal contains two terminal steps: `iterate()` and `next()`. Our traversal also processes several steps (`cap("x").next()`) after our `iterate()` step. We violated both of these rules! Because this doesn't work, let's take a moment and consider how to approach this differently.

Think about what we know about graph traversals and see if you can think of a different approach to this problem. Were you able to come up with some ideas? What do you think would happen if we switched the steps in our traversal to include the `drop()` step in the `sideEffect()` step and the `count()` step in the main traversal?

```
Long vertexCount = g.V().
  has("person","first_name", name).    <-- Finds a person by first_name
  sideEffect(__.drop()).               <-- Deletes everyone with this first_name
  count().                             <-- Returns the count
  next();                              <-- Runs a terminal step

  return vertexCount.toString();
```

Try it out and see if this code works. Well, the good news is that it works; the bad news is that this is a strange way to make this work. If you're confused as to why this second one works when the first one doesn't, don't despair. This confusion is something that happens in Gremlin from time to time as you're learning to think about these problems from a traversal perspective. Let's take a moment to remember what we're doing in terms of traversing our graph:

1. Find all the person vertices with a given name.
2. For each of those vertices, `drop()` it from our graph.
3. For all of the original vertices, return the `count()`.

When we think about what this traversal is doing, we can see why this new approach works. Because we are now using a side effect to perform the deletion, we are able to retrieve the original count. Coming from a relational background, though, it seems confusing and backward.

When you run into this sort of frustration or confusion, we recommend taking a step back and thinking about what you're trying to accomplish and how you would approach this problem by traversing through the graph structure. We find that this approach frequently allows us to break our learned bias for handling issues from the relational perspective.

6.5 Translating our list and path traversals

Whew, that took some work! We now have a framework and some of the basic functionality of our social network. Now let's circle back to the requirements of the DiningByFriends social network from chapter 2 and write some methods to retrieve this data. As you may recall, the social network use case for DiningByFriends had three questions that need to be answered:

- Who are my friends?
- Who are the friends of my friends?
- How is person *X* connected to person *Y*?

In this section, let's take the traversals we previously wrote and turn these into methods in our application. We aren't going to go into this in depth, however, but we'll highlight a few key items as we move through each question.

When we have a series of traversals to implement, we like to see if there's a natural order in which to approach those, building on one another in progression and enabling some copy and pasting of the code. Looking at these questions, it seems like the second question—"Who are the friends of my friends?"—is likely an extension of "Who are my friends?" As a result, it looks like implementing "Who are my friends?" would be an excellent place to begin.

6.5.1 Getting a list of results

We start by answering the question, "Who are my friends?" Let's review our process of building a method in our application:

1. Write our traversal.
2. Translate our traversal into the Java equivalent.
3. Process the results.

Back in chapter 3, we wrote the traversal to retrieve all the friends of Ted:

```
g.V().has('person', 'first_name', 'Ted').
  out('friends').dedup().
  values('first_name').next()
```

Using this traversal as a starting point, we obtain a Java equivalent of

```
List<Object> friends = g.V().has("person", "first_name", name).
            out("friends").dedup().
            values("first_name").
            toList();
```

Happily, this completes the first two steps, leaving us with processing the results. As we see, the traversal only requires changing the terminal step to `toList()` instead of `next()` for our traversal to work in the application. One question remains: Why are we are returning a `List<Object>` instead of some stronger typed class?

We return a more ambiguous class because, unlike a SQL query, a graph traversal isn't required to return properties of the same data type. This ability to return different data types provides quite a bit of flexibility for the traversal, but has the downside of ambiguity in the nature of data returned. Taking this traversal and combining what we've learned about how to run traversals and return results in Java yields this `getFriends()` method:

```
public static String getFriends(GraphTraversalSource g){
    Scanner keyboard = new Scanner(System.in);
    System.out.println("Enter the name for the person " +
        "to find the friends of:");
    String name = keyboard.nextLine();

    //Returns a list of objects representing
    //the friend person vertex properties
    List<Object> friends = g.V().                      // Expects a list of objects
            has("person", "first_name", name).
            out("friends").dedup().
            values("first_name").
            toList();                                  // Returns a list from our traversal

    return StringUtils.join(friends, System.lineSeparator());
}
```

6.5.2 Implementing recursive traversals

You've learned a lot so far in this chapter, so we think it's time for you, once again, to practice what you've learned and tackle another exercise on your own. In this case, let's extend the `getFriends()` method we built in the last section to answer, "Who are the friends of my friends?" To solve this, you may want to refer to the traversal we wrote to answer this question for Ted:

```
g.V().has('person', 'first_name', 'Ted').
    repeat(
      out('friends')
    ).times(2).dedup().values().next()
```

> **EXERCISE** Create a `getFriendsOfFriends()` method that returns the answer for, "Who are the friends of my friends?" How does your code look? Does it return the expected results?

Let's look at our answer:

```
public static String getFriendsOfFriends(GraphTraversalSource g){
    Scanner keyboard = new Scanner(System.in);
    System.out.println("Enter the name for the person " +
        "to find the friends of:");
    String name = keyboard.nextLine();

    // Returns a list of objects representing the vertex properties
    // of the friend of a friend person
    List<Object> foff = g.V().                          <-- Expects a list of objects
            has("person", "first_name", name).
            repeat(
                    out("friends")                      <-- This requires an additional import.
            ).times(2).dedup().
            values("first_name").                       <-- Returns only the first_name property
            toList();                                   <-- Returns a list from our traversal

    return StringUtils.join(foff, System.lineSeparator());
}
```

If you came up with something similar, then it's also likely that your IDE prompted you to add an import for the `out()` method:

```
import static org.apache.tinkerpop.gremlin.process.traversal.dsl.graph.__.out;
```

This is because the `out()` method requires a starting traversal source. If we didn't want to include the import statement, we'd use an anonymous traversal like this:

```
List<Object> foff = g.V().has("person", "first_name", name).
        repeat(
                __.out("friends")                      <-- Requires an anonymous traversal when not adding a static import for out()
        ).times(2).dedup().
        values("first_name").
        toList();
```

The choice of which one you use is up to you. Both provide the expected results.

6.5.3 Implementing paths

For the final section of this chapter, we look at how to implement the last question of our social network use case, "How is person *X* connected to person *Y*?" To do that, this section demonstrates the construction of the `findPathBetweenPeople()` method. Let's start by taking a look at where we ended with this traversal in chapter 4:

```
g.V().has('person', 'first_name', 'Ted').
  until(has('person', 'first_name', 'Denise')).
  repeat(
    both('friends').simplePath()
  ).path().next()
```

Using what we learned, we can translate this into the following Java code:

```
List<Path> friends = g.V().has("person", "first_name", fromName).
    until(has("person", "first_name", toName)).
    repeat(
      both("friends").simplePath()
    ).path().toList();
```

This code looks similar to what we've done to date, with the exception that we return a list of `Path` objects instead of the more generic `Object` type used previously. We could also point out that we use `both()` instead of `out()` like we did for `getFriends()` and `get-FriendsOfFriends()`. This is a simple illustration of how we can use the direction of the edges in one context (friend-finding) and ignore it in another (pathfinding between two people), all with the same data set within the same set of use cases.

Remember, `path()` returns not just a single vertex or edge, but all the intermediate steps. Because we know what the return type from our traversal is, we can use `List<Path>` instead of the more generic `List<Object>`. The `Path` class is a particular class within the TinkerPop Java driver that contains an ordered list of objects; each object represents an individual vertex or edge in the path. With this understood, we finalize our method to get one that returns the paths between people as shown here:

```
public static String findPathBetweenPeople(GraphTraversalSource g){
    Scanner keyboard = new Scanner(System.in);
    System.out.println("Enter the name for the person " +
        "to start the edge at:");
    String fromName = keyboard.nextLine();
    System.out.println("Enter the name for the person " +
        "to end the edge at:");
    String toName = keyboard.nextLine();

    // Returns a List of Path objects which represent
    // the path between the two person vertices
    List<Path> friends = g.V().                          <-- Expects a list of paths
            has("person", "first_name", fromName).
            until(has("person", "first_name", toName)).
            repeat(
```

```
            both("friends").simplePath()      <-- Only finds simple paths
        ).path().      <-- Yields the Path objects
toList();      <-- Returns a list from the traversal

    return StringUtils.join(friends, System.lineSeparator());
}
```

Congratulations! You've reached a milestone: you now have a fully functioning console-based application for the DiningByFriends social networking use case, all backed by a graph database. You know how to add, edit, and remove users and edges, as well as how to answer the three requirements for the social networking part of the app.

In the next chapter, we'll start adding more features to our model to handle both our restaurant recommendation and the personalization of our recommendation results. Along the way, you'll learn how to manage some of the more complicated graph data modeling scenarios that we didn't cover previously.

Summary

- Setting up a graph-backed application is similar to any data-backed application and entails selecting our tools, setting up our project, obtaining the correct driver, and preparing our database server.
- Connecting to a graph database involves configuring the appropriate network client and setting the `GraphTraversalSource` for TinkerPop databases only.
- Apache TinkerPop provides several Gremlin Language Variants (GLVs). These enable us to translate our previously written Gremlin traversals into native language functional calls. This means that we are able to use functions in Java, .NET, Python, or JavaScript to write our traversals instead of using string manipulations.
- Using GLV leverages the IDE's code completion functionality and also utilizes strongly-typed results.
- Terminal steps are required when writing application code. Unlike Gremlin Console, application drivers do not automatically add these.
- Translating traversals for use in our application is usually as easy as replacing each Gremlin step with a method call of the same name.
- Solving problems in a graph requires a mind shift to approach these from a perspective of how to traverse the graph to answer the problems.

Part 2

Building on Graph Databases

In part 2 of this book, we continue our journey into working with connected data and graph databases. Having become familiar with the basics of graph data modeling and building graph-backed applications, we're going to stretch your new-found skills by tackling two common graph data patterns—known walks and subgraphs. Along the way, you'll learn how to construct more complex data models and traversals.

As in part 1, we start by creating a data modeling problem to demonstrate these graph patterns. In chapter 7, we extend our existing data model to introduce the challenge of working with multiple types of vertices and edges. This new challenge requires us to learn and apply several additional data modeling concepts to graphs. In chapter 8, we introduce the known-walk traversing pattern, which we apply to construct the recommendation functionality of our graph-backed DiningByFriends application. We then wrap up this part in chapter 9, where we introduce the concept of subgraphs as we address the personalization use case for DiningByFriends.

7 Advanced data modeling techniques

This chapter covers

- Applying the data modeling process to more complex use cases
- Improving performance by using generic labels
- Denormalizing data for more efficient graph traversals
- Moving properties to edges to simplify traversals

So far, we've walked through the entire process of building a simple graph application. We went from data modeling to traversal construction to coding a Java application for the recursive and pathfinding traversals that we used in our social network. Although the model for our social network was simplistic, it allowed us to demonstrate the patterns and processes required to build graph-backed applications, while showing some of the powerful tools, such as recursive and pathfinding traversals, that graphs bring to the table.

These basic modeling steps provided us with a strong foundation and worked well on the relatively simple social network data model. But as the complexity of our data model increases, these basic steps need to be combined with additional techniques in order to create a logical data model capable of handling any scale of data. Most real-life applications, like recommendation engines or personalization applications, are more complicated than the one-vertex, one-edge data model required by our social network. In this chapter, we'll construct the data models for the recommendation and personalization use cases of DiningByFriends to explore three advanced data modeling techniques frequently used in more complex models:

- Increasing performance with generic data labels
- Simplifying traversals by moving properties to edges
- Creating more efficient traversals with data denormalization

We'll show you not only how to apply each of these techniques, but when to apply them and how they help to improve your model. We'll then demonstrate these techniques by applying them to the process of extending our existing model with the vertices, edges, and properties required for our restaurant recommendation engine. Finally, we'll wrap up this chapter with a challenge, and ask you to take on the task of extending the data model for the personalization use case.

By the end of this chapter, we'll complete the working data model for DiningByFriends, and you'll learn additional skills for building better, more scalable data models. However, before we jump into the recommendations and personalization use cases, let's take a moment to review the current state of our conceptual and logical data models.

7.1 Reviewing our current data models

Our focus this chapter is on demonstrating advanced data modeling techniques, but we won't start from scratch. We start with our current data model as defined in chapter 2, and then extend it for the recommendation and personalization use cases. The logical place to begin is to take stock of our current data model before making changes. Remember, we built this data model with a four-step process:

1. Defining the problem
2. Creating the conceptual data model
3. Creating the logical data model
4. Testing the model

Looking back to the work we did in chapter 2, we see that we completed the first two steps for the social network use case and also started on the recommendation engine and personalization use cases. This resulted in the conceptual model shown in figure 7.1.

We also completed the third step, creating the logical data model, but only for our social network use case. Figure 7.2 shows the resulting logical data model thus far.

This logical data model was the basis for all of the traversals written in chapters 3 through 5 and the implementation we completed in chapter 6. We got a lot of code out of that one little picture, and we hope you appreciate how helpful it is to think through the essential design points up front. However, that is as far as this data model can take us by itself. In order to model more complex scenarios, we'll need to extend

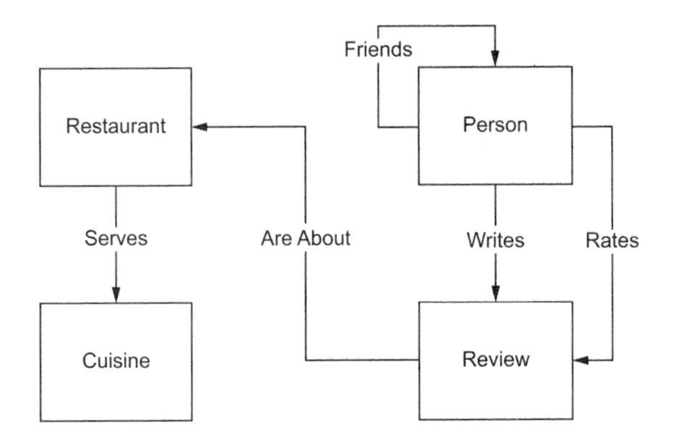

Figure 7.1 Conceptual data model showing the entities (boxes) and relationships (arrows) for the DiningByFriends app

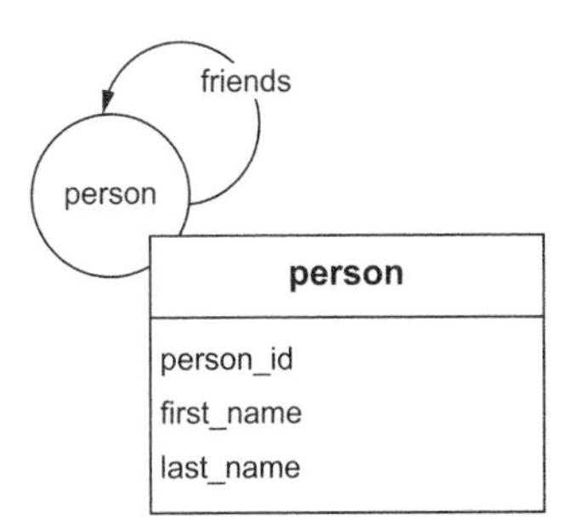

Figure 7.2 Logical data model for the DiningByFriends social network from chapter 2

this logical model to include the different entities required for the recommendation and personalization use cases.

7.2 Extending our logical data model

A model with one vertex and one edge isn't sufficient for complex work. So the question is how we go about extending this model. To remind you, the process of turning a conceptual model into a logical model involves four operations:

1. Translating entities to vertices
2. Translating relationships to edges
3. Assigning properties to those vertices and edges
4. Testing the model

These basic modeling steps provide a strong foundation and work well on the relatively simple social network data model. But as the complexity of our data model increases, these basic steps need to be combined with additional techniques to create a logical data model designed to handle any scale of data. Let's begin extending our data model by reviewing our recommendation engine use case requirements.

If you remember, the recommendation engine provides restaurant recommendations based on reviews, location, and cuisine type. For the initial version of the recommendation engine, the application needs to support answering the following questions for a user:

- What restaurant near me with a specific cuisine is the highest rated?
- What are the ten highest-rated restaurants near me?
- What are the newest reviews for this restaurant?

Just as we did for our social network in section 2.4, let's start by identifing the portions of our conceptual model required for the recommendation engine. Figure 7.3 shows the entities and relationships for this use case.

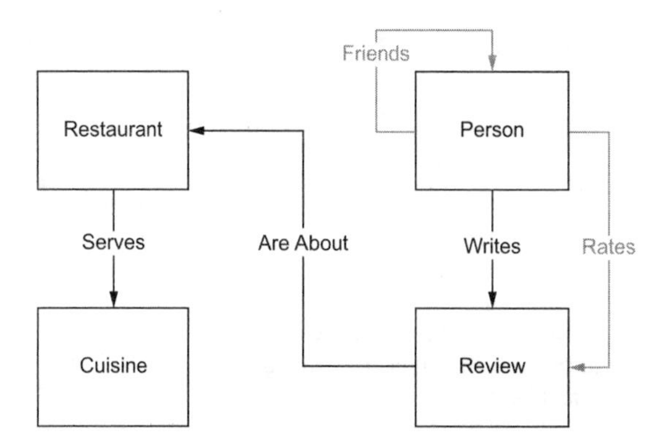

Figure 7.3 Conceptual data model with the relevant portions highlighted for the recommendation use case of DiningByFriends

While our recommendation engine will focus only on these three questions for its initial version, the requirements are noticeably more complicated than the social networking use case. Although our social networking use case also needed to answer three questions, it only required one entity, Person, and one relationship, Friends, to answer those questions. For our recommendation use case, we require four entities (Person, Review, Cuisine, and Restaurant) and three relationships among those (Writes, Are About, and Serves). This more closely represents our real-life scenario.

It's rather rare that any application or use case of an application only traverses a single entity and relationship. More often than not, we are required to traverse multiple vertices and edges to arrive at the desired result. The pathfinding and recursive traversals developed for our social network did not require traversing different entity types, but this is actually quite rare in complex domains. Most common graph patterns rely heavily on the connections between data that involve multiple types of vertex and edge labels. This is true of recursive and pathfinding traversals as well as another common graph pattern, called a *known walk.*

To see how these questions apply in a real scenario, let's look at each question for our restaurant recommendation engine. Then we can determine what makes it a known walk.

Known walk

In graph theory, a *walk* is a sequence of edges and vertices. If this sounds like the definition for a path, then you're correct. A *path* is a specific type of walk, which only contains distinct vertices. A *known walk* is a pattern in graph applications where we have prior knowledge of the exact series of vertices and edges to traverse to get our answer.

You might think that this is like the pathfinding algorithms we did for our social network, and you're correct except for one crucial difference. In pathfinding problems, we know the series of vertices and edges to traverse, but we do not know the number of times we must traverse these. For example, in our social network, when we were trying to find how person *X* is connected to person *Y*, we knew that we needed to traverse the `friends` edge, but we did not know if these were connected or how many hops it would take to traverse them.

In known-walk problems, on the other hand, we know the series of vertices and edges to traverse and the number of times we need to traverse them. For example, in our social network, when we wanted to find the friends-of-friends, we knew that we needed to traverse the `friends` edge to the `person` vertex and then traverse the `friends` edge to the `person` vertex a second time. The fact that we knew both the path and number of repetitions for that path allowed us to optimize both our data model and traversal.

Prior knowledge about the depth of the traversal is one of the factors differentiating between the two patterns, pathfinding and known walks. To decide if a traversal is a known walk, we use the following two questions:

- Do we know the exact definition of the steps (e.g., entities and relationships) needed to traverse from the starting vertex to the ending vertex?
- Do we know the number of times we need to traverse these steps to get our result?

If we are able to answer Yes to both questions, then this traversal is a known walk. Generally, known walk traversals are preferred over recursive traversals because the number of steps required to traverse the graph is well known. This often results in more consistent traversal execution times than in recursive traversals with their unknown number of iterations.

"*What restaurant near me with a specific cuisine is the highest rated?*" Do we know the exact definition of the steps we need to traverse to get from the starting vertex to the ending vertex? Yes. Looking at our conceptual model, we know we need to use the following entities:

- *Restaurant*—Entity being returned
- *Review*—Contains the rating for this restaurant
- *Cuisine*—Needed to filter on a specific cuisine

Each of these entities is connected to the others via a single relationship, meaning that we have a clearly defined path to traverse between these. If there were multiple relationships among them, we would need additional criteria to define which of the relationships to use in the traversals.

Do we know the number of times we need to traverse between the entities to get our result? Yes, we know that we need to traverse the Serves and Are About relationships once to retrieve the entities associated with the restaurant. This is unlike our pathfinding and recursive traversals because we would not know how many times to traverse some edges until executing the traversal.

"*What are the ten highest-rated restaurants near me?*" Are the required entities and relationships between these entities well defined? Yes. Examining our conceptual model, we see that answering this question requires the following entities:

- *Restaurant*—Entity being returned
- *Review*—Contains the rating for this restaurant

And because Restaurant and Review only contain a single relationship between them, the path is defined. Continuing with our known-walk assessment, do we know the number of times we need to traverse between entities to get our result? Yes. For each Restaurant, we know that this requires one iteration of the Are About relationship to get the Review that's needed to calculate the highest rating.

"*What are the newest reviews for this restaurant?*" Are the required entities and relationships between these entities well defined? Yes. As with the last question, answering this requires the following entities:

- *Restaurant*—Entity being reviewed
- *Review*—Entity being returned

Restaurant and Review only contain a single relationship between them, so the path is defined. Do we know the number of times we need to traverse between entities to get our result? Yes. This question only requires one iteration over the Are About relationship to get the Review entities for a Restaurant.

As we can see, applying these two selection criteria helps us not only determine that we have a known walk traversal pattern, but also gives us a jump start on our data modeling by forcing us to think about the entities and relationships required to answer each question. Now that we have reviewed the requirements of our recommendation engine use case, let's return to the process of creating our logical data model.

7.3 Translating entities to vertices

Let's begin extending our logical data model by translating entities in our conceptual model into vertex labels. Remember that to translate our entities into vertices, we need to

- Find the entities required in our conceptual data model.
- Create corresponding vertex types in our data model.
- Assign descriptive label names to those vertex types.

This time, we ask first that you complete the steps on your own. Then we'll walk you through them. (If you need a refresher, refer to section 2.3.1.)

> **EXERCISE** Write down what you think are the appropriate vertex types and label names for the entities in our system. In the review of our conceptual data model, we identified four entities: Person, Review, Restaurant, and Cuisine. What label did you give to each?

In chapter 2, we stated that it was a best practice to use generic labels for vertices and edges, but we did not discuss why. Before transitioning to the next step, let's investigate the reasoning behind this recommendation.

7.3.1 Using generic labels

The primary goal of a label on a vertex or edge is to provide a name for a category of similar items. As we previously mentioned, it is generally best practice to use more general terms, such as `user` or `person`, instead of specific terms, such as `reviewer` or `restaurant_patron`. When we made this recommendation, we didn't go into why this is the case. Let's examine why we recommend generic over specific label types.

Generic labels allow us to group related items together, which simplifies both the model and the writing of traversals by having fewer entity types. This frequently creates more performant traversals as well. Generic labels also allow us to group similar entities into the broadest categories, while still being able to differentiate between them.

While this sounds easy in concept, the devil is in the details: if we make the label too generic, such as grouping all entities under a single label—`item` or `entity`, for example—then these labels no longer provide insight about the entity represented. If we make it too specific, such as a vertex label for each entity, we no longer gain the advantage that grouping entities provides. What we want to find is the "Goldilocks Zone," the perfect balance of labels generic enough to simplify our traversals but specific enough to provide insight about its represented entity. Now for the million-dollar question: "How do we decide the right level of specificity for label names?"

LABELING WITH CONTACT TYPES

Let's walk through an example graph for tracking a person's contact information and take a look at what makes an ideal specificity of a label. In this system, we define a `person`, an `email`, a `phone`, and a `fax` vertice, illustrated in figure 7.4.

> **NOTE** Code examples are provided in this section to illustrate how data model decisions impact the code we write. All of the traversals are functional, but no data initialization is provided, aside from the figures of the example graphs. We leave it as an exercise for enterprising readers to apply the skills gained from chapter 4, if they so desire, to create these graphs and try out the example code.

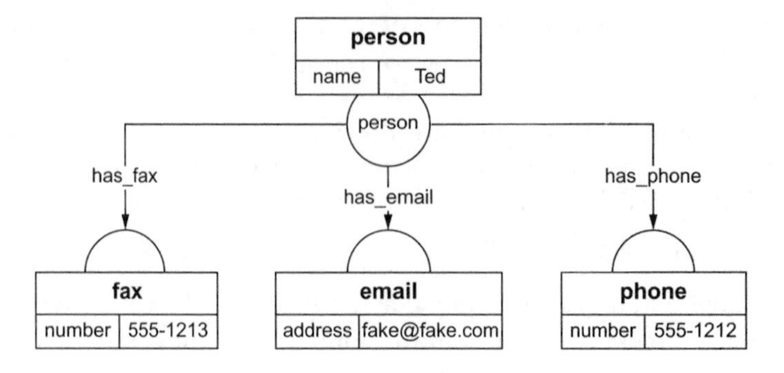

Figure 7.4 Contact graph showing each entity with a specific vertex label

Because categorizing our entities to simplify and optimize our traversals is one of the major reasons we recommend using generic labels, let's examine what a few traversals would look like using this label per entity model:

- What is the phone number for Ted?

```
g.V().has('person', 'name', 'Ted').
  out('has_phone').              <-- A single edge
  values('number')                   to traverse
```

- What are all the contact methods for Ted?

```
g.V().has('person', 'name', 'Ted').
  union(                                  <-- A union of three
    out('has_phone' ).values('number'),       edge traversals
    out('has_email').values('address'),
    out('has_fax').values('number')
  )
```

- What is all the contact information for people in the system?

```
g.V().                   A union of three
  union(            <--  edge traversals
    out('has_phone').values('number'),
    out('has_email').values('address'),
    out('has_fax').values('number' )
  )
```

To make these traversals work, we introduce a new step here, the `union()` step:

- `union(traversal, traversal, ...)`—Processes each traversal separately and outputs the combined results as a single result set.

It is an oversimplification to say that because a traversal is easier to read that it is also better-performing; however, this rule of thumb generally applies.

NOTE In the context of this chapter, we use the readability of a traversal as a proxy for the relative performance of that traversal. In chapter 10, we'll dive into several other factors affecting traversal performance and how to quantify them.

Examining our traversals, the first one ("What is the phone number for Ted?") appears reasonably concise. It starts by filtering to a single vertex (`person`) and then traversing a single edge (`has_phone`). However, the other two traversals are not as pretty because these require a `union()` step to join the results. Though this may seem reasonably straightforward, the `union()` step creates some complexity that we can improve on.

A `union()` step is a branching step that requires that the current traverser be copied to each branch of the `union()` step in order to run. This means our final two traversals require three copies of the traverser in order to continue processing. While not a tremendous burden on our small sample graph, this becomes a large amount of additional overhead in larger graphs.

The first adjustment is to combine the `email`, `phone`, and `fax` labels into a generic label, `contact`, as each represents the same logical construct: a contact method. This introduces a new complication—we just lost the type of contact. Loss of information defining a subtype of an entity is a common side effect of changing from specific to generic labels, but it can be quickly remedied by adding a `type` property to the `contact` vertex, highlighted in figure 7.5.

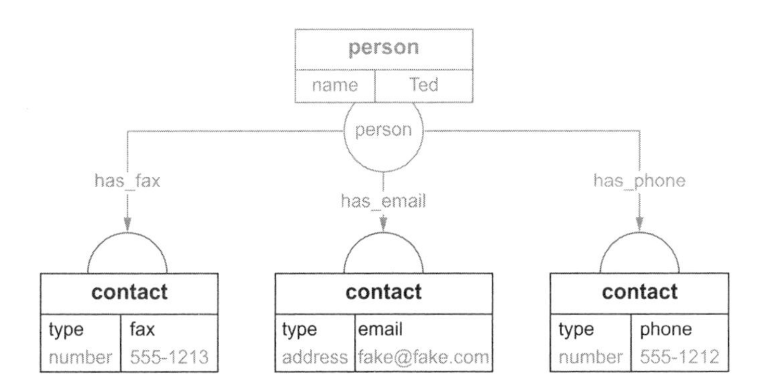

Figure 7.5 Contact graph with the vertex labels and a type property (changes highlighted)

Let's reexamine our traversals after this first adjustment and see how they've changed:

- What is the phone number for Ted?

```
g.V().has('person', 'name', 'Ted').
  out('has_phone').
  values('number')
```

Same; a single edge to traverse

- What are all the contact methods for Ted?

```
g.V().has('person', 'name', 'Ted').
  union(
    out('has_phone' ).values('number'),
    out('has_email').values('address'),
    out('has_fax').values('number')
   )
```

Same; a union of three edge traversals

- What is all the contact information for people in the system?

```
g.V().
  hasLabel('contact').
  values('number', 'address', 'type')
```

Now only a single step to traverse

Looking at our first two traversals, we see that there was no change. However, in our third traversal, we see that changing to a generic vertex label `contact` makes the traversal shorter and removes the need to copy the traversers in `union()`. Are there any other locations where we can apply generic labels to simplify our traversal?

Looking at the traversals as they are now written, both the first and third traversal seem rather straightforward and tidy. As written, the second traversal seems a bit lengthy and complex, and requires copying the current traversal state three times to fulfill the `union()` step. Because we use readability and length as a proxy for performance, let's see if we can simplify that. What if we replaced the `has_fax`, `has_email`, and `has_phone` edge labels with a generic `contact_by` edge. How does this affect our model? Let's investigate figure 7.6 and see what our model looks like with this adjustment.

NOTE While it might be tempting to create a simple edge label called `has` or `has_a` to simplify our model, the reality is that labels such as these are too generic to provide any useful information. When we review data models, this is one of the first things we look for as a code smell.

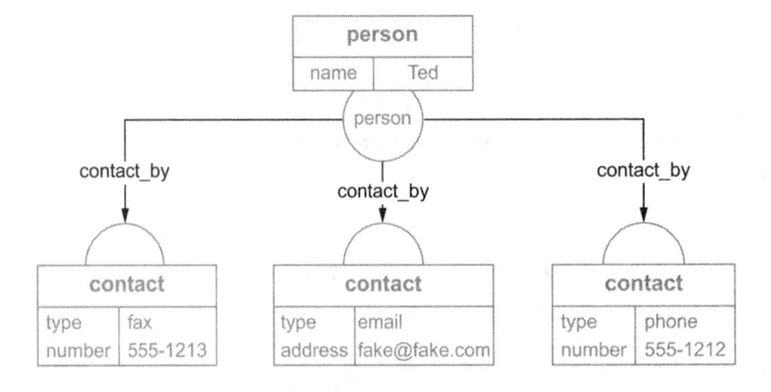

Figure 7.6 Contact graph with generic edge labels highlighted

We see that even with these two changes, from specific to generic labels, we are still able to easily understand what this graph is showing us. Let's see how this adjustment impacts our traversals:

- What is the phone number for Ted?

```
g.V().has('person', 'name', 'Ted').
  out('contact_by').                      <-- Same; a single edge
  has('contact', 'type', 'phone').        <-- Adds an additional filter
  values('number', 'type')
```

- What are all the contact methods for Ted?

```
g.V().has('person', 'name', 'Ted').
  out('contact_by').                      <-- Changed to a single edge
  values('number', 'address', 'type')
```

- What is all the contact information for people in the system?

```
g.V().hasLabel('contact').                <-- Same; a single step to traverse
  values('number', 'address', 'type')
```

With this adjustment to the model, we see that our first traversal becomes a little more complex, while the other two are simplified:

- The first traversal added a filter on the `contact` vertices to find the vertices that have a type of `phone`.
- The second traversal was simplified to traverse out the `contact_by` edge instead of the `has_fax`, `has_email`, and `has_phone` edges.
- The third traversal is unaffected by the change.

Commonly, changing your data model positively affects some traversals while negatively impacting others. Although this might seem like a step backward, we believe that the tradeoff in making the first traversal a bit more complex is outweighed by the simplification achieved with the second traversal. Data model optimization, as with most things in life, is about tradeoffs. Understanding these tradeoffs and balancing them is necessary when building a data model that preferentially optimizes for the most common traversal patterns.

LABELING FOR RECOMMENDATION ENGINE VERTICES

Returning to the data model for our recommendation use case, let's examine how we can apply these principles to creating generic labels for our recommendation engine. Looking at our first entity, Person, we see that our data model has a `person` vertex from our social network. Because these vertices represent the same conceptual entity, we can reuse this vertex label for our recommendation engine. The ability to reuse vertices is one of the key benefits of generic labels. If we had made a specific vertex label for our friends—say, a `friend` vertex—we could not have reused it. Let's look at how to add the three new entities (Review, Restaurant, and Cuisine) to our model.

Applying our process from chapter 2, we convert each entity into a unique vertex. Grouping these entities doesn't make sense because they represent fundamentally different concepts in the domain. We can validate that these represent different concepts by looking at their properties. Because there is no overlap of properties among these three vertices, they all have different relationships.

Now that we have our entity types, our next task is to determine a descriptive label for each type. Utilizing best practices, we label our entities `review`, `restaurant`, and `cuisine`. Figure 7.7 shows the changes to our data model resulting from adding these entities.

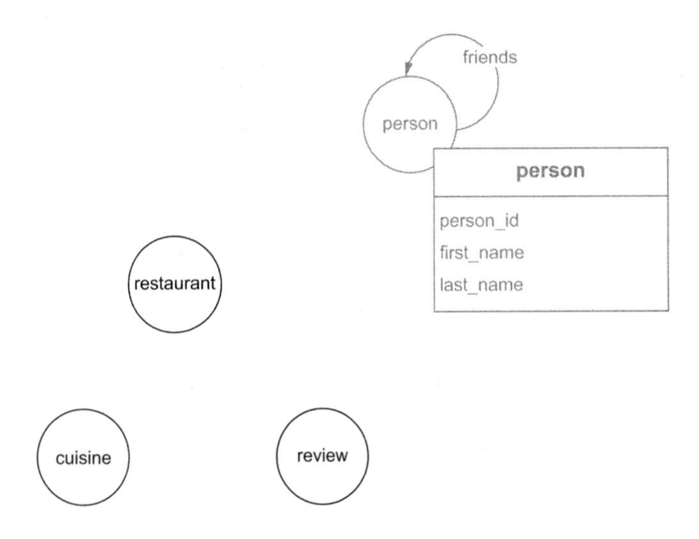

Figure 7.7 The logical graph model with the highlighted `restaurant`, `cuisine`, and `review` labels added to the schema

Now that we've added the vertices from our conceptual data model, let's look once again at the recommendation engine requirements. That helps us to ensure that we include all the necessary entities:

- What restaurant near me with a specific cuisine is the highest rated?

  To answer this question, we have the essential entities: `restaurant`, `cuisine`, and `review` vertex types. Aside from the address on `restaurant`, we do not have any location elements in our data model, elements required for the "near me" part of the question. This means that our model is incomplete, so we should evaluate our need to add the missing piece.
- What are the ten highest-rated restaurants near me?

  To answer this question, we have the `restaurant` and `review` entities. But as with the last question, we can't handle the "near me" geographical aspect of the question.

- What are the newest reviews for this restaurant?

  The `restaurant` and `review` entities do answer this question.

We answered the third question confidently, but the first two have an unresolved issue; namely, how do we handle the restaurant's geographical location? In data modeling, graph or otherwise, we have several options for modeling geography. The two most common approaches are either as separate entities (such as `city` and `state`) or denormalized as a property (such as an `address` property for `restaurant`).

We should also mention that for more functionality (e.g., rendering on street maps) and complexity, geospatial coordinates are often used. That level of complexity isn't required for DiningByFriends, so we'll tackle denormalization instead of geospatial data. What do we mean when we talk about denormalizing data in a graph?

7.3.2 Denormalizing graph data

Denormalization in a graph data model is similar to denormalization in a relational data model; both involve copying data into multiple locations at write time to increase performance at read time. As with relational systems, data denormalization presents several downsides:

- *Increased disk usage*—Because the data is written to more than one location, the data size increases. While the cost of disk space is rarely an issue anymore, these costs are something to consider, especially for large or cost-sensitive projects.
- *Data synchronization issues*—As we write data to multiple locations, each location must be updated every time a change is made. If any of these locations are out-of-sync, different traversals can return different results.
- *Reduced write performance*— Because we must update the value in multiple locations, more write operations are required. This is commonly called *write amplification.*

With these downsides, when should we use data denormalization? We should denormalize data when a normalized data model is unable to retrieve (read) data fast enough. Poor read performance is most often due to the number of operations required to retrieve the information. This is especially problematic in any sort of distributed system where data access can also require additional network access, in addition to memory and disk access. In a graph database, denormalization is all about reducing the length of the traversal required to get from our starting point to the ending data.

Because of the downsides, denormalization should not be the first technique you choose to solve a performance problem. Rather, it should be considered after proper data modeling and traversal optimization fail to achieve the desired performance.

NOTE The discerning reader will recognize that indexing is a form of denormalization, though we don't always think of it as such. Regardless of which data engine you use, if you add an index, you utilize a form of denormalization.

The engine's simple syntax for doing so does not change the fact that it is a duplication of existing data for read performance purposes.

While there are many different forms of data denormalization, we'll only discuss the two most common ones: precalculated fields and duplicated data. Let's look at those next.

USING PRECALCULATED FIELDS

If a traversal performs an aggregation (e.g., sum, average, count), and the traversal is executed significantly more often than we update the vertices, then it is a strong candidate for adding precalculated fields into a vertex or edge. *Precalculated fields* are properties of a vertex or edge that store the result of performing a calculation at write time to allow quick retrieval of the data at read time. To examine how precalculation works, let's return to our *Gremlins* movie example graph from chapter 2, which is shown in figure 7.8.

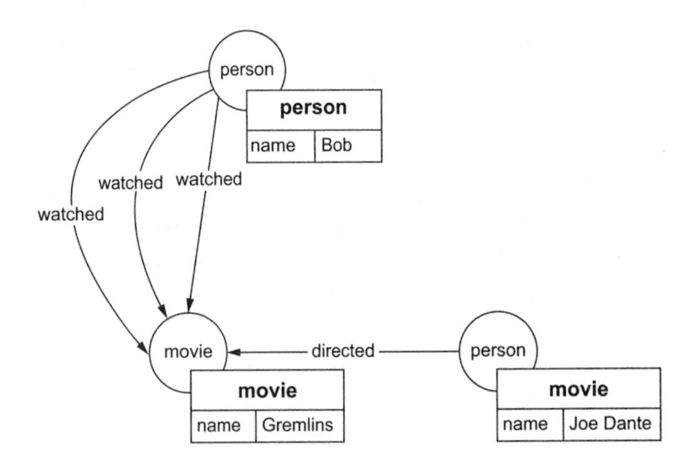

Figure 7.8 Our *Gremlins* movie graph from chapter 2

Let's say we want to know, "How many people have watched *Gremlins*?" With the current model, we count the number of `watched` edges incident to the `Gremlins movie` vertex:

```
g.V().has('movie', 'name', 'Gremlins').
  bothE('watched').
  count()
```

This traversal returns the correct count of `watched` edges but has a lurking issue: the number of incident edges is directly proportional to the time it takes to return the count. This means that the performance degrades as more `watched` edges are added to the `Gremlins movie` vertex, so popular movies take longer to retrieve than unpopular ones. This is especially detrimental as popular movies are viewed more frequently than unpopular ones. This type of performance problem is common in graphs because the vertices with the most connections are usually the vertices that are touched most often in traversals.

To alleviate the issue, we precalculate this value by placing a `watched_count` property on the `movie` vertex, illustrated in figure 7.9. Then we update the count when adding, updating, or deleting a `watched` edge.

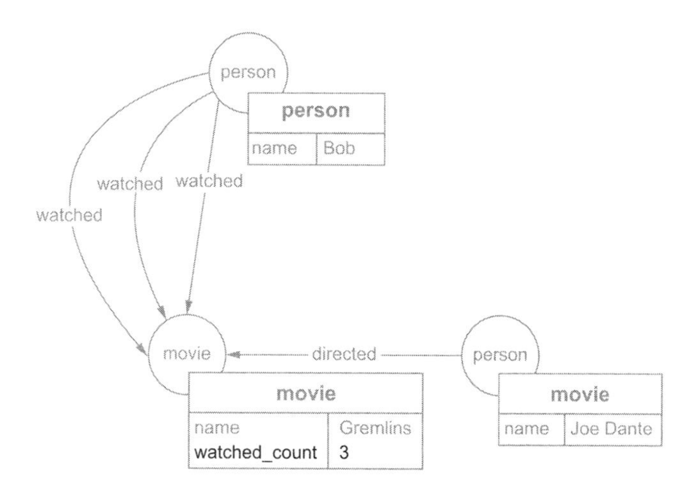

Figure 7.9 ***Gremlins* movie graph with a precalculated `watched_count` property highlighted on the `movie` vertex**

With the `watched_count` property, the question—"How many people have watched *Gremlins*?"—only requires retrieving the `Gremlins` `movie` vertex to access the `watched_count` property. Precalculated fields do not suffer the same read-performance degradation over time. Performance does not degrade because retrieving the `watched_count` requires getting a single vertex instead of accessing an unknown number of edges. By precalculating this value, we now have a constant time lookup for the `watched_count`, regardless of how popular it is.

USING DUPLICATE DATA

The second use of data denormalization involves duplicating properties from one vertex or edge to another vertex or edge. Copying properties into more than one location in our graph allows us to optimize for multiple, different traversal paths at the expense of keeping data in sync.

Like precalculated values, duplicating data is another example where write performance suffers, due to writing data to multiple places, to optimize read performance. Returning to our order-processing system from chapter 5, shown in figure 7.10, let's see how we might apply data duplication to answer those questions more efficiently.

Figure 7.10 Order-processing system graph from chapter 5 with only the `order_date` on the `placed` edge

To answer, "What date did I place order 123?", our traversal needs to take two steps; first, it needs to find the `order` vertex for `order_id 123`, and then traverse out to the `placed` edge to get the `order_date`:

```
g.V().has('order', 'order_id', '123').
  outE('placed').
  values('order_date')
```

If we expect to retrieve orders by ID frequently, and we probably do, this additional overhead is a potential bottleneck. To alleviate this overhead, we can copy the `order_date` onto both the `placed` edge and the `order` vertex as figure 7.11 illustrates.

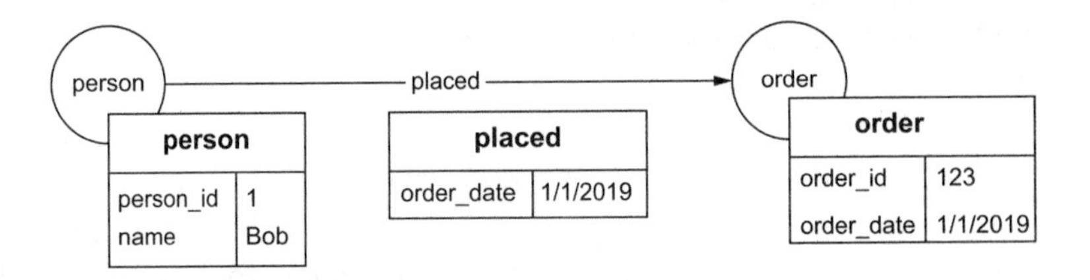

Figure 7.11 Order-processing system with `order_date` duplicated on the `placed` edge and `order` vertex

With this change, we can answer, "What is the date I placed order 123?" in a single step. And we don't need to traverse to any unnecessary vertices or edges.

With these different types of data denormalization available to us, how do we know which one to choose? Well, the answer depends on a few specific factors. Precalculating fields is a great choice when you have an aggregation (sum, average, count) value, and the traversal is executed significantly more often than the vertices are updated. Duplicating data into multiple places in your graph is an excellent choice when you have multiple traversal patterns, each of which needs to be optimized by moving the desired data earlier in the traversal steps.

EVALUATING THE DININGBYFRIENDS LOGICAL MODEL FOR DENORMALIZATION

Now that we know a bit about what it means to denormalize data in a graph, let's decide if this approach is right for our restaurant recommendation use case. Let's start by reminding ourselves which questions require geographical information:

- What restaurant near me with a specific cuisine is the highest rated?
- What are the ten highest-rated restaurants near me?

Now, let's compare our two potential approaches, adding separate vertices for `city` and `state` or adding these as properties on the `restaurant` vertex. First, let's discuss what happens if we make separate vertex types for `city` and `state` as figure 7.12 shows.

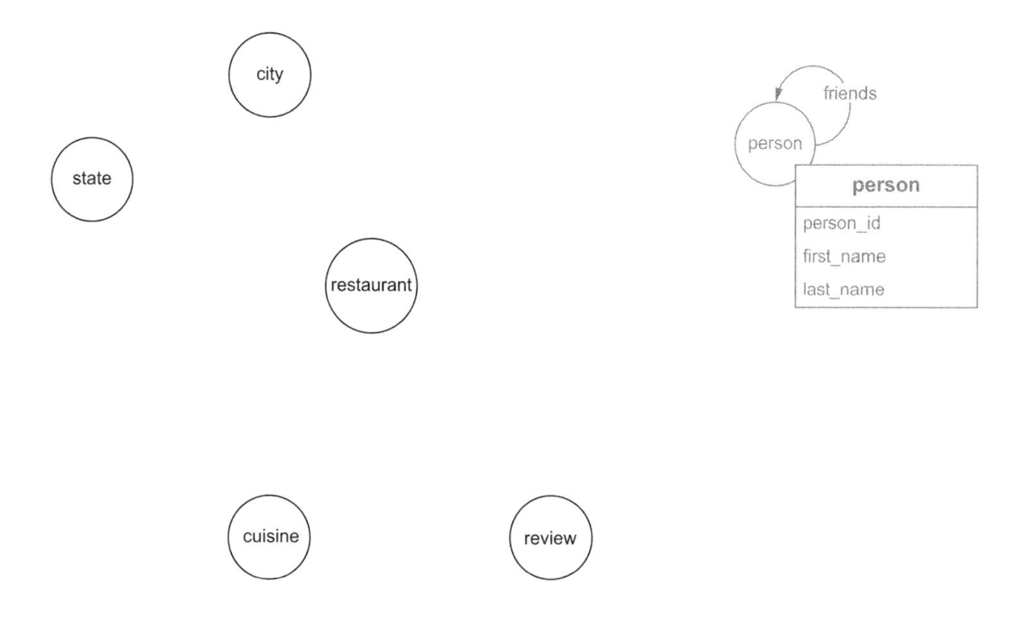

Figure 7.12 Logical graph data model with the `city` and `state` vertex types added

If we assume that both the `person` and `restaurant` vertices are connected to the `city` vertex, then this model allows us to use the `city` vertex to quickly traverse from a person to all the restaurants "near me." Because knowing which restaurants are "near me" is a crucial piece of the puzzle for both restaurant questions, this seems like a good fit. The downside we see is that answering a question such as "What is the location of this restaurant?" requires an additional traversal from a `restaurant` to the `city` and `state` vertices.

The second approach is to denormalize the `city` and `state` information onto both the `restaurant` and the `person` vertices. By adding these properties, we are able to answer, "What is the location of this restaurant?" by returning the `restaurant` vertex.

The tradeoff is that "What restaurants are near me?" requires us to find the `person` vertex of the user and then scan all of the `restaurant` vertices to find the restaurants in that location. While different databases have some optimizations to help with this (such as geospatial data indexes), the availability and functionality of these are implementation-specific.

Comparing these two methods with what is required, we do not see any advantages to denormalizing the `city` and `state` data in on our model. In fact, it is likely that this would make our model much worse. However, if at some point in the future we want to answer questions like, "What is the location of this restaurant?", we need to revisit this decision. The good news is that denormalizing data is an easy change to make after we quantify the performance of the application. Now that we've added `city` and `state` vertices to our logical data model, let's revisit the questions requiring geographic information, and decide if all the necessary information exists:

- What restaurant near me with a specific cuisine is the highest rated?
- What are the ten highest-rated restaurants near me?

Yes, with the addition of the `city` and `state` vertex types, we are now able to answer the "near me" aspect of these questions. Our logical data model now contains all the required vertex types, so our next step is to define the edges.

7.3.3 Translating relationships to edges

To define edges for our recomendation engine use case, we begin by identifying the relationships in our conceptual model. Figure 7.13 shows these relationships.

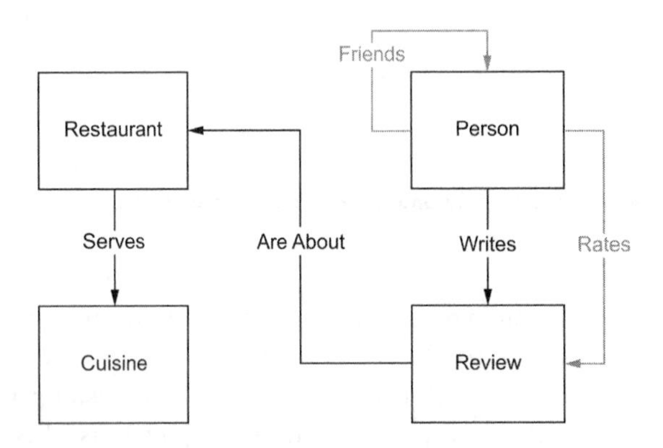

Figure 7.13 The conceptual data model for the recommendation engine use case of DiningByFriends with the required portions highlighted

In our conceptual data model, we observe the following relationships:

- Person–Writes–Review
- Review(s)-Are About–Restaurant
- Restaurant–Serves–Cuisine

In our logical data model, we also added the `city` and `state` vertex types, so we need to consider the relationship between `restaurant` and `city` as well as `city` and `state`. But one requirement that we've not touched on is how to determine a user's location, which is needed to decide which restaurants are "near me."

To represent this relationship, we add a relationship from the `person` vertex to the `city` vertex to denote the city and state where a person `lives`. Based on this information, we compile a list of required relationships:

- Person–Writes–Review
- Person–Lives In–City
- Review–Are About–Restaurant
- Restaurant–Serves–Cuisine
- Restaurant–Within–City
- City–Within–State

After reviewing the list, we can see that the last two relationships share the same name, yet another example of generic labels. Let's add these six relationships to our model, illustrated in figure 7.14.

NOTE We drop the prepositions from our relationship descriptions when we translate these into edge labels to keep the graph data model simple.

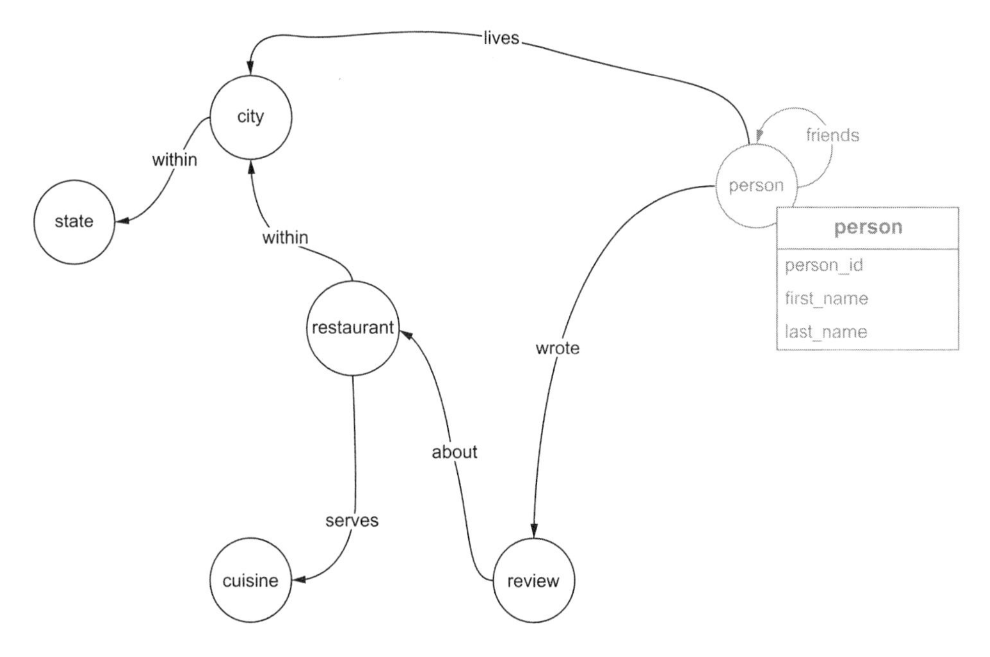

Figure 7.14 DiningByFriends graph data model with the edges added for the recommendation engine

With our edges identified and labeled, we need to decide on the uniqueness of each edge. The following list describes each edge:

- `wrote`—It is reasonable to expect that a specific person can only write a particular review once, so the edge uniqueness is *single*.
- `about`—A specific review is about a single restaurant, so the edge uniqueness is *single*.
- `serves`—While a restaurant can serve many types of cuisine, the expectation is that a specific restaurant is only associated with a particular cuisine once, so the edge uniqueness is *single*. (Remember, this designation does not exclude a restaurant from being associated with more than one cuisine. The designation of *single* means that you can only have an edge from a specific restaurant to a specific cuisine once (e.g., El Rey Taqueria restaurant can only have one edge to Mexican cuisine.)
- `lives`—A `person` lives in a single city, so the edge uniqueness would be *single*.
- `within`—A particular restaurant can only exist in a single city, and a city can exist in a single state, so the edge uniqueness would be single.

All of the edge labels have a uniqueness of single, and that shouldn't surprise us. This uniqueness specification is the most common and the safest starting point.

7.3.4 Finding and assigning properties

With the structural elements of our model in place, it is time to start adding properties.

> **EXERCISE** Referring to the conceptual model in figure 7.13, what properties do we need to add to the model to answer the three questions for the recommendation engine? (What restaurant near me with a specific cuisine is the highest rated? What are the ten highest rated restaurants near me? What are the newest reviews for this restaurant?).

Once you have decided the properties you believe are required, take a look at the properties we chose. Table 7.1 shows the vertex properties; no properties are required on the edge labels.

Table 7.1 Properties for each vertex label

Person	Review	Restaurant	Cuisine	City	State
`person_id`	`rating`	`restaurant_id`	`name`	`name`	`name`
`first_name`	`body`	`name`			
`last_name`	`created_date`	`address`			

How did your choices compare with ours? Hopefully they are close, but if not, that's fine. There's more than one way to construct a data model. If we do not match, that does not mean either is right or wrong. The important outcome is that the logical data model answers the questions.

Before we finalize these properties in our model, we need to check the location of the properties. As we discussed when talking about data denormalization, it is sometimes possible to improve the performance of traversals by moving data so that it is reached in the fewest number of steps. One common way to do this is to move properties from a vertex to the incident edge.

7.3.5 Moving properties to edges

The concept behind this optimization is straightforward. Move a property from a vertex to the incident edge to reduce the number of steps your traversal needs to process. However, in practice this process is more nuanced.

Let's return to the example of a simple order-processing system for a retail website. In this retail site, a person can place one or more orders, and each order has a date. Coming from a relational world, our instinct is to build a model with a `person` vertex, an `order` vertex, and an edge named `placed` between the vertices. We also need to add properties for `person_id` and the `name` of a person to the `person` vertex, as well as the `order_id` and `order_date` to the `order` vertex. Figure 7.15 demonstrates this.

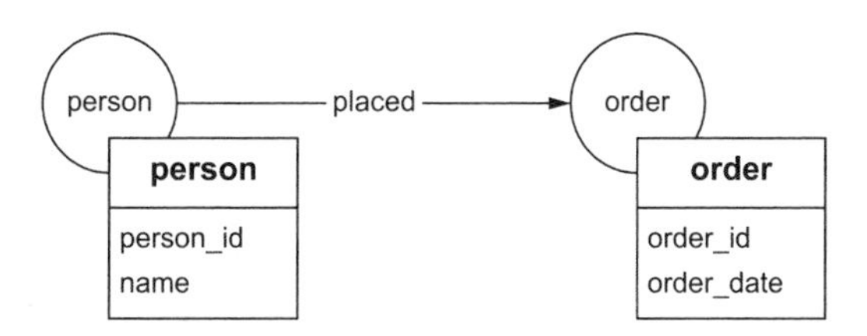

Figure 7.15 An order-processing system showing properties on vertices only

Using this model to answer the question, "Show me all my orders in the last three months?" requires traversing from the `person` vertex to the `order` vertex. When we use a single `out()` step to do this, we think of it as a single operation. But in reality, the `out()` step is effectively an alias for the `outE().inV()` combination of steps. This means that there are always two operations (memory reads, cache hits, disk reads, or even network access in distributed systems) that happen to traverse an edge: moving onto the edge and moving from the edge to the destination vertex. Figure 7.16 illustrates this two-step traversal.

There is nothing wrong with this model. Overall, it works well, but to answer our question, it is less than ideal because it requires a second read operation before we filter on `order_date`. As we've learned, the sooner we filter our traversal, the better that traversal performs. In the relational world, the model is analogous to filtering a table after performing a join, instead of before the join operation. To help mitigate this concern, let's move the `order_date` from the `order` vertex to the `placed` edge, as figure 7.17 shows.

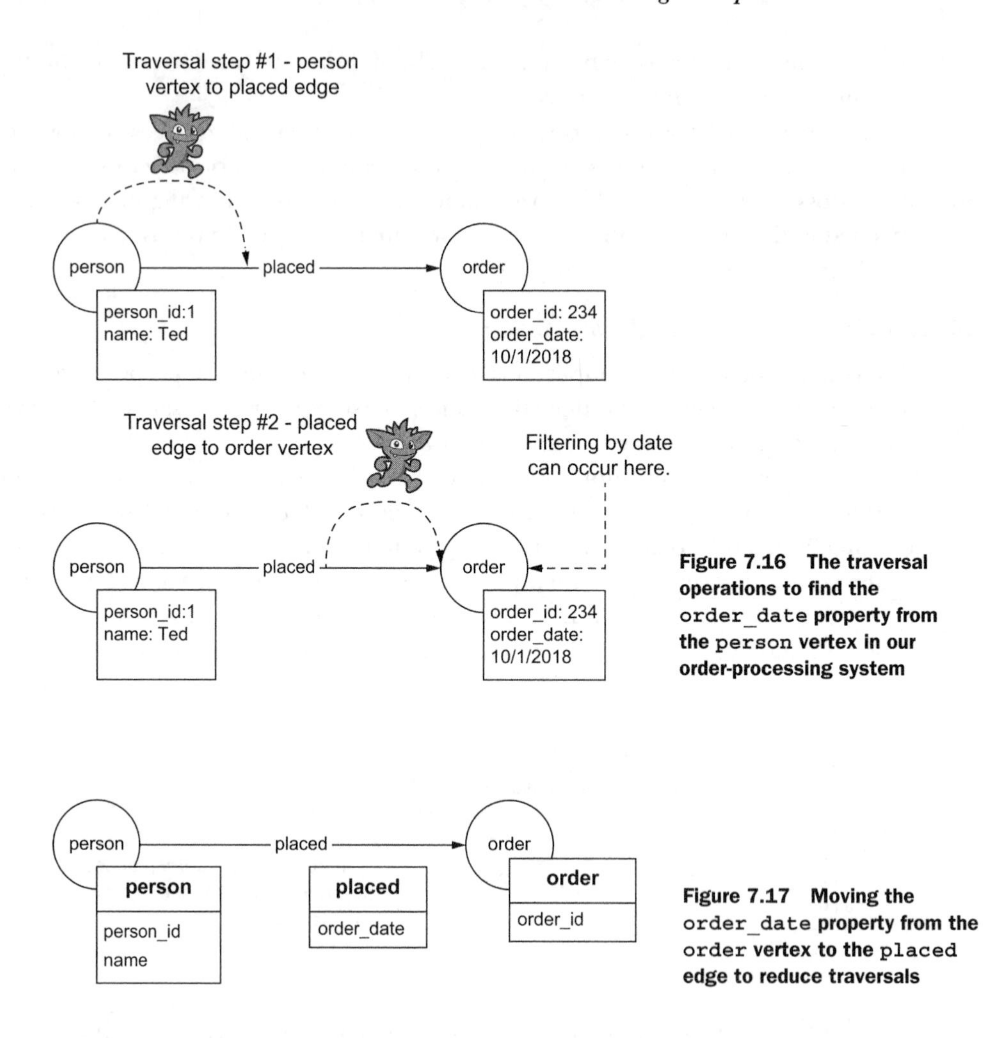

Figure 7.16 The traversal operations to find the `order_date` property from the `person` vertex in our order-processing system

Figure 7.17 Moving the `order_date` property from the `order` vertex to the `placed` edge to reduce traversals

Why make this change? This rearrangement of the property location enables us to filter our traversal after only a single operation from the `person` vertex to the `placed` edge, reducing the computation required and increasing the traversal speed. (This is not true in all databases. Some databases, such as Neo4j and AWS Neptune, have optimized their underlying data model to collocate the edge information on the associated vertices in their on-disk representations. This means that moving properties to edges is not an optimization in these systems.) Figure 7.18 demonstrates this change.

While this sort of single-operation optimization can seem insignificant on a single traverser, it becomes a huge performance gain when running multiple traversers. In fact, it is one of the best ways to increase overall traversal performance, just as with relational databases where, if we filter the data before the join operations, then performance improves.

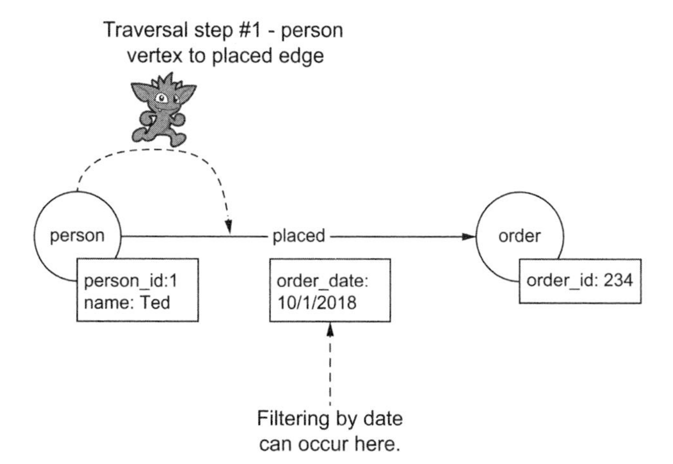

Figure 7.18 The traversal steps in our order-processing system to find the `order_date` property now that it has moved to the `placed` edge

Looking at our model and the requirements for our recommendation engine, we do not see a need for moving any property to an edge. Still, this technique is an excellent tool to keep in mind later when we look more at performance optimizations. Figure 7.19 shows the result of adding the properties to our data model.

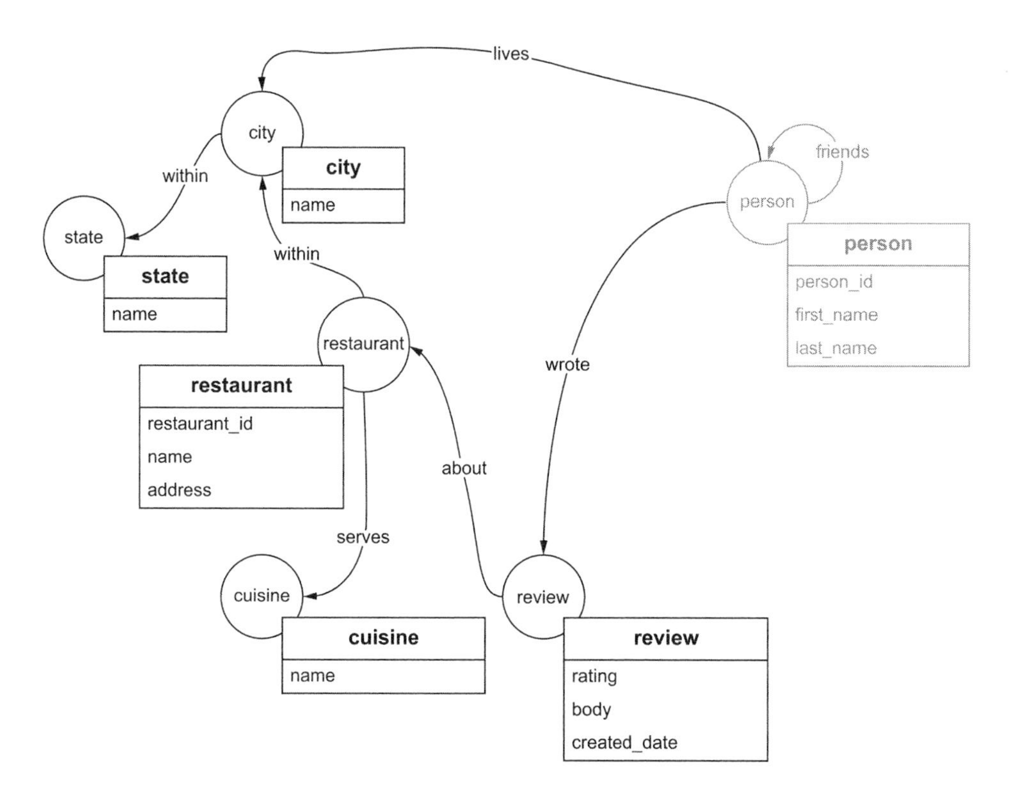

Figure 7.19 The recommendation engine logical data model for DiningByFriends with the added properties

7.3.6 Checking our model

The last step for completing our model is to validate its construction. We let you do this in the form of an exercise.

> **EXERCISE** Reflect on the following questions and check the validity of our model:
>
> - Do the vertices and relations read like a sentence?
> - Do we have different vertex or edge labels with the same properties?
> - Does the model make sense?

Do you think we have a valid model? Reviewing these questions, we feel confident that our model works for our use case.

7.4 Extending our data model for personalization

Now, it is time to add the personalization use cases. Looking at our conceptual model, we highlight the relevant portions in figure 7.20.

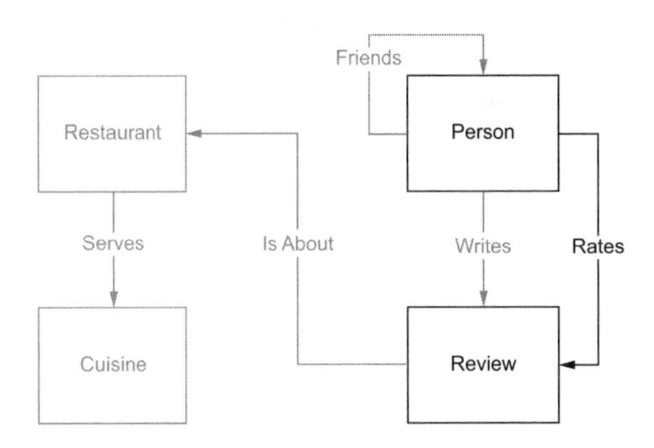

Figure 7.20 The relevant portions of our conceptual data model highlighted for the personalization use case of DiningByFriends.

For the last two feature sets, we demonstrated the process step by step. For this use case, we perform the four-step process independently. The personalization use case provides a mechanism to individualize recommendations based on the circle of friends and how the friends rate other reviews. As a reminder, the questions for the personalization use case follow:

- What restaurants do my friends recommend?
- Based on the review ratings from my friends, what are the best restaurants for me?
- What restaurants have my friends reviewed or rated a review for in the past *X* days?

> **EXERCISE** Run through the graph data modeling process yourself, following the process we completed for the recommendation engine. Once you finish, compare your results with ours.

Now that you have (hopefully) taken the time to walk through the process independently, it is time to see what we developed. Figure 7.21 shows our model.

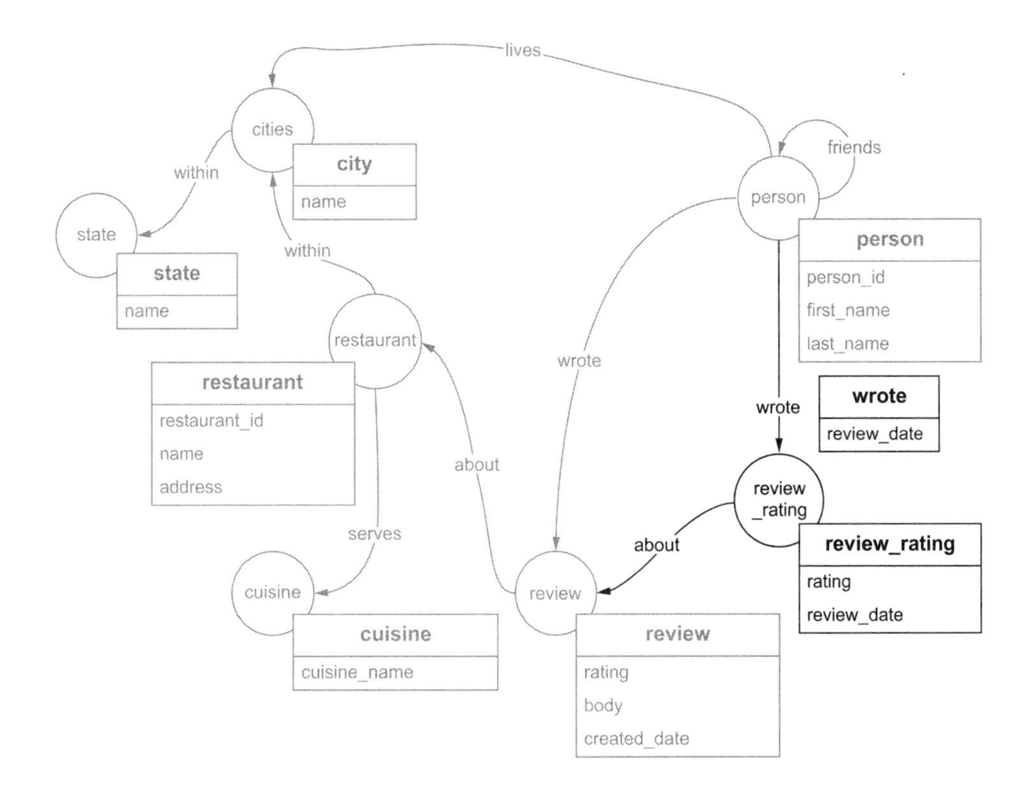

Figure 7.21 The logical data model we developed for our personalization use case in DiningByFriends

Compare what you designed with what we produced. Are these similar? Where do these differ? Specifically, did you add a single vertex like we did? What did you name your edges? And, finally, how did your model handle the review date and review rating?

In our model, we added a single new vertex labeled `review_rating` and two edges, `wrote` and `about`, to represent the entities and relationships for this use case. To arrive at this model, we followed the same process we used previously, but we applied one technique that we discussed but had not put into practice yet: edge properties.

We chose to denormalize the `review_date` property to both the `review` vertex and the `wrote` edge. Denormalizing this property allows us to efficiently answer, "What restaurants have my friends reviewed or rated a review for in the past *X* days?" It enables us to filter on the edge as well as returning the `review_date` with the `review_rating` vertex. While filtering on edges may not seem like it has a significant impact, when we think about this optimization being hit hundreds or thousands of times in a single traversal, the summation of these micro-optimizations adds up to significant performance gains.

7.5 Comparing the results

Now that we've completed our graph data model for DiningByFriends, let's see what our final logical data model looks. Figure 7.22 shows this data model.

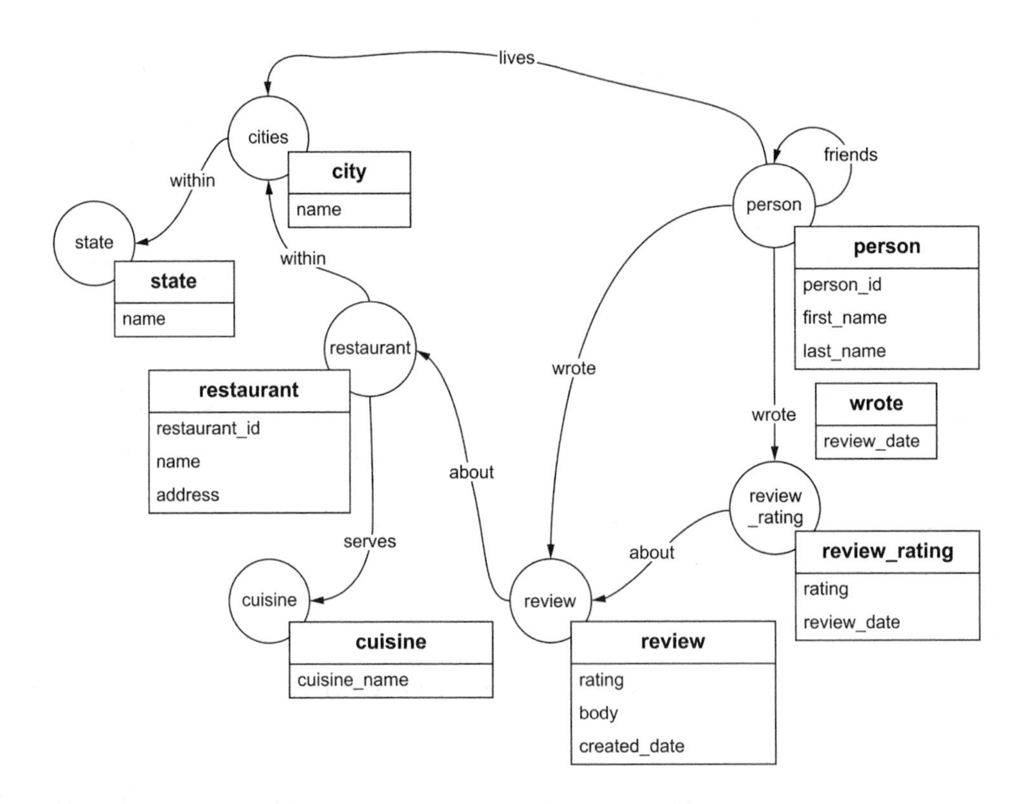

Figure 7.22 The final graph showing the logical data model for DiningByFriends

Reviewing the logical data model, we notice it's similar to our conceptual data model. The vertex and edge names form human-readable sentences, and the relationships between the entities are understandable. This is one of the great benefits of graph data modeling. Unlike relational data models, both technical and non-technical users can understand graph data models.

We'll use this logical data model throughout the remainder of this book, although it will likely change a bit as we work through the remaining chapters. As we said before, no data model survives first contact with code or with actual data. But this model provides a solid foundation to begin work.

Summary

- Generic data labels allow us to reuse labels to simplify our traversals by enabling us to group similar entities so that we create more performant and scalable traversals.
- Denormalizing data through precalculating fields or duplicating data reduces the complexity of our traversals by making the data available earlier in our traversal.
- Precalculating fields is a great choice when the field to be calculated is read much more frequently than written.
- Data duplication involves copying properties into multiple locations in our graph to optimize for multiple different traversal paths at the expense of keeping data in sync.
- Moving properties from vertices to edges can reduce the complexity of our traversals by reducing the number of steps our traversal has to perform.
- Applying these advanced modeling techniques allows us to create complex data models for real-world situations, such as recommendation and personalization use cases.

8 Building traversals using known walks

This chapter covers

- Creating known-walk traversals
- Translating business questions into graph traversals
- Prioritizing strategies for traversal development
- Paginating results in a graph traversal

Denise is one of the users of DiningByFriends. She recently traveled to Cincinnati, Ohio, for work and is looking for a recommendation for an excellent restaurant for dinner. From the work we did in the last chapter, we know that our data model contains all the necessary information to answer this type of question. We also know that to answer this question, we need to develop traversals that

- Traverse a specified set of vertices and edges
- Traverse those elements in a set order
- Traverse those elements a specific number of times

In the last chapter, we learned that traversing a graph with the attributes from this list is a pattern described as a known walk. Although we introduced the concept of known walks in chapter 7, this chapter dives into how we develop traversals for this pattern. To demonstrate this, we use the tangible target of our recommendation engine use case. We start by revisiting the requirements of the restaurant recommendation engine. We then identify the vertices and edges needed for our known-walk traversals. We follow that with developing the traversals for the use case. Finally, we incorporate the traversals into our application.

In previous chapters, we developed our traversals separately from the Java code and then added these to our application. This separated traversal drafting and testing and application development into two distinct steps. We did this to avoid intermingling developing traversals with that of constructing an application. The reality is that most developers do both at the same time, regardless of database engine.

In this chapter, we'll follow a more standard developer workflow and combine the creation of traversals with adding these to our application. Along the way, we'll provide a variety of tips and best practices to aid in our process of developing graph-backed applications. By the end of this chapter, you'll have a simple recommendation engine for our DiningByFriends app. You'll also learn how to develop applications using known walks to solve substantive business problems. Let's get started.

8.1 Preparing to develop our traversals

Before we begin our development process, we need to gather two critical pieces of information. First, we need the requirements of our use case. Second, we need our graph data model. Revisiting the requirements and data model for our recommendation engine, these questions make up the requirements for our recommendation engine use case:

- What restaurant near me with a specific cuisine is the highest rated?
- What are the ten highest-rated restaurants near me?
- What are the newest reviews for this restaurant?

To develop the traversals that answers these questions, we'll use our logical data model. Figure 8.1 shows this model.

8.1.1 Identifying the required elements

With the questions and logical data model, we can begin writing our traversals by identifying the necessary vertex labels and edge labels for this use case as we did in chapters 3, 4, and 5. One addition to our process is that, because we are working with a more complex logical model, one with multiple vertices and edges, we need to perform a few steps in preparation for writing our traversals. We go through these steps for each of the three questions in our use case:

1. Examine each requirement and break it into the components needed to answer the question.
2. Identify the required vertex labels.
3. Identify the required edge labels.

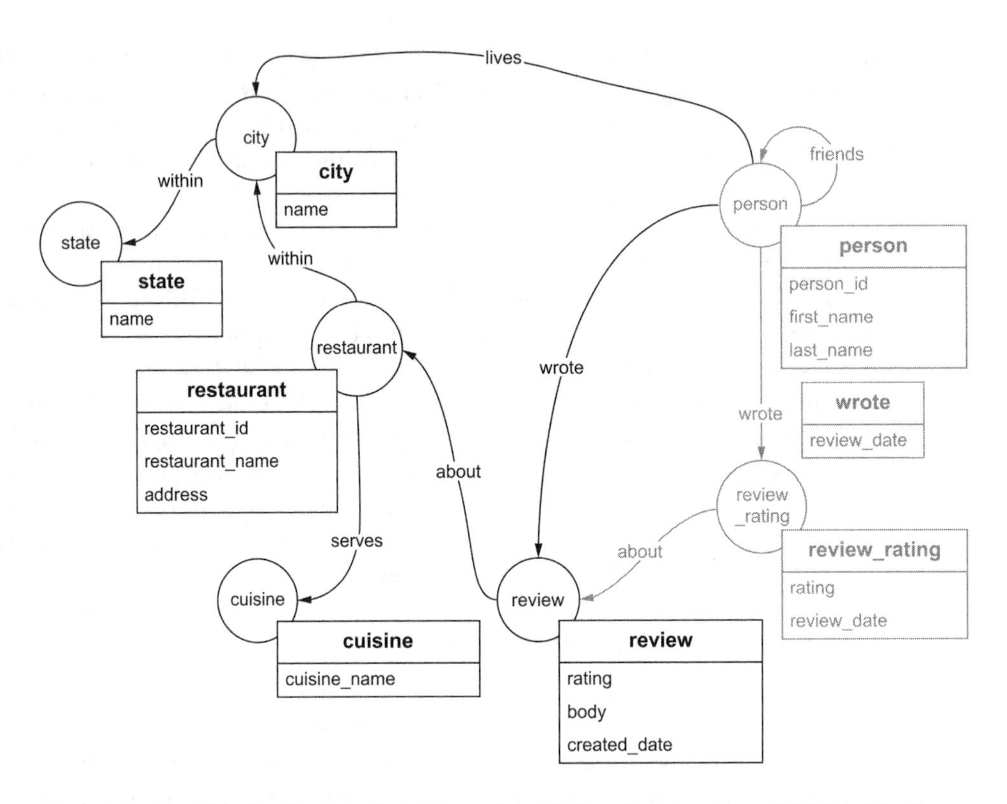

Figure 8.1 The DiningByFriends logical data model with the relevant portions highlighted for the recommendation engine use case

Although these steps seem logical and straightforward, the devil is in the details. In our experience, we find that this process is better understood by working through a few examples. Let's use our restaurant recommendation engine requirements to show how the preparation is different when working with a more involved model.

Our first question in the use case is, "What restaurant near me with a specific cuisine is the highest rated?" Breaking this question into the required actions, we see three pieces are needed:

- What restaurant near me . . . (locates the restaurants in a geographic area "near me")
- . . . with a specific cuisine . . . (filters by cuisine to find a specific cuisine)
- . . . is the highest rated? (calculates the average rating for each restaurant to find the highest rated)

Looking at our data model, let's see which vertex labels provide the information needed to satisfy these requirements. When looking for the required vertex labels, a good place to start is by looking for the nouns in the requirements. After we find the nouns, we look in the data model for the corresponding vertex labels for those nouns. While finding the corresponding labels can be a straightforward lookup, it frequently involves synonyms or additional concepts. We already did much of this work during our data modeling process, so we are able to leverage that work to quickly identify the required elements.

Applying this process to our requirements, we can identify the following required vertex labels. Note that the order of these doesn't matter. We are in a design phase and simply want to make sure that we have what we need to answer the overall question:

- `restaurant`—This is the core piece of information we need to return.
- `city`—We need to find restaurants "near me," so we use `city` to represent the geographical location of a restaurant as we explained in section 7.2.1.
- `cuisine`—This allows us to filter by cuisine.
- `review`—We need to calculate the average rating of a restaurant, and `rating` is a property of this entity.

What do you think? Have we listed all of the key vertices to return an answer for this traversal?

After reviewing, we find a subtle element missing. Did you notice that we need a `person` vertex? Because we need to find a restaurant "near me," we need to know what city to search. This is contingent on where the person lives. From our data model, we know that the current user's location is represented by the `lives` edge from the `person` vertex, so we need to include the `person` vertex.

> **IMPORTANT** Pronouns are easily overlooked when translating business questions into technical requirements and implementation. These tend to hide additional and more subtle requirements. Pay attention to pronouns when identifying required elements. In the current example, note how we called out the phrase "near me" (with an emphasis on "me" as an example of a potentially hidden element).

Now that we've identified the vertex labels, our next step is to look for the required edges. In our current model, this step is simplified because, for our four vertex labels (`restaurant`, `city`, `cuisine`, and `review`), we are limited in the edges we can use. We could just use the edge labels connecting those vertex labels and then proceed with a fair chance of success. But sometimes, there are multiple edge labels between vertices, or we might have missed an important part of the use case by only looking at the nouns. We should also remember to look for edge labels by examining the requirements of the use case and looking at the verbs (actions) in the questions.

Once we identify the verbs in our requirements, we can look for the corresponding edge names in our data model, much as we did with our nouns. As with that step, we are able to reuse much of the work we did while data modeling to quickly identify the required edge labels. Analyzing our requirements, we find the corresponding edges in our data model:

- `restaurant within city`—Determines the location of a restaurant to use as a filter
- `restaurant serves cuisine`—Categorizes the cuisine a restaurant serves to filter on it
- `review about restaurant`—Provides all the reviews for a restaurant to calculate the average rating for that restaurant
- `person lives city`—Provides a person's location to find restaurants "near me." (This is one of the more subtle or hidden requirements from our data model.)

> **NOTE** Two of these edges contain a preposition (`within`, `about`) implying a verb, rather than a verb itself. For the grammarians among our readers, in these cases in our data model, we decided to omit the verb *is* for simplicity; *within* is simpler to write than *is_within*. As we mentioned, naming things is hard, and naming the edges is no different.

Turns out that these are the only options for connecting the four vertex labels, so these are the edges we use, given our schema. It is possible that we missed something in our modeling work in chapter 7. If so, we would discover that at this point because there would be a disconnect between our use case requirements and the schema. Alternatively, if we skip the modeling work altogether, we must do that now so that the design satisfies the requirements. But because these edges from the logical model align nicely with the use case requirements, our efforts in chapter 7 were sufficient.

In our example, we found nine total elements for the first question: five vertex labels and four edges. Let's move on to see what's required for the other two questions that make up the recommendation engine. For each question, we follow the same process: extract the information needed to answer the question into a series of steps, find the nouns (or pronouns) in these steps to locate the corresponding vertex labels, and then find the verbs in our actions to identify the required edges. Let's see how this looks for our second question, "What are the ten highest-rated restaurants near me?" We start by extracting the requirements needed to answer the question:

- Locate restaurants in a geographic area.
- Determine a user's location to filter on restaurants in that area.
- Calculate the average restaurant rating in order to sort the restaurants and return only the top 10.

As with the previous example, we find the vertices by looking for the nouns or their synonyms, which provide the corresponding vertex labels in our data model:

- `restaurant`—This is the core piece of information we need to return.
- `person`—Locates a user, which is needed to satisfy the "near me" requirements of our question.

- `city`—Defines the location of both the user and the restaurant.
- `review`—Required to calculate the average rating for a restaurant because `rating` is a property.

Moving on to our next step, reviewing the logical data model to locate the verbs, we need the following edge labels:

- `restaurant within city`—Defines the location of a restaurant "near me" to filter on
- `person lives city`—Defines where the user lives
- `review about restaurant`—Provides all the reviews for a restaurant needed to calculate the average rating

Following this process, we identify seven required elements for this question: four vertex labels and three edges. While this is fewer than the previous question, that's not a problem. Because we follow the same process, we have a high degree of confidence that we did not miss anything. We can also compare the two questions. Doing so, we see that the main difference is that this question does not include a reference to cuisine. Otherwise, these questions are quite similar. This gives us the confidence that we're on the right track.

> **EXERCISE** For the third question—"What are the newest reviews for this restaurant?"—go through the process on your own and compile a list of vertices and edges that you think are required.

How many elements did you come up with? When we looked at that question, we found three. Our two vertex labels include

- `restaurant`—Required to find appropriate reviews for the current restaurant
- `review`—This is the core information being returned. For this question, we also assume that "newest" refers to the date the review was written, so we also need the `created_date` property, which allows us to find the newest reviews.

And our sole edge label is

- `about` (connecting review and restaurant)—Required to associate a set of reviews to the appropriate restaurant. This is also where our `created_date` property lives, which allows us to find the newest reviews.

In the three questions for our restaurant recommendation engine, identifying some of the required elements, such as restaurant, was fairly straightforward. However, identifying other elements, such the elements required for "near me," are less obvious and require us to leverage our experience when creating the logical data model to answer these. We are also aided in that our logical model has only a single edge label among vertex labels in all cases. We are now ready to get started writing our traversals, but where do we begin?

8.1.2 Selecting a starting place

Before writing our traversals, we need to make a crucial decision: Where do we begin our development work? We can't build three traversals all at the same time, so which use case should we address first? For this, we see two reasonable approaches.

One approach is to pick what we think is the most challenging problem and start there. This approach works well when there are compelling unknowns or project risks, such as introducing new technologies or processes into the development ecosystem. This approach allows us to fail fast and is the right choice when a quick decision of some sort is needed—perhaps to make a "go/no-go" decision or to determine whether the technology is the right choice for the problem.

Another approach is to start with the most straightforward or the least complicated question and use it as a building block for the rest of our work. This path allows for the progressive development of the code base and provides an excellent way to avoid biting off more than you can chew. The idea here is to get a quick win or success with a smaller, simpler problem before tackling more complex issues.

Let's look at the questions for our recommendation use case. Then we can decide which approach and questions look like the best place to start.

Question	Vertex Labels	Edge Labels

What restaurant near me with a specific cuisine is the highest rated?	`person`
	`restaurant`
	`city`
	`cuisine`
	`review`
		`lives` (connecting person → city)
		`within` (connecting restaurant → city)
		`serves` (connecting restaurant → cuisine)
		`about` (connecting review → restaurant)

What are the ten highest-rated restaurants near me?	`restaurant`
	`city`
	`review`
	`person`
		`lives` (connecting person → city)
		`within` (connecting restaurant → city)
		`about` (connectingreview → restaurant)

What are the newest reviews for this restaurant?	`restaurant` `review`	`about` (connecting review → restaurant)

The questions seem to become progressively simpler, with the last question having the fewest required elements. Which of the approaches would you choose?

If we were to choose the first approach, starting with the most challenging, we'd want to start with the first or second question. These questions are more complicated and involved than the last one. If we wanted to prove our graph technologies, choosing one of these questions would provide us with the most knowledge about the problem with a single effort. On the other hand, we could choose the second approach, starting with least complicated, which allows us to get a win or success with a smaller, simpler problem before tackling more complex issues.

For our present work, let's choose the second approach: start with the simplest problem and use that as a building block for the harder questions. For this project, we don't have a go/no-go point, and we don't have significant budget constraints. Without these constraints, we'd rather start small and get a quick win. However, before we begin developing in earnest, there's one more task to accomplish: setting up some test data for our work.

8.1.3 Setting up test data

Before we begin writing our traversals, the last step in our preparation is to load some test data. As with any database development, work goes much faster with data, preferably real data. Thus, a useful test data set should cover our core use cases at a minimum and, ideally, the known edge cases. As we work with the code and with the data, we expect to uncover additional edge cases. These make excellent candidates for adding to our test data set, as well as using these additional edge cases for unit and integration tests.

For this book, we included a set of test data with the code for this chapter, which may be found here: https://github.com/bechbd/graph-databases-in-action. Note that this script works a bit differently than the previous scripts. Instead of using individual commands to independently create vertices and edges, this script loads data from a JSON file. To see the details of how this works, see http://mng.bz/jVV9 to look at the `io()` step in Gremlin. The downside of this approach is that we need to update the reference location of the data file in the loading script, `8.1-restaurant-review-network-io.groovy`. To update the script, open it in a text editor of your choice and edit the line shown here to point to the full path and file location of chapter08/scripts/restaurant-review-network.json:

```
full_path_and_filename = "/path/to/restaurant-review-network.json"
```

Once that is done, if you set up the Gremlin Console according to the instructions in appendix A, then you can start the Gremlin Console and load the data for this chapter with this single command on MacOS or Linux:

```
bin/gremlin.sh -i $BASE_DIR/chapter08/scripts/8.1-restaurant-review-network-
    io.groovy
```

Or use this command on Windows:

```
bin\gremlin.bat -i $BASE_DIR\chapter08\scripts\8.1-restaurant-review-network-
    io.groovy
```

Once this script completes, our graph now contains test data we will use throughout the remainder of this chapter for testing during our traversal writing process. We can quickly verify that the data set is loaded correctly in the Gremlin Console by typing `g` and pressing Enter:

```
         \,,,/
         (o o)
-----oOOo-(3)-oOOo-----
```

```
plugin activated: tinkerpop.server
plugin activated: tinkerpop.utilities
plugin activated: tinkerpop.tinkergraph
gremlin> g
==>graphtraversalsource[tinkergraph[vertices:185 edges:318], standard]
gremlin>
```

8.2 Writing our first traversal

Now that we've decided where to start and have loaded our test data, it's time to begin writing the first traversal. At this point, we like to break down the question into parts and progressively build out the code in a sequential, or at a least systematic way, using these steps:

1. Identify the vertex labels and edge labels required to answer the question.
2. Find the starting location for the traversal.
3. Find the end location for the traversal.
4. Write out the steps in plain English (or in your preferred language). The first step is your input step; the last step is the output or return step.
5. Code each step, one at a time, with Gremlin and verify the results against the test data after each step.

As you probably noticed, this sounds like the process we went through to develop the traversals for our social network. This similarity is not an accident. Although we followed this process, we did not formalize it.

In chapters 3, 4, and 5, we chose to focus on the basics of writing a traversal instead of getting bogged down in the process. These were also (by design) much simpler traversals than what we build in this chapter. Now that you've learned the basics and are familiar with both thinking about and traversing a graph, as well as the Gremlin syntax, it is time to formalize our process as we take on more involved traversals. For the next section, we'll employ this process consistently. However, for section 8.2.2, we'll streamline the instructions and focus on how the execution looks when we are in the flow of writing code.

8.2.1 Designing our traversal

Now we are ready to begin designing our traversal to answer, "What are the newest reviews for this restaurant?" We start this process by identifying the required portions of the schema, based on the work from section 8.1.1. Figure 8.2 highlights the part of the logical data model (schema) that we'll use to answer this question.

We use this part of the schema to choose a starting point for our traversal. As we learned in chapter 3, choosing a starting point is about looking for the criteria that quickly narrows down the number of starting vertices. This minimizes the number of traversers required and results in faster performance overall.

Looking at the question, we see that it revolves around, "this restaurant," so that is a great place to start our traversal. Because we are looking for a single starting vertex

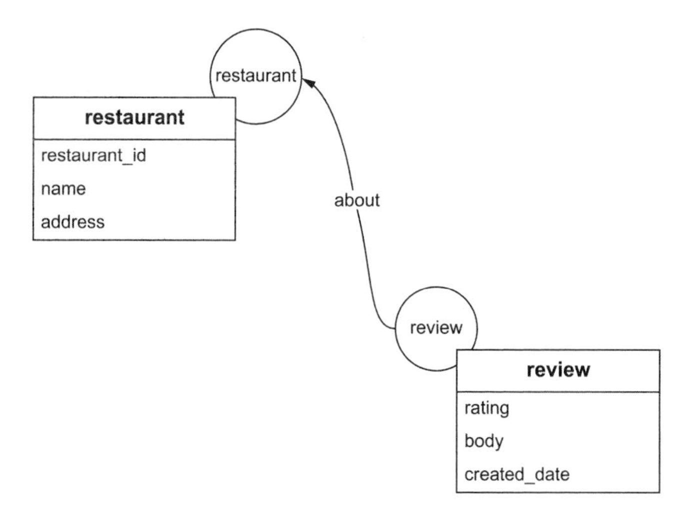

Figure 8.2 The logical data model showing the information required to answer the question, "What are the newest reviews for this restaurant?"

("this restaurant"), we start by filtering our traversal. This leads to a dilemma: Which property do we use to identify a restaurant, `restaurant_id`, `name`, or `address`?

- `name`—This is problematic as restaurants can belong to chains, all of which have the same name. Using this as a filter could potentially return many restaurants.
- `address`—Works well for standalone buildings, but a user might be in a food court, multi-tenant building, or at a mobile food truck. In any case, the address might not be apparent or readily available.
- `restaurant_id`—This property more precisely describes a particular restaurant, so we can use this unique identifier to ensure that we get the single starting vertex.

Given these factors, let's make a simplifying assumption that our traversal takes a `restaurant_id` as an input. Because the `restaurant_id` is a unique identifier of the restaurant, using this ID ensures that we get the single starting vertex we desire. With our starting vertex and filtering criteria identified, let's move on to the next step of our process: determining the end point. We find the end point by looking for the nouns and qualifiers that describe what the traversal needs to return; what type of thing will the answer be?

Examining the question, we see that we want to return "the reviews" for the restaurant. This means that our traversal needs to traverse from our starting point, the `restaurant` vertex, to the `review` vertex to retrieve the reviews. This is the only additional information the question asks for, so this makes the `review` vertex our traversal's endpoint. However, we don't want to return the entire vertex. Users don't want to get vertices back. An end user wants the text of the review, so we need to return the body property. Because we want the "newest" reviews, we also need to order by when a review was made. For that, we return the `created_date` property as well.

Now that we know that we start with the `restaurant` identified by its `restaurant_id` and return the `created_date` and body properties of the `review` vertices, our process gives us the following steps that our traversal must perform (we sort for newest later in this process):

- Get the `restaurant` based on the `restaurant_id`.

  ...

- Return the `created_date` and body properties of the `review` vertices.

In between these two steps is one or more steps that we'll fill in as we complete this section. We have denoted these middle steps with an ellipsis (. . .) because we don't know exactly how many steps there will be to get from our starting point to the final response. But we do know that we have some steps in between the starting point and the return data because we are not simply returning the starting vertex.

As a point of comparison, in a relational database these starting and ending objects would be two tables, and we need to determine the join conditions necessary to return the correct rows, which may require additional tables. In our graph database, we have identified two vertex labels, and we must construct the appropriate traversal steps to craft the desired response data. Some of this involves traversing through the graph with the steps we first introduced in chapter 3. Other parts of this involve formatting the response using the steps introduced in chapter 5.

Having identified our starting point and return data, let's examine our schema (figure 8.3). This allows us to find the appropriate vertices and edges we need to traverse from one to another.

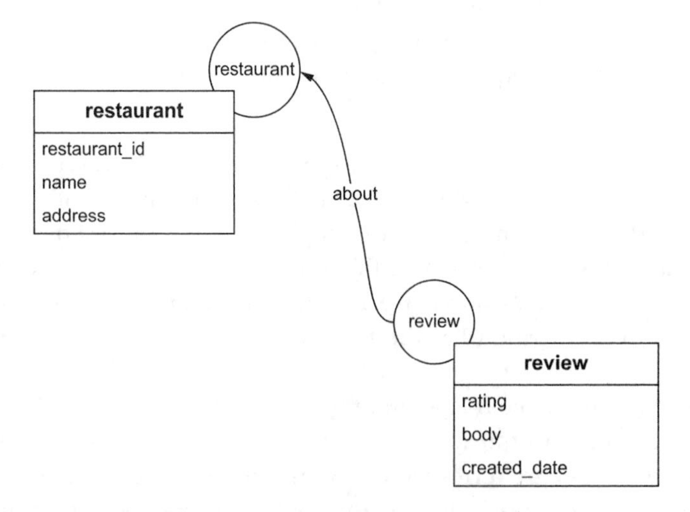

Figure 8.3 The graph schema shows that the starting point (`restaurant_id`) and ending point (`created_date` and body properties of the `review` vertices) are directly connected.

Because the starting and ending points are directly connected, this keeps things simple. Now we know we need to

- Get the `restaurant` based on the `restaurant_id`.
- Traverse the `about` edges to the `review` vertices.

  . . .
- Return the `created_date` and `body` properties of the `review` vertices.

That's closer, but the output needs to be ordered a specific way, namely, by the most recent reviews. Our `review` vertex has a `created_date` property that is well-suited for this purpose. Because we want the newest reviews, we sort the `review` vertices by `created_date` in descending order. Using properties such as a created date or an updated date is a widespread pattern in graph databases, as with other databases. That allows us to sort items chronologically. With this, we have a total of four steps in our path:

- Get the `restaurant` based on the `restaurant_id`.
- Traverse the `about` edges to the `review` vertices.
- Sort `review` vertices by `created_date` in descending order.
- Return the `created_date` and `body` properties of the `review` vertices.

Upon first glance, we might think that we're done, but there's another factor we should consider. There could be hundreds or thousands of reviews going back years for restaurants in the system. There's a chance that the result set could be extensive, so let's limit the results. At the end of this section, we throw in some bonus material on how to handle pagination, but for now let's only return the top three. This wasn't directly specified in the use case, but as experienced developers, we know that this is a reasonable implementation approach, one that we can later validate with user testing. Bringing everything all together, we now see that our traversal needs to perform the following actions:

- Get the `restaurant` based on the `restaurant_id` .
- Traverse the `about` edges to the `review` vertices.
- Sort `review` vertices by `created_date` in descending order.
- Limit the results to the first three returned.
- Return the `created_date` and `body` properties of the `review` vertices.

Comparing these steps against the requirements of our question, we see that we've defined the traversal in sufficient detail to get started writing. The old saying, "The proof of the pudding is in the eating of it," applies here too. We won't know for certain that we have covered the requirements of the use case until we code the traversal. But now we have a clear set of steps that should make short work of writing the code.

The set of steps needed for this traversal looks similar, though not exactly identical, to the recursive and path-based traversals we developed for our social network. When we developed our recursive traversals, we had a known series of vertices and edges to traverse but an unknown number of times with which to traverse these. With those path-based traversals, we were interested in *how* our start and end points were connected (the path). In contrast, the known-walk or known-path traversals we work with now have a defined set of vertices and edges to traverse and a known number of times we must traverse these. We use known-path traversals when we are interested in finding out if our starting and ending points are related, not in *how* these are related.

In our present example, we need to traverse a single type of edge label. The variable won't be in the number of edges labels traversed (there's only the single `about` label), but in the number of instances of `about` edges. In relational databases, this is akin to the difference between the number of joins we have to use verses the number of rows actually returned.

8.2.2 Developing the traversal code

Now that we've designed our traversal approach, we're ready to start developing it using our test data. This is the process of taking the bulleted points we designed in the previous section and writing the corresponding code. Looking at the actions we need to accomplish, we notice that we've already performed each of these steps in previous chapters when building pathfinding traversals. Because we already know the steps required, let's develop this traversal to learn an iterative approach to constructing complex traversals.

This process is straightforward. We begin by writing the code for the first step and progressively adding code for each additional step, one step at a time. Then we verify that we are returning the expected results after each one.

Looking at our test data, let's use that great greasy eatery, that hopping Houston mainstay, Dave's Big Deluxe, as our example, which has a `restaurant_id = 31`. We should probably mention that this restaurant, as with all of the people and other restaurants in the examples, is entirely fictitious.

To allow us to easily test against multiple restaurants, let's create a variable named `rid` and assign the `restaurant_id` of Dave's Big Deluxe, `31`. We add the variable using the following command in the Gremlin Console:

```
rid = 31
```

NOTE As previously mentioned, the Gremlin Console allows for creating variables using Groovy syntax. In this case, we create a variable simply to ease the development of the traversal, so it's not required. But using a variable makes it easy to test our traversal on multiple restaurants by merely changing the variable value instead of the traversal.

We already know how to create the traversals to find our starting point. Let's filter on restaurants that match the `rid` variable:

```
g.V().has('restaurant','restaurant_id',rid)      <-- Gets the restaurant based
                                                     on the restaurant_id = rid
==>v[288]
```

It returned a result! That's great, but it'd be nice to know if it's the correct result, so let's check our results. While not specifically required to construct our traversal, we find it helpful to add values steps to traversals while constructing these for immediate validation. In this case, we add the `valueMap()` step:

```
g.V().has('restaurant','restaurant_id',rid).
  valueMap(true)                              <-- Returns all properties

==> {id=288, label=restaurant, address=[490 Ivan Cape],
➥ restaurant_id=[31], name=[Dave's Big Deluxe]}
```

Perfect! We got back Dave's Big Deluxe, so we confirmed that we have the right `restaurant` vertex. Let's move on to the next step: traversing the `about` edges.

Examining the schema, we see it is an inbound edge, meaning we use an `in()` step. (Don't worry that the body of a review looks like gibberish. The data set we loaded contained some autogenerated text values, so these are not meant to be understandable.)

```
g.V().has('restaurant','restaurant_id',rid).
  in('about').                     <-- Traverses the about edges
  valueMap(true)                       to the review vertices

==>{id=894, label=review, rating=[5], created_date=[Wed Sep 26 18:30:16 CDT
➥ 2018], body=[Soluta velit quasi explicabo ut atque ratione nisi. ...]}
...                                <-- Results truncated to improve readability.
==>{id=666, label=review, rating=[5], created_date=[Wed May 01 07:37:44 CDT
➥ 2019], body=[Quo et non aut ipsam qui autem aut. Voluptatem id. ...]}
```

Whoa! Perhaps that `valueMap()` step is too helpful. If you're following along, you should have returned a wall of text for the eight review results. We need a way to see those results, but perhaps without all of the noise. Referring back to our plan for this traversal, we know we want to return the `created_date` and `body` properties, so let's update our traversal to only return those properties:

```
g.V().has('restaurant','restaurant_id',rid).
  in('about').
  valueMap('created_date', 'body')     <-- Returns only the created_date
                                           and body properties

==>{created_date=[Sun Jul 19 01:43:31 CDT 2015],
➥ body=[Dolorem  ...                          <--
==>{created_date=[Wed Sep 26 18:30:16 CDT 2018],    Results
➥ body=[Soluta    ...                         <--  returned
==>{created_date=[Wed Jul 27 07:30:46 CDT 2016],    unsorted
➥ body=[Officiis ...                          <--
...
```

NOTE Within Gremlin Server, dates are stored in Coordinated Universal Time (UTC) formats; but for display purposes, Gremlin Console automatically translates dates to the local time zone.

That's better; at least we only have a pair of properties. Looking back at our plan, the next step is to add logic to sort the results by the `created_date` value:

```
g.V().has('restaurant','restaurant_id',rid).
  in('about').
  order().by('created_date').          <-- Orders reviews by the created_date
  valueMap('created_date', 'body')

==>{created_date=[Sun Jul 19 01:43:31 CDT 2015],
➥ body=[Dolorem ...
==>{created_date=[Wed Jul 27 07:30:46 CDT 2016],
➥ body=[Officiis ...
==>{created_date=[Thu Mar 09 03:37:52 CST 2017],
➥ body=[Rerum omnis ...
...
```

Results ordered in ascending order by created_date

We're almost there. The results are sorted by date, but the wrong way! Recall that the `order()` step defaults to ascending order, so let's use descending order:

```
g.V().has('restaurant','restaurant_id',rid).
  in('about').
  order().by('created_date', desc).    <-- Adds desc for descending order so the ordering is in the expected direction
  valueMap('created_date', 'body')

==>{created_date=[Wed May 01 07:37:44 CDT 2019],
➥ body=[Quo et ...
==>{created_date=[Tue Mar 12 20:33:43 CDT 2019],
➥ body=[Ducimus ...
==>{created_date=[Wed Sep 26 18:30:16 CDT 2018],
➥ body=[Soluta ...
...
```

Results sorted in descending order by created_date

We accomplished all the actions our traversal requires except for limiting the results. Let's add that functionality now:

```
g.V().has('restaurant','restaurant_id',rid).
  in('about').
  order().by('created_date', desc).
  limit(3).                            <-- Limits results to three
  valueMap('created_date', 'body')

==>{created_date=[Wed May 01 07:37:44 CDT 2019],
➥ body=[Quo et ...
==>{created_date=[Tue Mar 12 20:33:43 CDT 2019],
➥ body=[Ducimus ...
==>{created_date=[Wed Sep 26 18:30:16 CDT 2018],
➥ body=[Soluta ...
```

Only receive three results.

That's great! It looks like we're done. We have the right data, returned in the proper order, and limited to the correct result size. What's next?

EXTENDING OUR TRAVERSAL WITH THE ID

While the previous traversal handles all the requirements of our question, our application is likely going to need more than just the body of the review text. Most applications work from domain objects, so we probably want to create a review object for our application, with a unique identifier tying our domain object back to the underlying database.

As discussed in section 4.1.1, we discourage the use of a database engine's internal ID values for any business logic; however, due to the ease of using that ID, it's common practice, which is why we discuss it here. Using the internal ID results in a leaky abstraction that tightly couples your application logic to the underlying database implementation. The best practice is to use either a natural key from your data or an application-generated synthetic key. Now that we've restated our "use ID's wisely" message, let's update our traversal to return the ID of the review vertex, along with the created_date and body properties:

```
g.V().has('restaurant','restaurant_id',rid).
  in('about').
  order().by('created_date', decr).
  limit(3).
  valueMap('created_date', 'body').
    with(WithOptions.tokens)                 <-- Returns the ID and label metadata of the vertex

==>{id=666, label=review,
➥ created_date=[Wed May 01 07:37:44 CDT 2019],
➥ body=[Quo et non aut ipsam qui autem aut...     <--
==>{id=564, label=review,
➥ created_date=[Tue Mar 12 20:33:43 CDT 2019],
➥ body=[Ducimus maxime corrupti et aut...         <-- Results now include id and label properties.
==>{id=894, label=review,
➥ created_date=[Wed Sep 26 18:30:16 CDT 2018],
➥ body=[Soluta velit quasi explicabo ut...        <--
```

> **valueMap() and the with() step**
>
> As of TinkerPop version 3.4, the `valueMap()` step takes an optional `with()` step for modifying the output, so it is a fairly recent addition to TinkerPop. Not all vendors may support it. Prior to TinkerPop version 3.4, the `valueMap()` step, including the IDs and labels, uses a Boolean parameter like `valueMap(true)`, or, more specific to our present case, `valueMap(true, 'created_date', 'body')`. That form should still work, at least up to TinkerPop version 3.5.
>
> TinkerPop is striving for more consistency in its implementation and now uses a `with()` step to provide configuration information for the `valueMap()` step. By specifying `with(WithOptions.tokens)`, we can include both Gremlin's internal ID value as well as the `label` of the vertex in our responses.

Adding this traversal to our application

Whew, it certainly took more work to develop this known-walk traversal than the social network ones, but the hard work is over. Luckily, the process of adding it to our actual application is the same as the one we used for adding our social network traversals in chapter 6. We won't go over the details again, but you can find the full Java code in the chapter08/java folder in a new method called `newestRestaurantReviews()`. We share the relevant parts of the method here with some call-outs for the minor differences from the Gremlin code we drafted. The following shows the `newestRestaurantReviews()` method:

```
private static String newestRestaurantReviews(GraphTraversalSource g) {
    Scanner keyboard = new Scanner(System.in);
    System.out.println("Enter the id for the restaurant:");
    Integer restaurantId = Integer.valueOf(keyboard.nextLine());

    List<Map<Object, Object>> reviews = g.V().
        has("restaurant",
➥ "restaurant_id", restaurantId).            <-- All strings in Java must use double quotes.
        in("about").
        order().
          by("created_date", Order.desc).     <-- Uses TinkerPop's Order enumeration
        limit(3).
        valueMap("rating", "created_date", "body").
          with(WithOptions.tokens).
        toList();                             <-- All traversals require a terminal step; here we use toList() when not using the Gremlin Console.

    return StringUtils.join(reviews, System.lineSeparator());
}
```

Congratulations, you built your first known-walk traversal and you now know the process of iteratively constructing a traversal! You can repeat this process of starting with a specification of steps, adding each Gremlin step one at a time, and testing the traversal against the test data to make sure you get the expected results until you've satisfied all the identified parts.

8.3 Pagination and graph databases

This traversal is also a great way to illustrate one of the more challenging patterns with graph databases—pagination.[1] As we noted, it's possible that our traversal can return more results than the requesting software wants or can handle. We addressed that concern in the last section by limiting our results to just the newest three reviews. What if the requesting software wanted access to all the results, just not all of those at the same time?

---

[1] We want to give a shout-out to Jason, an early MEAP reader that requested we address the subject of pagination. Jason, thanks for being an early reader of our work and for reminding us to discuss this everyday use case.

Before we investigate how graph databases handle pagination, let's take a quick look at how relational databases handle pagination to use that as a comparison. Wait—we can't take a *quick* look because it seems that every relational database engine does it slightly differently, each with its own semantics. What *is* common is that most pagination implementations take two inputs:

- `offset`—The number of records to skip. To start at the beginning of the data set is to have `offset = 0`. Offset values are multiples of the page size. With a page size of 10, possible offset values would include `0`, `10`, `20`, `30`, and so forth.
- `limit`—The page size or the maximum number of items to return. We stress that the limit is the maximum number of items because result sets won't always return the page size (e.g., the final page can contain fewer objects than the specified page size).

The general pattern for dealing with pagination in a relational database, and in a graph database it is the same:

1. Retrieve the values of the traversal.
2. Start with the record at the `offset` index.
3. Return the `limit` number of values.
4. Repeat the process with a new offset value equal to `offset + limit`.

To handle this in Gremlin, we use the `range()` step. Although we briefly introduced the `range()` step in chapter 5, here we expand the definition slightly:

- `range(startNumber, endNumber)`—Passes through the objects, starting with and including those indexed at the `startNumber`, continuing up to but not including those indexed with `endNumber`. So `startNumber` is *inclusive*, and `endNumber` is *exclusive*.

A keen observer will notice that while `startNumber` is the same as the offset, `endNumber` isn't the same as the `limit` used in relational databases. Instead, `endNumber = startNumber + limit`. A pagination function that take the usual inputs of `offset` and `limit` must compute the `endNumber`.

The `startNumber` and `endNumber` values apply to an element index returned by the traversal. This element index value is somewhat analogous to the results of a SQL `ROW_NUMBER()` function. This value is a zero-based element index number assigned to each of the elements based on the order those are passed to the `range()` step.

Importance of ordering the inputs before calling range()

For pagination to work as expected, the order of objects passed to the `range()` step matters. This means we need to order elements prior to paginating them. Without ordering, the results can arrive at the `range()` step in any sequence, and subsequent calls to the same traversal can result in a different ordering of the objects as these enter that step. Not sure what we mean by this? Well, let's take a look.

For example, assume that a graph has five vertices: ([v[0], v[1], v[2], v[3], v[4]]). Then let's assume that we request vertices two at a time, and that we make that request three times. This looks like the three following calls:

```
g.V().range(0,2)
g.V().range(2,4)
g.V().range(4,6)
```

Running this, we expect to obtain the following output:

```
v[0], v[1]
v[2], v[3]
v[4]
```

This output is only true if g.V() returns the same order of vertices each time. What would happen if it changed the ordering each time?

In theory, if a database provides a randomized return order, then we get back a seemingly random pair of vertices for each call. Because our call returns the values based on an index, if the element in that index changes between runs, so does the value that is returned. TinkerPop guarantees that it returns elements in the order these enter a step, but it is up to the actual underlying engine to specify that order. In other words, there's no guarantee about the order unless we specify an order.

This is no different than relational databases: the database engine determines the order of the results according to its own internal logic. This means that to provide a consistent experience to our users, we must sort things before calling the range() step.

ORDERING IS AN EXPENSIVE OPERATION

We must note that ordering results is an expensive operation in any database, particularly with large data sets. To order the results of a traversal, the database must first return all the results and then sort them. This cost is the same in both relational and graph databases.

In TinkerPop, the order() step is categorized as a "collecting barrier step." Unlike most TinkerPop steps, which are lazy, meaning that the data is processed opportunistically as new values enter the step, the order() step (and other collecting barrier steps) first collect *all* incoming values before ordering these and then send the results to the following steps. Yet again, this is no different than any relational database, because to provide a sorted set of values, we must first know all the values we need to sort.

Let's take a look at what our traversal from the last section looks like when we update it to include pagination. To reduce the output text, we use the rating property instead of the body property. When adding pagination, we need to make the following changes:

1. Replace the limit() step with a range() step.
2. Define a limit variable with a value of 3.
3. Define an offset variable and increment it by the limit value for each call.

NOTE Because the Gremlin Console converts timestamps to local time zones and we can update our sample data after publication, your results might not match our results exactly.

Implementing these changes, our traversal now looks like this:

```
limit = 3                                  Sets the number of
==>3                                       results to return
offset = 0                                 Sets our initial offset
==>0
g.V().has('restaurant','restaurant_id',rid).
  in('about').
  order().by('created_date', decr).        Replaces the limit() step
  range(offset, offset + limit).           with the range() step
  valueMap('rating','created_date')
==>{rating=[2],
   created_date=[Sun May 26 00:53:56 AKDT 2019]}
==>{rating=[1],                                          Newest
   created_date=[Thu Mar 28 21:56:30 AKDT 2019]}         three results
==>{rating=[5],                                          returned
   created_date=[Fri Nov 09 20:09:49 AKST 2018]}
```

We see from this code example that the newest three reviews are returned. Let's see what happens if we try to get the next page of results:

```
offset = offset + limit                    Updates our offset to get
==>3                                       the next page of results
g.V().has('restaurant','restaurant_id',rid).
  in('about').
  order().by('created_date', decr).
  range(offset, offset + limit).
  valueMap('rating','created_date')
==>{rating=[5],
   created_date=[Tue Sep 11 14:39:05 AKDT 2018]}
==>{rating=[3],                                          Returns the
   created_date=[Tue Oct 24 07:38:21 AKDT 2017]}         next three
==>{rating=[2],                                          results
   created_date=[Tue Mar 28 18:10:00 AKDT 2017]}
```

We retrieve three results, but these are older than the results from the previous run of this traversal. To continue paging our results, we only need to update the `offset` each time by calculating a new `endNumber` using the `offset + limit` calculation. Let's update our `offset` again and do this one more time:

```
offset = offset + limit                    Updates our
==>6                                       offset to 6
g.V().has('restaurant','restaurant_id',rid).
  in('about').
  order().by('created_date', decr).
  range(offset, offset + limit).
  valueMap('rating','created_date')
```

```
==>{rating=[2],
➥ created_date=[Thu Jun 09 08:58:35 AKDT 2016]}
==>{rating=[2],
➥ created_date=[Sun Sep 27 10:21:17 AKDT 2015]}
```

Returns only two results instead of three

Well, that is a bit strange. Why did we only get two results back instead of three? This is because we only had eight reviews for this restaurant. This leads us to one final question to address: How do we know when to stop paging?

One approach is to keep running the traversal and incrementing the offset until it returns an empty result set. This approach is useful when we expect a large number of results and don't need to know the total number, or when we want to avoid the cost of counting all of the results up front. A second approach is to initialize a count of the possible results and then use that as an upper bound to the `offset + limit` value. This latter approach is particularly useful when the application needs to know the total number of possible results it is required to display.

8.4 Recommending the highest-rated restaurants

We've finished writing the first known-walk traversal for our recommendation engine. Let's move on to answering our second question, "What are the ten highest-rated restaurants near me?" For this question, we follow the same methodology as we did in the last section:

1. Identify the vertex labels and edge labels required to answer the question.
2. Find the starting location for the traversal.
3. Find the end location for the traversal.
4. Write out the steps in plain English (or in your preferred language). The first step is your input step; the last step is the output or return step.
5. Code each step with Gremlin, one at a time, verifying the results against the test data after each step.

8.4.1 Designing our traversal

Now that we know the process for constructing a traversal run through, let's repeat the process for this traversal. However, unlike the detailed walk-through we did in the last section, we progress through the steps quickly, only stopping to dive into the details where they differ from the previous process. In section 8.1, we specified that we need the following elements to answer this question.

Question	Vertex Labels	Edge Labels

What are the ten highest-rated restaurants near me?	`restaurant`
	`city`
	`review`
	`person`
		`lives` (connecting person → city)
		`within` (connecting restaurant → city)
		`about` (connecting review → restaurant)

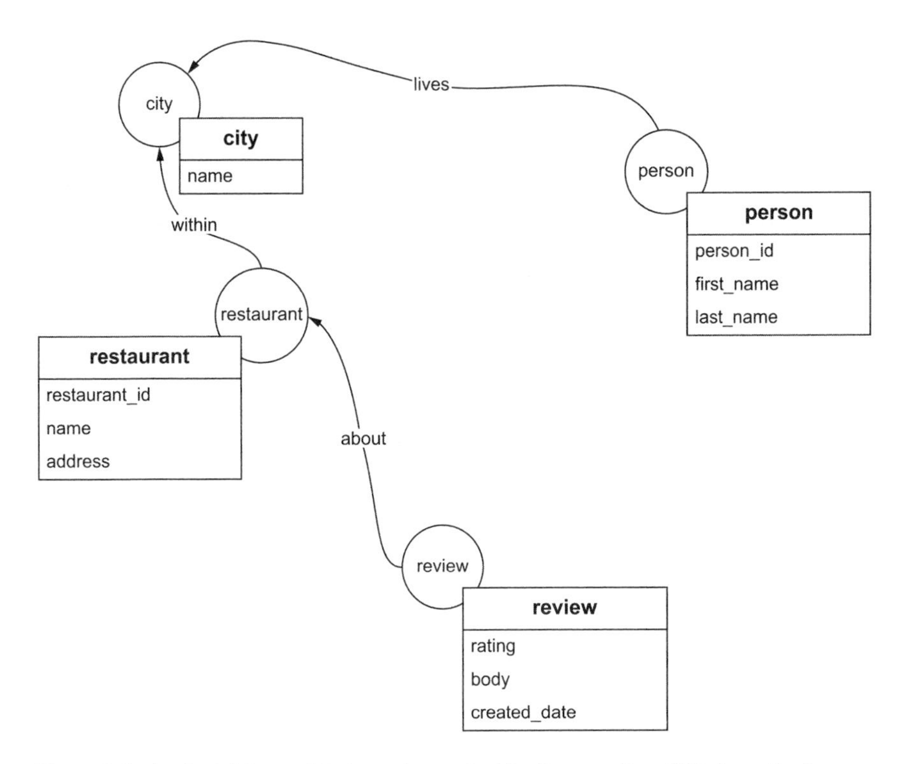

Figure 8.4 Logical data model elements required for the question, "What are the ten highest-rated restaurants near me?"

Let's highlight the relevant elements of the schema. Figure 8.4 shows the data model for this use case.

Because we've identified the relevant schema elements for our use case, we can move on to the first step in finding our starting step. Remember that locating the starting point is all about finding the portion of the question that narrows down our traversal to the minimum number of starting vertices. Looking at the question, "What are the ten highest-rated restaurants near me?", we see that a user wants all the restaurants located "near me." This means that the `person` vertex filtered by `person_id` needs to narrow down our starting vertex to a single instance, "me."

Next, we move to the second step in our process—finding the end point. Examining the question tells us that what the user wants is a list of nearby restaurants, meaning our endpoint should be the `restaurant` vertex. While not explicit, it seems likely that a user wants to get back several properties of the restaurant, such as the `name`, `address`, and the average `rating` for that restaurant. With these two pieces of information, we can start our step specification as

- Start with the current `person`, identified by their `person_id`.

  . . .

- Return the name, address, and average rating for the `restaurant` vertices.

Next, we move on to figuring out the series of vertices and edges needed to traverse from our starting and ending points. As we did in the last question, we are going to assume that "near me" means the city in which the user lives. To get the city for that person, we traverse from the person to the `lives` edge to find the city. Now that we are on the correct `city` vertex, we need the `restaurant` vertices in the same city. Examining our schema, we see that the `within` edge connects the two.

As with our last traversal, we need to do some ordering and limit the results; in this case, 10 is specified before we return the `restaurant` vertices. This leaves us with the following set of actions that our traversal needs to accomplish:

- Start with the current `person`, identified by their `person_id`.
- Traverse the `lives` edge to get their `city`.
- Traverse the `within` edge to get the `restaurant` vertices of the `city`.
- Calculate and perform a descending sort by `average_rating`.
- Limit to the first 10 results.
- Return the name, address, and average rating for the `restaurant` vertices.

With these actions identified, let's go on to the next step. This step is where we iteratively develop our traversal.

8.4.2 Developing the traversal code

Just as we did in the last section, we'll develop this traversal in an iterative fashion. In this section, we'll write the code one step at a time, testing our traversal after each step to ensure that we get the results we expect. Along the way, we'll revisit some of the concepts we learned in previous sections and show how these can be applied to more complex, real-world scenarios.

Before we begin, as with the last example, let's make it easy to test multiple people by using a variable to represent the `person_id` we use for filtering. Let's use Denise in Cincinnati this time and set `pid = 8`:

```
pid = 8
==>8
```

As always, we start our traversal filtering on the starting point (in this case, Denise). We learned in chapter 3 how to use Gremlin to filter the `person` vertex based on a `person_id` value, so let's apply that here and add a `valueMap()` step to return the properties so we can verify our results

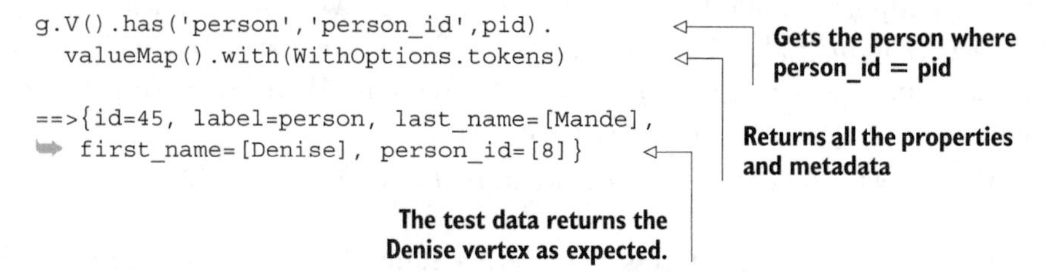

We're off to a good start! Let's now traverse out from the `Denise` vertex to find the `city`. From the schema, we see this is an outbound edge, so we use an `out()` step:

```
g.V().has('person','person_id',pid).
  out('lives').                             <- Traverses the lives edge to get the city
  valueMap().with(WithOptions.tokens)

==>{id=7, label=city, name=[Cincinnati]}    <- The test data returns Cincinnati as expected.
```

Because Denise lives in Cincinnati, our next action is to traverse the `within` edge, which is an inbound edge. With this traversal, we can find the nearby restaurants in that fair urban conclave:

```
g.V().has('person','person_id',pid).
  out('lives').
  in('within').                             <- Traverses the within edge to get the restaurants in Cincinnati
  valueMap().with(WithOptions.tokens)

==>{id=60, label=restaurant, address=[600 Bergnaum Locks],
➥ restaurant_id=[1], name=[Rare Bull]}
==>{id=192, label=restaurant, address=[102 Kuhlman Point],
➥ restaurant_id=[18], name=[Without Heat]}
...
```

We've truncated the results to only a subset, but it's clear that we're dealing with test data. How else can we explain the complete lack of chili restaurants in Cincinnati? (Although what they call "chili" in Cincinnati would never be considered authentic chili in other parts of the world.)

Now that we've successfully returned the restaurants in the city of our user, Denise, we're ready to order this list of restaurants by their average ratings. Here is where it gets a little bit trickier. In the previous examples, all the values we needed to answer our question resided on the end point vertex. For this example, this is not the case. The `rating` value we need to use is located on the `review` vertex, so we need to traverse to the `review` vertices for a restaurant and compute the average rating.

Recall section 5.3, where we discussed ordering and grouping? There, we stated that not only can we pass a property into the `order()` step's `by()` modulator, but we can also pass in a traversal. In this case, we want to pass in a traversal, which traverses to the `review` vertex, and average the `rating` values for a restaurant. While we know how to traverse to the `review` vertex, we need a new step to calculate the average:

- `mean()`—Aggregation to compute the mean or average of a set of values; used most commonly in the `group().by().by()` step pattern

Let's combine this step with what we know about traversing and give this a try:

```
g.V().has('person','person_id',pid).
  out('lives').
  in('within').
```

```
  order().
    by(__.in('about').values('rating').mean()).    <-- Orders results by the average rating
  valueMap().with(WithOptions.tokens)

The provided traverser does not map to a value:
➥ v[232]->[VertexStep(IN,[about],vertex),
➥ PropertiesStep([rating],value), MeanGlobalStep]
➥ Type ':help' or ':h' for help.
➥ Display stack trace? [yN].    <-- Standard Gremlin Server error statement
```

Uh-oh. Whoops! Something certainly didn't go as expected there. Let's see if we can parse this unexpected response and sort out the problem.

TROUBLESHOOTING ERRORS WHILE DEVELOPING A TRAVERSAL

We'll start by saying that we rarely look at the stack trace for an error. We find that if we are following the development process of adding one action at a time to the traverser, then we already know where to begin debugging. Combining that with the details in the error and our knowledge of how graph traversals work is usually sufficient to troubleshoot. Now, about the actual troubleshooting. Because we iteratively add steps to our traversal, we know that the issue is likely with the last step:

```
order().
  by(__.in('about').values('rating').mean())
```

Let's take a look at the error message to see what details it might give us about the problem:

```
The provided traverser does not map to a value.
```

Just as with any debugging problem, we need to combine the information in this error message with what we know about how graph traversals work to postulate a likely reason for the error. Recall that a traverser is a process that performs a specific task. The traversal, which is the whole string of steps, is broken down into multiple traversers along the way. In this case, we can see that one of these traversers ran into a problem: it didn't return a result. The question is, what was the traverser trying to do when it failed to get a result? Because we know that the issue was with our ordering step, we already know the problematic step, but how do we check our suspicion?

How exactly we check for this information is different for each database. Luckily, for those using Gremlin, this is easy because the next part of the error statement tells us:

```
v[232]->[VertexStep(IN,[about],vertex), PropertiesStep([rating],value),
     MeanGlobalStep]
```

The information you are looking at is Gremlin bytecode. Generally, as application developers, we don't worry about the bytecode. It is an implementation detail within the Gremlin Server and in the TinkerPop drivers. However, it is usually what is displayed when there is an error and points out a problem. While there is not a one-to-one mapping of bytecode steps to Gremlin steps, it is usually easy enough to map between the two (more on this in chapter 10). In this case, we see that one vertex (`v[232]`) generated the error. Which traversal step caused the error can be inferred from knowing which steps were added last; in this case

```
order().
    by(__.in('about').values('rating').mean())
```

More specifically, we see that the bytecode identifies the `MeanGlobalStep`, so we can infer that our issue is with the `in('about').values('rating').mean()` part of the traversal. This is an important feature of our step-by-step approach to traversal development (or any complex software, for that matter). By testing after each incremental change, when we run into an error, we know exactly what caused it!

Because vertex `v[232]` didn't like that part of the traversal, let's investigate why. To answer these types of questions about a specific vertex, we start by looking at the data for that vertex:

```
g.V(232).valueMap().with(WithOptions.tokens)      <- Traversal gets the full details about
                                                     the vertex with the ID 232

==>{id=232, label=restaurant, address=[212 Lorraine Court],
➥ restaurant_id=[23], name=[With Sauce]}    <- Response with full details about
                                                the vertex with the ID 232
```

Nothing looks out of order with the details about the vertex. But then again, that wasn't where there was a problem. The vertex made it through the traversal just fine, up to the point where we sorted these by the average ratings. Because the ratings are located on the `review` vertex and not the `restaurant` vertex, let's look at the `about` edges that take us to the `review` vertices for this restaurant:

```
g.V(232).inE('about')      <- Traversal displays the about edges
                              for the vertex with the ID 232
==>          <- Response listing a within
                edge and a serves edge
```

Well, that's it! There are no reviews for this restaurant. By combining our iterative development process with what we know about traversing a graph, as well as the details of our error message, we quickly pinpointed the issue with our traversal.

In locating the source of the problem in this example, we found that we had made an incorrect assumption about the data. But we could just as easily have made a typo in our script, or there could have been a problem with the logic. We can fix this problem by adding a review, but the problem wasn't that the data was wrong. The real problem is that we had an incorrect assumption about the data. We were fortunate enough to discover that in our development process, not in production, and so we're able to write the necessary defensive code to handle a valid state of the data.

In other situations with incorrect assumptions about the data, adding to the sample data might be the best approach. Depending on your specific circumstances, you might want to investigate whether there is some validation process in place with production data that isn't reflected in a test data set. Or perhaps the test data isn't crafted well enough to fit the actual shape and scope of the real-world data. Maybe the solution is to add a validation process and clean the data before this traversal runs on it. There are multiple possibilities governed by your environment and the nature of the software you develop. There are, however, some approaches to troubleshooting that can be generally helpful.

In our example, we knew with certainty that the traversal was working as expected before we added the `order()` step. Additionally, we were able to parse the error statement and get helpful clues. We could then investigate the problem vertex, its properties, and its edges. And we were able to lay out all of that information and reason back to the true cause of the error.

Finally, when troubleshooting these types of errors, there are other resources at your disposal. Sometimes just talking about the problem with a colleague can help you get to a solution. Another possible approach is to attempt to duplicate the conditions in a more controlled way, perhaps with a smaller set of data. Last but not least, there are also online resources like the Gremlin Users email group and the Stack Overflow website, or a vendor's support team, or even consulting services for researching, asking questions, or getting paid support. The final point is that you are not alone.

MID-TRAVERSAL FILTERING

Returning to our traversal, we have now pinpointed our issue as being the lack of reviews for restaurant `v[232]`. The question is, what are we going to do about it?

We could say that a lack of reviews is a problem with the data, and as authors, we were sorely tempted to just add one more review to the sample data and avoid the little trouble-shooting digression in the last section. But we felt that the teachable moment was too good to pass up. A lot of times, we make assumptions about the data that are later exposed in our traversal development. In our recommendation use case, we assumed that all the restaurants had ratings. However, this was a bad assumption. Having a restaurant without a rating is actually a reasonable expectation and should be handled. So how do we handle this?

In this example, we filter out restaurants that don't have ratings, which is to say, do not have any `about` edges. For this filtering, we introduce a new step, the `where()` step:

- `where(traversal)`—Filters incoming objects based on a traversal, and only passes through objects when the traversal returns a result

The `has()` step, however, is the primary go-to filtering step and is best for filtering logic based on properties. The `where()` step is generally for all other filtering use cases, where we filter based on a more complex set of logic beyond simple property matching.

In SQL terms, using a `where()` step is similar to writing a subquery in the WHERE clause such as this:

```
SELECT
  FirstName,
  LastName
FROM
  Person.Person
WHERE
  BusinessEntityID =
  (
    SELECT BusinessEntityID
    FROM HumanResources.Employee
    WHERE ID_Number = 123
  );
```

For our traversal, we want to traverse only from restaurants that have reviews, so we filter based on the existence of an `about` edge. To accomplish this, let's insert a `where()` step that checks the presence of an `about` edge before our `order()` step:

```
g.V().has('person','person_id',pid).
  out('lives').
  in('within').
  where(__.inE('about')).
  order().
    by(__.in('about').values('rating').mean()).
  valueMap().with(WithOptions.tokens)

==>{id=224, label=restaurant, address=[3134 Keenan Stravenue],
➥ restaurant_id=[22], name=[With Shell]}
==>{id=108, label=restaurant, address=[2419 Pouros Garden],
➥ restaurant_id=[7], name=[Eastern Winds]}
...
```

Adds a where() step to filter out vertices without an inbound about edge

Lists the restaurants

Well, that gives us a result, but not exactly the one we are looking for. We got back a list of ordered vertices, but the ratings aren't visible, so we aren't sure if these were ordered correctly. To get this list with the computed averages, we need to switch our approach. Before we order the data, we need to compute the mean rating for each restaurant and associate it with the `restaurant` vertex. Let's group vertices into key-value pairs with the `restaurant` vertex as the key and the mean of the `rating` values as the value.

In section 5.3.2, we showed you how to use a `group().by().by()` series of steps to produce a collection of key-value pairs. The first `by()` modulator specifies the keys; the second `by()` modulator specifies the values. To create key-value pairs for this use case, we use the `group()` step, but how do we return the `restaurant` vertex as the key? To return this key, we need another Gremlin step:

- `identity()`—Takes the element entering the step and returns that same element unaltered

Now that we know how to calculate both the key and value parts of our key-value pair, we can add these steps with the `group().by().by()` like this:

```
g.V().has('person','person_id',pid).
  out('lives').
  in('within').
  where(__.inE('about')).
  group().                                        <-- Groups vertices to create our key-value pair
    by(__.identity()).                            <-- Assigns the current element as the key
    by(__.in('about').values('rating').mean())    <-- Traverses the about edge and returns the average rating as the value

==>{v[192]=1.5, v[324]=4.0, v[262]=3.3333333333333335,
➥ v[200]=3.25, v[330]=2.25, v[270]=2.0, v[208]=4.0,
➥ v[336]=4.0, v[146]=3.5, v[84]=1.75, v[276]=2.0,
➥ v[342]=5.0, v[216]=3.6666666666666665, v[154]=3.5,
➥ v[282]=3.5, v[92]=2.5, v[224]=1.0,
➥ v[162]=3.6666666666666665, v[100]=4.0, v[294]=4.5,
➥ v[108]=1.3333333333333333, v[176]=2.5, v[306]=4.0,
➥ [246]=3.0, v[60]=4.333333333333333              <-- Results now contain key-value pairs.
```

Whoo-hoo! The `group().by().by()` gave us the key-value pairs we wanted, with the key being the `restaurant` vertex and the value being the average rating for all reviews of that restaurant. Key-value pairs are convenient to work with, as long as we get these ordered by their values. Let's add our `order()` step back in and look at our results:

```
g.V().has('person','person_id',pid).
  out('lives').
  in('within').
  where(__.inE('about')).
  group().
    by(identity()).
    by(__.in('about').values('rating').mean()).
  order().                 <-- Adds the order step
    by(values, desc)       <-- Orders results in descending order by the values of our key-value pair

==>{v[193]=3.0, v[289]=2.75, v[163]=4.0,
➥ v[69]=3.3333333333333335, v[133]=1.3333333333333333,
➥ v[139]=3.0, v[331]=4.0, v[109]=2.25, v[301]=4.0,
➥ v[177]=4.0, v[209]=2.6666666666666665,
➥ v[147]=2.6666666666666665, v[307]=3.0, v[117]=3.0,
➥ [277]=3.0, v[247]=3.0, v[185]=3.0, v[313]=5.0,
➥ v[155]=3.0, v[61]=4.333333333333333, v[125]=2.5}    <-- Results are not ordered as expected.
```

Wait a minute. If we look at the results, these don't appear to be ordered correctly. What is going on? Looking at the results before we added our `order()` step, we might see the answer:

```
==>{v[193]=3.0, v[289]=2.75, v[163]=4.0,
➥ v[69]=3.3333333333333335, v[133]=1.333333333333333,
➥ v[139]=3.0, v[331]=4.0, v[109]=2.25, v[301]=4.0,
➥ v[177]=4.0, v[209]=2.6666666666666665,
```

```
➥ v[147]=2.6666666666666665, v[307]=3.0, v[117]=3.0,
➥ v[277]=3.0, v[247]=3.0, v[185]=3.0, v[313]=5.0,
➥ v[155]=3.0, v[61]=4.333333333333333, v[125]=2.5}
```

Examining this code snippet closely, we see that what we are getting back is not a collection of key-value pairs as one might expect, but a single object. Note the starting and ending braces: { and }. This is an unexpected result, but one that we know how to remedy. Back in section 5.3.2, we dealt with this same issue with our grouped results. There, we used a step to explode an object into its individual properties using the `unfold()` step. Let's add this step before our `order()` step and see if we get a properly ordered set of results:

```
g.V().has('person','person_id',pid).
  out('lives').
  in('within').
  where(__.inE('about')).
  group().
    by(identity()).
    by(__.in('about').values('rating').mean()).
  unfold().                    <-- Unwinds the incoming object into individual key-value pairs
  order().
    by(values, desc)
==> v[342]=5.0
==> v[294]=4.5
...
==> v[224]=1.0                  Results are ordered as expected.
```

While the results are truncated, it is pretty clear that we are getting back the results in descending order by average rating as expected. Whew, that was a lot of work, but it looks much better. We returned our aggregation, associated it with the vertex, and got our ordering. That only leaves limiting our results:

```
g.V().has('person','person_id',pid).
  out('lives').
  in('within').
  where(__.inE('about')).
  group().
    by(identity()).
    by(__.in('about').values('rating').mean()).
  unfold().
  order().
    by(values, desc).
  limit(10)                    <-- Limits results to 10

==> v[342]=5.0
==> v[294]=4.5
...
==> v[162]=3.6666666666666665   Top 10 results ordered by descending average rating
```

Now that we have the right number of results in the correct order, and we have all of the data we need, we're almost at the end of our process. All that is left is to format our results to return the name, address, and average rating for each restaurant. This task is more complicated than the previous examples because it requires that we combine both the `project()` and the `select()` steps to create our object.

PROJECTING KEY-VALUE PAIRS

Our traversal returns key-value pairs containing the `restaurant` vertex and the average rating. But what we really want to return is a map of properties containing the name, address, and average rating. To generate this new object from the current key-value pairs, we need to revisit what we learned in section 5.2.1 about formatting results.

We know that we have two options for formatting results: selection and projection. Because we want to create our result object from the our current location (instead of selecting data from earlier in our traversal), we use a `project()` step. We start by creating three property names for the returned object and add a `by()` modulator for each one:

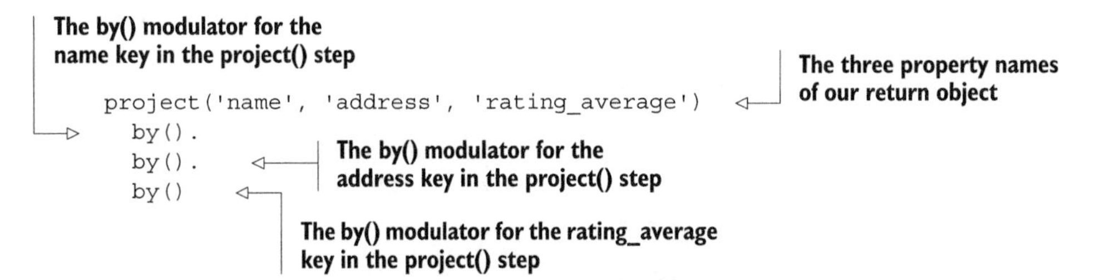

Before we jump into projecting our traversal, we need to take a minute and talk about how to use the `project()` step with key-value pair data. Using the `project()` step with a key-value pair instead of a graph element is more complicated. We need to pull data from either the key or the value portion of the incoming key-value pair.

Remember, at this point in the traversal, we are working a collection of key-value pairs, the first of which looks like this: `v[313]=5.0`. The key part is the `v[313]`, which represents a vertex with an ID value of `313`. The value part is `5.0`, which is the average rating we computed in the `group()` step.

When working with key-value pairs, we choose the key part or value part using a special overload of the Gremlin `select()` step. This overload takes a token, either `keys` or `values`, to specify whether we want to choose the key (`select(keys)`) or the value (`select(values)`) portion of the key-value pair.

> **IMPORTANT** The token `values` is different from the `values()` step. The token `values` refers to the value portion of a key-value pair, while the `values()` step specifies the properties to return from an element. We wanted to call this to your attention because we use both in the same traversal. We know this is confusing, but we didn't name this, so please don't shoot the messenger.

Returning from that brief aside, let's apply our knowledge to this traversal. We know that we want the `name` and `address` properties from the key, containing the `restaurant` vertex, and the `rating_average` from the value of our key-value pair. Combining this with our knowledge of how to select portions of our key-value pair, we get the following traversal:

```
project('name', 'address', 'rating_average')
  by(select(keys).values('name')).              <-- Selects the name from the restaurant vertex in the key
  by(select(keys).values('address')).           <-- Selects the address from the restaurant vertex in the key
  by(select(values))                            <-- Selects the rating_average from the value
```

Applying this to the end of our earlier traversal, we get

```
g.V().has('person','person_id',pid).
  out('lives').
  in('within').
  where(__.inE('about')).
  group().
    by(identity()).
    by(__.in('about').values('rating').mean()).
  unfold().
  order().
    by(values, desc).
  limit(10).
  unfold().
  project('name', 'address', 'rating_average').
    by(select(keys).values('name')).
    by(select(keys).values('address')).
    by(select(values))
                                     Adds project() step to return results with all needed properties

==>{name=Lonely Grape, address=09418 Torphy Cape,
➥ rating_average=5.0}
==>{name=Perryman's, address=644 Reta Stream,
➥ rating_average=4.5}                Results with all properties are ordered as expected.
...          <-- Results truncated for brevity.
```

That's all, folks. We've successfully written a traversal to answer, "What are the ten highest-rated restaurants near me?" While this wasn't as straightforward as we might have hoped, it denotes the complexity we can encounter when writing graph traversals. It also allowed us to demonstrate some common graph concepts, such as how to deal with key-value pairs and how to construct complex result objects, not to mention our little excursion into mid-traversal development troubleshooting. Now that the hard work of finishing this use case is complete, the only thing left to do is to add it into our application.

ADDING THE TRAVERSAL TO OUR APPLICATION

As in the last section, in this section, we'll follow the same process as we did in chapter 6. In our example app, we'll add a new method called `highestRatedRestaurants`. In Java, the traversal code looks like this:

```
List<Map<String, Object>> restaurants = g.V().
    has("person", "person_id", personId).
```

```
out("lives").
in("within").
where(inE("about")).
group().
    by(identity()).
    by(in("about").values("rating").mean()).
unfold().
order().
    by(values, Order.desc).
limit(10).
project("name", "address", "rating_average").
    by(select(keys).values("name")).
    by(select(keys).values("address")).
    by(select(values)).
toList();
```

An anonymous traversal is not required due to imports.

Enum values (Column.keys, Column.values) included in imports

8.5 Writing the last recommendation engine traversal

We've worked our way back to the first question, "What restaurant near me with a specific cuisine is the highest rated?" Let's make this section a little more concrete. Let's choose a random person from our test data.

Say we're Kelly Gorman. Because we are Kelly, and because Kelly is a social catalyst, we're out with our friends. As everyone knows, everywhere Kelly goes, there's always a group of people eating and hanging out. The group is getting a little unruly because they're hungry and thirsty, but they can't decide where to go, debating between a diner or just going to a bar. Naturally, Kelly pulls up DiningByFriends and gets everyone to agree that they'll go to the top local diner or bar as returned by the app. Your job now is to use what you have learned to build the traversal and return that restaurant. Doing so saves Kelly from her hangry friends.[2]

We'll give you a couple of hints to get started, provide a quick review of the process used in the two previous sections, and then offer some space to sort out the answer on your own. Then we'll close the chapter by revealing our traversal and walking through our thinking. To start, as Kelly Gorman, we assume that we're logged into DiningByFriends with

```
pid = 5
==>5
```

We need to enter two cuisines for our search. To do so, we create a list variable in the Gremlin Console like this:

```
cuisine_list = ['diner','bar']
==>diner
==>bar
```

[2] The word *hangry* is an English portmanteau of "hungry" and "angry."

Finally, the answer we expect to display from DiningByFriends is this:

```
{name=Without Chaser, address=01511 Casper Fall,
rating_average=3.5, cuisine=bar}
```

We invite you to go through the process we used with the first two questions:

1. Identify the vertex labels and edge labels required to answer the question. Perhaps make a small schema drawing to help you process what to use in the graph.
2. Find the starting location for the traversal.
3. Find the ending location for the traversal.
4. Write out the steps in plain English (or in your preferred language). The first step is your input step, and the last step is the output step.
5. Code each step with Gremlin, one at a time, verifying the results against the test data after each step.

We broke down the relevant question for you to answer at the start of the chapter, but we repeat that in the following table for your convenience. To help with the first step, figure 8.5 provides a picture of the schema.

Question	Vertex Labels	Edge Labels

What restaurant near me with a specific cuisine is the highest rated?	`person`
	`city`
	`restaurant`
	`cuisine`
	`review`
		`lives` (connecting person → city)
		`within` (connecting restaurant → city)
		`serves` (connecting restaurant → cuisine)
		`about` (connecting review → restaurant)

Now take a few minutes and record the steps the traversal requires to answer the question. We think you'll need eight or nine steps, including the initial one with the starting and the ending return steps. Hint: many of the steps are identical to the traversal used in the previous section. Given all of that, feel free to plan out your approach with the following bullet list:

- (Starting Point)
- (Listing of steps to traverse)

  . . .
- (Ending Point)

We encourage you to refer back to the two previous examples in this chapter, as well as the last chapter's content as you work through this exercise on your own. When you have outlined your approach with bulleted points, go ahead and code it and test it with the data. We also suggest that you look at the `within()` predicate to help you filter by cuisine (see http://mng.bz/wpp7). Aside from that, the other parts of our solution have been discussed previously in this book, many in this chapter. When you're ready, take a look at our solution in the next section.

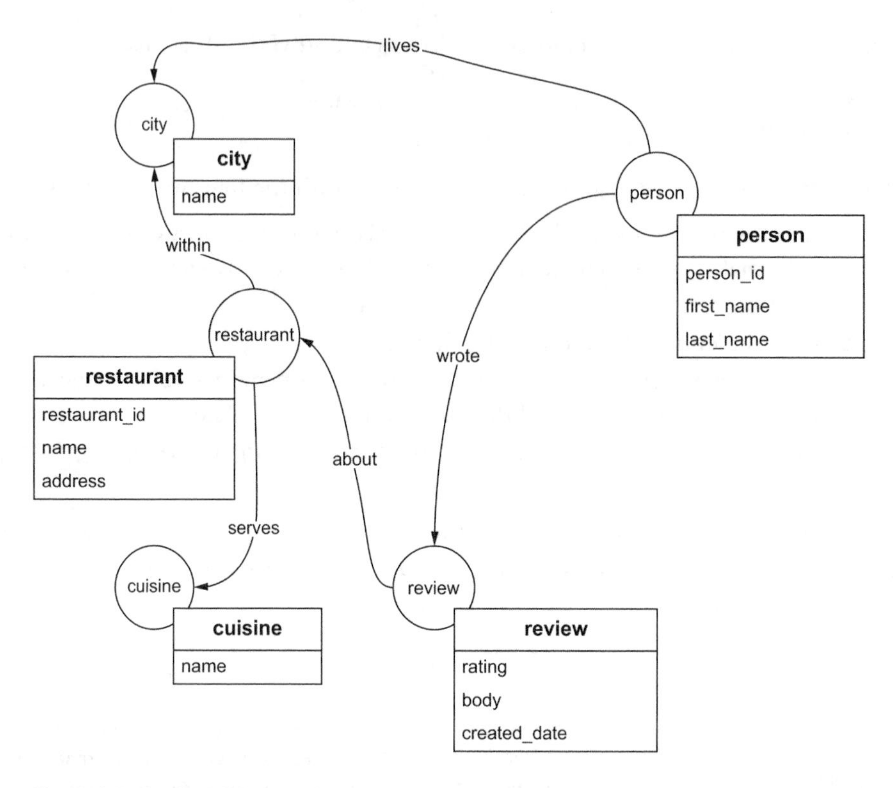

Figure 8.5 Logical data model elements required for the question, "What restaurant near me with a specific cuisine is the highest rated?"

8.5.1 Designing our traversal

Our first thought is that this traversal is similar to the traversal from the last section. Much like that one, we look for restaurants near the current user, so the best starting point here is a `person` vertex filtered by the `person_id`. The ending point is also going to be a list of restaurant vertices with the restaurant's name, address, average rating, and type of cuisine served. Here is the list of actions we came up with:

1. Get the `person` based on the `person_id` input with the `pid` variable.
2. Traverse the `lives` edge to get their `city`.
3. Traverse the `within` edge to get the `restaurant` vertices for the `city`.
4. Filter the `restaurant` vertices based on the adjacent `cuisine` vertex to show only those that offer one of the cuisines specified in the `cuisine_list` variable.
5. Group the vertices with computed `rating_average`, including a filter to ensure each `restaurant` vertex has a review.
6. Order by `average_rating`, descending.
7. Limit to one result.
8. Return name, address, and average rating for the `restaurant` vertex.

With those parts laid out, we know the steps we need to accomplish this traversal. With the exception of the number of items returned and filtering on the `cuisine`, we can reuse much of the traversal we constructed in section 8.3. Let's start by copying that traversal here and limiting our results to one:

```
g.V().has('person','person_id',pid).
  out('lives').
  in('within').
  where(inE('about')).
  group().
    by(identity()).
    by(__.in('about').values('rating').mean()).
  unfold().
  order().
    by(values, desc).
  limit(1).                          <-- Limits results to just one element
  project('name', 'address', 'rating_average').
    by(select(keys).values('name')).
    by(select(keys).values('address')).
    by(select(values))
==>{name=Dave's Big Deluxe, address=490 Ivan Cape, rating_average=4.0}
```

The next missing piece of this traversal is to filter for the input cuisine(s). For this, we want to add a simple filter using `has()`:

```
has('cuisine',within(cuisine_list))
```

However, we can't. That's not how the schema is designed. In our model, cuisine is a separate vertex, so we need to filter on a traversal instead of directly on a property. That traversal needs to reach out to the `cuisine` vertex in order to filter by the cuisine type.

In Gremlin, this can't be accomplished using a `has()` step because it filters just on properties, not on the presence of an edge. Recalling our work with the previous traversal, the `where()` step does what we need. This step takes a traversal as a parameter and filters any traversers that do not return a result. We do this by creating this statement:

```
where(out('serves').has('name',within(cuisine_list)))
```

For each `restaurant`, this statement traverses out to its incident `cuisine` vertex to check if its `name` is in our cuisine list. We have a filter within a filter. The outer filter is the `where()` step, which only passes through results if the inner traversal completes. The inner filter is at the end of the traversal and uses a `has()` step to filter on the `name` property of the `cuisine` vertex.

Refactoring your model

This use of `cuisine` is a candidate for refactoring. Do you recall what we said about escape rooms back in chapter 3? We used this analogy to show that being on a vertex in a graph is like being in a room for that vertex.

(continued)

Within our immediate access are a series of drawers that represent the properties on the vertex and a series of doors that lead to other vertices. There's little to no cost to look in a drawer for the value of a property, because we already incurred the expense of loading the vertex with its properties into memory. There's also no additional cost to look at the edges to note their values. But checking the cuisine requires traversing to the `cuisine` vertex, or walking down the hall to the room representing that vertex in our analogy. We'll discuss refactoring in chapter 10 when we cover performance issues.

Plugging this into our traversal and making a minor update to our `project()` step to include the cuisine, we get

```
g.V().has('person','person_id',pid).
  out('lives').
  in('within').
  where(out('serves').has('name',
➥ within(cuisine_list))).             <- Adds a where() step
  where(inE('about')).                    to filter by cuisine
  group().
    by(identity()).
    by(__.in('about').values('rating').mean()).
  unfold().
  order().
    by(values, desc).
  limit(1).
  project('name', 'address',
➥ 'rating_average', 'cuisine').
    by(select(keys).values('name')).
    by(select(keys).values('address'
    by(select(values)).
    by(select(keys).out('serves').values('name'))
==> {name=Without Chaser, address=01511 Casper Fall,
➥ rating_average=3.5, cuisine=bar}
```

Updates the project() step to include the cuisine value in the result

Returns our result for the highest rated bar or diner near Kelly Gorman

This result is precisely the response we expected. If you got the same answer, then your approach was successful as well. If you didn't, compare your steps with ours and find where in your traversal the results start to vary from our approach.

8.5.2 Adding this traversal to our application

We finish by adding a new method called `highestRatedByCuisine` to our application. In Java, the traversal code is as follows:

```
List<Map<String, Object>> restaurants = g.V().
    has("person", "person_id", personId).
    out("lives").
    in("within").
    where(out("serves").has("name",
```

```
➥ P.within(cuisineList))).         <-- Predicates use P.within or
    where(inE("about")).               a static import statement.
    group().
        by(identity()).
        by(in("about").values("rating").mean()).
    unfold().
    order().
        by(values, Order.desc).
    limit(1).
    project("restaurant_name", "address",
➥ "rating_average", "cuisine").
        by(select(keys).values("name")).
        by(select(keys).values("address")).
        by(select(values)).
        by(select(keys).out("serves").values("name")).
    toList();
```

Whew, we did it! We built the traversals for our three recommendation engine use cases! We started by determining the required graph elements (vertices and edges) for each question. Identifying this information helped us to organize our thinking and prioritize our work. We decided to start with the simplest of the traversals and work our way through to the more involved ones. This order worked to our benefit because we were able to reuse much of the code from the traversal for our use case (number two) in section 8.3 to write the traversal for our use case (number one) in section 8.4.

We also practiced using some of the more practical aspects of software development with graph databases throughout this chapter, such as always having a good picture of the schema available. We drafted our traversals, first in plain English and then with Gremlin steps. And as we went through the process of implementing the Gremlin steps, we encountered a number of the usual challenges of building software: incorrect assumptions about the data, unexpected bugs, steps missing from the original plan, and new uses for familiar steps.

Throughout this process, our iterative "one-step-and-test" approach served us well in completing the work. In the next chapter, we'll use subgraphs to allow users to personalize the recommendations they receive.

Summary

- Start developing traversals for a use case by identifying the vertices and edges required to answer the business question.
- Developing known-walk traversals starts with identifying the relevant portions of the schema, finding the starting and ending points for the traversal, identifying the series of vertices and edges needed to traverse from the starting to the ending points, and finally, composing the traversal by iteratively adding steps while validating the results against test data as we build each step.
- A good starting point for a traversal minimizes the number of starting vertices, ideally to a single vertex. To do this, apply all possible filtering at the start of the traversal.

- Prioritization of which use case questions to start on can be done by either choosing the hardest question if we want to de-risk the use of graph technology, or the simplest question if we want an early win as a building block for future success.
- Using a systematic, step-by-step approach to building traversals makes it easier to identify the source of a problem if an error is encountered. Troubleshooting any errors can involve multiple steps, including investigating the data, changing the approach to the traversal, and consulting with other staff or online resources.
- Paging results in a graph traversal requires an ordered result set and the use of limits and offsets to specify the desired subset of results.
- Grouping and ordering traversals create results that are key-value pairs. Further processing key-value pairs uses special overloads of the `select()` step to work with the key and value portions of the pair.

9 Working with subgraphs

This chapter covers

- Defining subgraphs using traversals
- Extracting subgraphs for future use
- Working with previously extracted subgraphs
- Using subgraphs to create modular, reusable code

Let's say we have two users who both use DiningByFriends to find great restaurants in Houston, TX. Nancy lives north of Houston, and Sam lives to the south. Although each is friends with many people in the app, they have never connected to one another. We are making two reasonable assumptions here: Nancy and Sam have distinct friend groups because they live in different parts of town and have never met, and they want to visit local restaurants that their friends rate highly. What happens when they pose this question: "Based on how my friends rate restaurants, what are the best local restaurants for me?" They each expect to get recommendations for restaurants in their local area (North Houston for Nancy; South Houston for Sam) based on their friends' ratings. How can we deliver results that are the most relevant to each of them?

Personalization is a process of filtering data based on the connections in the data in order to serve the most relevant content. In DiningByFriends, we can personalize the recommendations a user receives based on that person's social network. For example, to answer the question for Nancy more appropriately, we can intentionally limit the data to the restaurants and reviews created by her friends. In other words, we create a way to focus on one subset of the data, her friends' recommendations, and ignore another subset, all of the other restaurants in Houston.

Because we ask our question on only a well-defined subset of the data in our graph, we want to work exclusively with that data set. The most efficient way to do this is to extract that subset of data from the global graph. This is a common operation, and this subset of data is known as a *subgraph*. Conceptually, a subgraph is a fairly simple thing: it's a subset of vertices and edges, usually closely connected, according to some rule or an understanding of the business domain.

In this chapter, you'll learn when and how to use a subgraph to filter results. Subgraphs are a natural fit for personalization problems, so we'll use a personalization question like "Based on the review ratings from my friends, what are the best restaurants for me?" to demonstrate the basic operations of creating and using subgraphs. Then, we'll walk through the process of developing a traversal to answer this question and use it to demonstrate how subgraphs enable personalized results for different users. Finally, we'll look at some of the differences in approach required when using subgraphs from within our application.

9.1 Working with subgraphs

Before we get too deep into the personalization use cases, let's use the social network graph we've used throughout this book to demonstrate the basics of subgraphs. If you'll recall, a *subgraph* is a graph in which all the vertices and edges are a subset of a larger graph. An example of a subgraph in a social network is a graph that contains you and all the people you are connected to via a `friends` edge. The fact that a subgraph is a graph itself is one of the things that makes subgraphs so useful: they work just like the larger graph, but with a smaller memory footprint.

9.1.1 Extracting a subgraph

Returning to our social network, let's say we want to retrieve a subgraph of Josh and his friends. In this case, we need to include his friends as well as those that have friended him. Figure 9.1 highlights this section for our subgraph.

To create this subgraph, we need to develop a traversal that defines the vertices and edges. We already know how to create a traversal to find someone's friends (see section 3.2), so the unknown part here is how we specify these friends as part of a subgraph. Depending on our choice of database engine, we can create subgraphs using one of two techniques: vertex-induced and edge-induced.

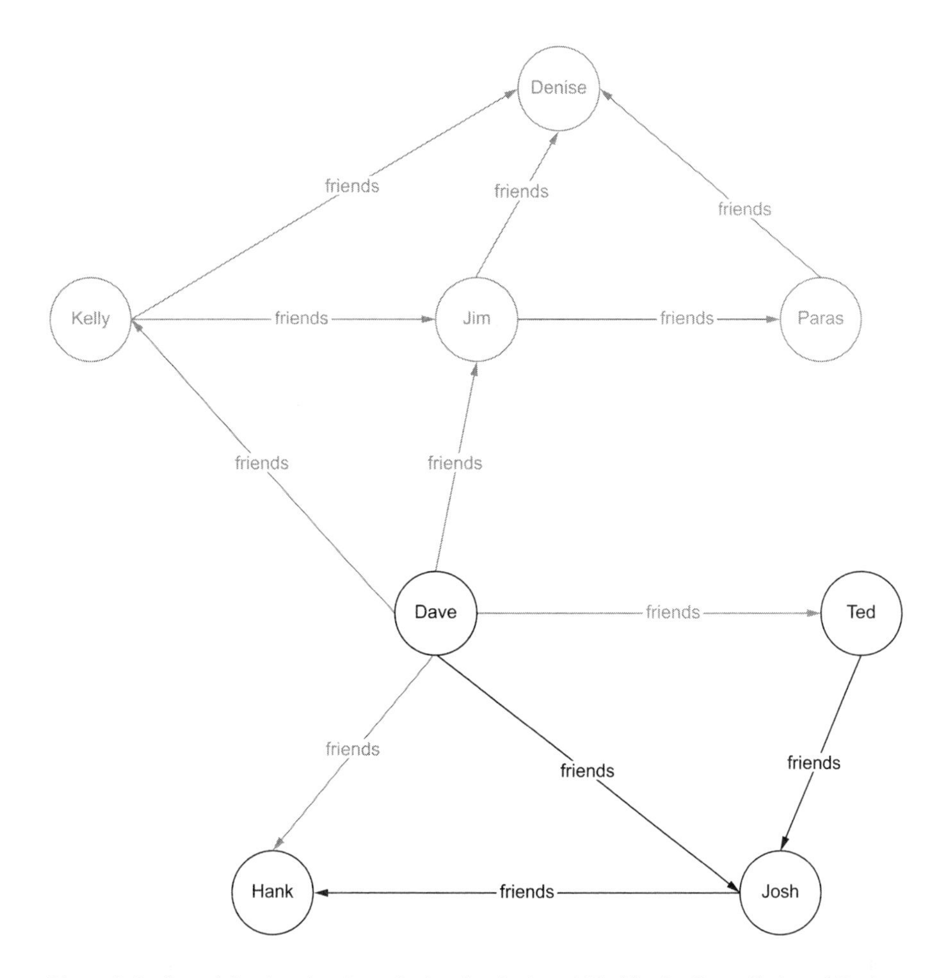

Figure 9.1 A social network subgraph showing Josh and his friends, Dave, Ted, and Hank.

VERTEX-INDUCED VERSUS EDGE-INDUCED SUBGRAPHS

A vertex-induced subgraph is defined by a set of vertices and any shared edges. For example, we can create a vertex-induced subgraph by specifying that we include only the even vertices in figure 9.2. Because this is a vertex-induced subgraph, it also includes any edges the vertices have in common, such as edges `H`, `I`, `J`, `K`, and `L` as highlighted in figure 9.2.

An *edge-induced* subgraph is also defined by a set of edges but includes the incident vertices. Figure 9.3 shows what this looks like, based on the edges connected to vertex `6`. In this case, we start with the edges `I`, `K`, and `L` and include the incident vertices `2`, `4`, `6`, and `8`.

Even though our subgraphs have the same vertices, these do not have identical edges. Comparing our two subgraphs, we can see that the vertex-induced subgraph

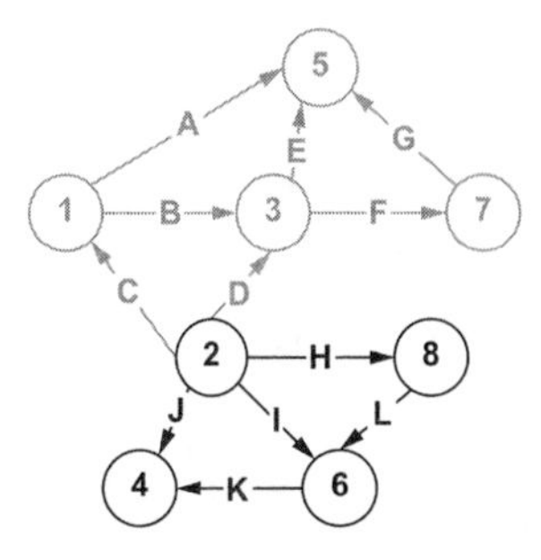

Figure 9.2 A vertex-induced subgraph based on choosing vertices `2`, `4`, `6`, and `8` also includes the shared edges `H`, `I`, `J`, `K`, and `L`.

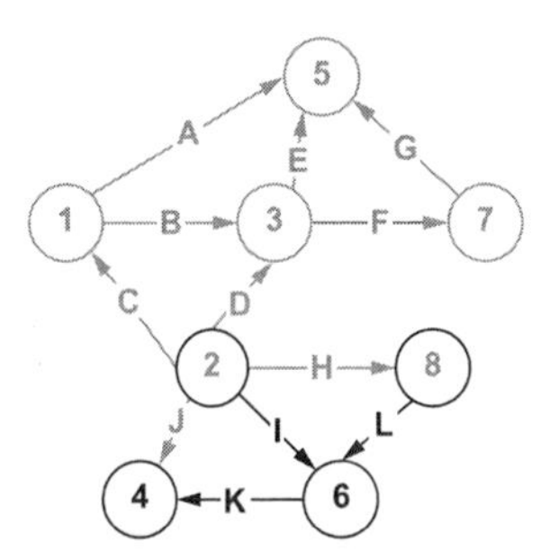

Figure 9.3 An edge-induced subgraph based on choosing the edges connected to vertex 6 includes the edges `I`, `K`, and `L` and the incident vertices `2`, `4`, `6`, and `8`.

also contains the `H` and `J` edges, which are not included in the edge-induced subgraph. The difference is because vertex-induced subgraphs include *all* shared edges, while edge-induced subgraphs include only a defined set of edges. These two approaches do not always achieve different results. Composition of the subgraph depends on which of these approaches you use and on which rules you use in the selection process.

Unfortunately, our approach is normally defined by the database vendor, and not all graph databases have explicit subgraph support. Defining boundaries using edges can seem counter-intuitive because, historically, we think about data with an *entity-first* if not *entity-only* mindset. However, the entity relationship of graphs allows us to use edges as first-class citizens to define the limits of a subgraph. This ends up being both safe and easy to properly delineate our subgraphs. TinkerPop's reference implementation of the Gremlin Server supports edge-induced subgraphs, so that is what we focus on for our personalization use case.

DEFINING OUR SUBGRAPH

Now that we understand the approaches used to define a subgraph, let's extend the actions required to find Josh's friends by returning the subgraph containing the Josh vertex, his friends' vertices, and the edges between those.

> **NOTE** This section demonstrates how to create edge-induced subgraphs using Gremlin. Different databases handle the creation of subgraphs using different

processes, but for TinkerPop-enabled databases, the process described here is standardized.

Setting up your local environment

Before you run traversals to retrieve a subgraph, you first need to set up your local graph with the appropriate data. As with previous chapters, we have provided a script (http://mng.bz/nz2g) to load a set of test data that we'll use throughout this chapter. And, as with the last chapter, we'll need to update our script to reference to location of the data file before running it.

To update this script, download it, then open it in a text editor of your choice and edit the following line to point to the file location from the downloaded source code, specifying the full path for chapter09/scripts/restaurant-review-network.json:

```
full_path_and_filename = "/path/to/restaurant-review-network.json"
```

If you set up the Gremlin Console according to the instructions in appendix A, you can start the Gremlin Console and load the data for this chapter with a single command. On MacOS and Linux systems, use

```
bin/gremlin.sh -i $BASE_DIR/chapter09/scripts/9.1-restaurant-review-
  network-io.groovy
```

For Windows, use

```
bin\gremlin.bat -i $BASE_DIR\chapter09\scripts\9.1-restaurant-review-
  network-io.groovy
```

Once this script completes, our graph contains the test data we'll reference throughout the chapter.

Using Gremlin, we create an edge-induced subgraph with the following approach:

1 Get the `Josh` vertex (`person_id =2`).
2 Traverse the `friends` edges in either direction.
3 Define a subgraph based on the edges traversed.
4 Extract the edges and vertices in the subgraph.
5 Return the results.

Examining these actions, we already know how to do the first two steps, as well as the last one. The novel actions are the middle two, where we define and extract the subgraph. Figure 9.4 shows how these actions map to the corresponding steps in Gremlin.

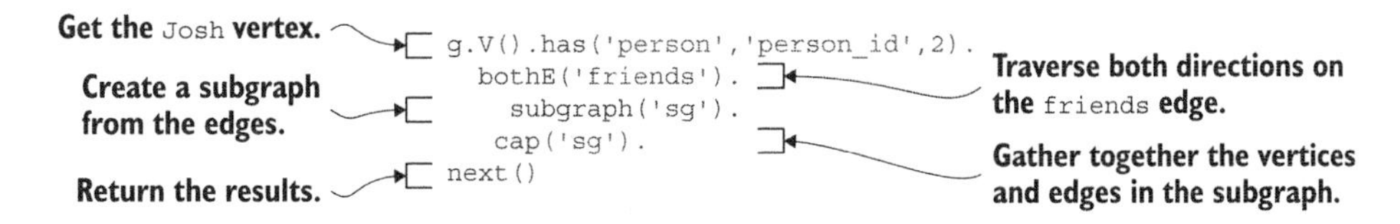

Figure 9.4 Mapping the plain text steps to the corresponding Gremlin steps to create a subgraph

This traversal primarily uses steps that are familiar, with the exception of the two steps required to define and extract the subgraph. These steps include

- `subgraph(sideEffectKey)`—Defines an edge-induced subgraph within a larger set of graph data. The `sideEffectKey` is a reference to the full results of the side effect.
- `cap(sideEffectKey)`—Iterates the traversal up to itself and emits the results of the side effect referenced by the `sideEffectKey`.

Side effects and Gremlin's rarely used general steps

By way of introduction to these steps, `subgraph()` and `cap()`, let's take a break to talk about side effects. Early in the TinkerPop documentation, we find a list of steps: `map`, `flatMap`, `filter`, `branch`, and `sideEffect`. Then, for the most part, the rest of the documentation proceeds to use every step *but* these five general steps.

Those familiar with functional programming should recognize some if not all of the steps as the staples of coding data transformations. All of the Gremlin steps, except for the ones that modulate or configure other steps, are essentially optimized versions of one of these five general steps, which are core concepts in programming. Side effects, however, may not be as obvious as the others. Let's boil it down to state: *side effects* are the way we change state.

We used several side effects in chapter 4 to mutate our graph by adding, removing, and updating elements. All of these mutations are a form of side-effect steps. When we call an `addV()` step and it returns a `Vertex` object, it changes the graph by adding that `Vertex` to the graph. The primary result of the `addV()` call returns a `Vertex` object, but the side effect changes what happens to the state of the graph: a new vertex is added in the data.

We tend to think about this as a single operation, which returns the vertex that was added to the graph, but there are two distinct parts. First, the data is added to the graph (the side-effect part of the operation). Second, a reference to the recently added data is retrieved and returned as the result of the operation.

The same is true with the `subgraph()` step. The primary effect of the `subgraph()` step is to return the edge that was its input. But, as a secondary operation, it adds those same edges and their incident vertices to an internal collection identified by a label.

There's a lot more that can be said about side effects and the other four steps mentioned earlier. We find this interesting in a fun, highly theoretical, but slightly impractical way. It's impractical because nearly every operation one would want to do in Gremlin can be done with the steps we've introduced throughout this book. In fact, all of those other steps are both better-performing and easier to read than the five general steps: `map`, `flatMap`, `filter`, `branch`, and `sideEffect`. This is why we've avoided the use of these general steps up to this point.

Let's look at running the traversal from figure 9.4 in the Gremlin Console:

Defines a variable called subgraph, which is equal to the result of this traversal

Traverses the friends edge in either direction

Assigns a subgraph with the key 'sg'

```
subgraph = g.V().has('person','person_id',2).
             bothE('friends').
             subgraph('sg').
             cap('sg').next()
==>tinkergraph[vertices:4 edges:3]
```

Iterates the traversal and emits the subgraph with key 'sg'

Returns a TinkerGraph object with four vertices and three edges

Well, that is interesting. We see that our result is not a list or a map as before, but a graph: we returned a graph that contains four vertices and three edges. Let's break down how that graph, our subgraph, was created. To begin, we start on the `Josh` vertex, as figure 9.5 shows.

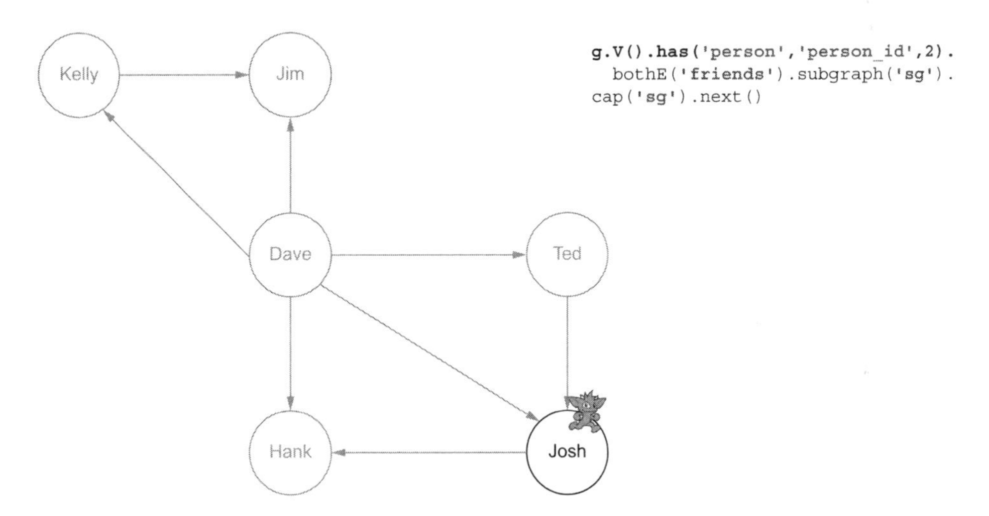

Figure 9.5 Our subgraph traverser begins on the `Josh` vertex.

Next, we traverse each of the `friends` edges adjacent to the `Josh` vertex with the `bothE()` step. Figure 9.6 shows this traversal.

With those three edges added to our subgraph, we call the `cap()` step to return the subgraph. Figure 9.7 shows this step.

Now we have a subgraph for Josh and his friends! This subgraph has all of the same graph capabilities as the larger graph it was taken from, although with slightly less data.

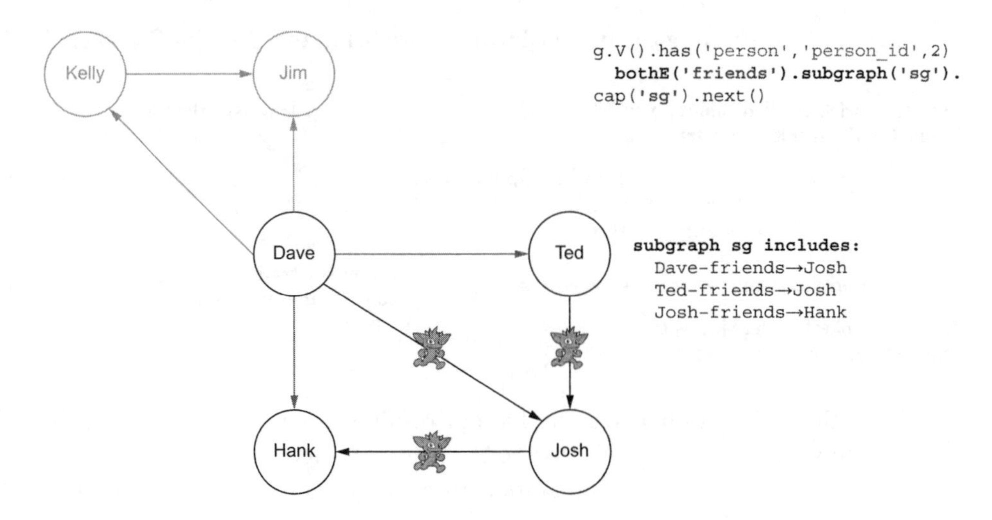

Figure 9.6 Our traverser branches onto the three `friends` edges and adds these and their corresponding vertices to our subgraph, `sg`.

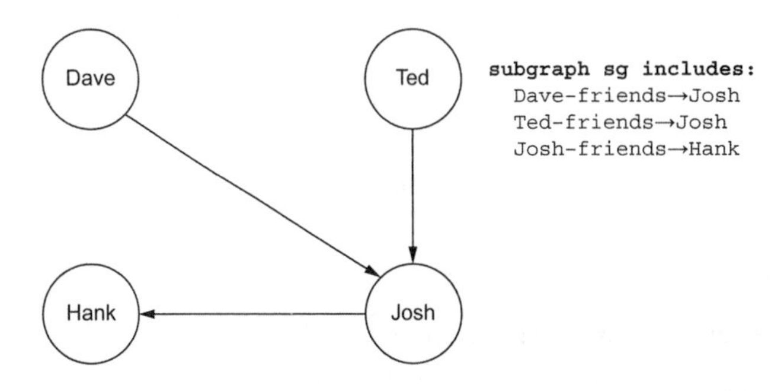

Figure 9.7 Returning the subgraph that our traversal extracted

9.1.2 Traversing a subgraph

Now that we've defined and isolated a subgraph, let's learn how to traverse it. In the last section, we assigned our `subgraph` variable a TinkerGraph object containing four vertices and three edges. Because our subgraph variable contains a TinkerGraph object, we can't continue working with or traversing through our subgraph until we create a `GraphTraversalSource` for this graph. In chapter 3, we introduced the difference between a `Graph` and a `GraphTraversalSource`. As a reminder

- `Graph` is a data store. It is simply a place to hold the data with no ability to access the data aside from the simplest of lookup operations.
- `GraphTraversalSource` is the base from which all traversals are written (the `g` in our traversals).

Having a `Graph` object without a `GraphTraversalSource` is akin to having a file system and its files without having any sort of file manager; that is, without any sort of tool for navigating the file system, reading the files and their attributes, or for moving the files around. This means that before we can work with our subgraph, we need to get a traversal source. In Gremlin, we do this by calling the `traversal()` method on the `Graph` object (in this case, our `subgraph` variable, `sg`):

```
sg = subgraph.traversal()
==> graphtraversalsource[tinkergraph[vertices:4 edges:3], standard]
```

Excellent. We now have our `GraphTraversalSource` in the `sg` variable and are ready to begin traversing our subgraph.

In relational database terms, extracting a subset of data and later reusing this data is akin to separating out a set of tables using a join, which then becomes a database of its own. This is like the concept of a view, sort of. Or maybe like a common table expression (CTE), kind of. Perhaps a set of temporary tables is a better analogy, although not really. The closest relational database equivalent might be to connect to the database with serializable isolation and run your operation without committing any changes back to the original database.

Sadly, there is no perfect analogy in the relational database world for this capability. Dynamically defining a functionally complete subset of the data in this manner is a skill somewhat unique to graphs.

Subgraphs for serial isolation

A subgraph can be thought of as a poor man's way of instituting serializable isolation in a graph database. Or, in other words, it can be deemed as *the* way to interact with graph data with a serializable isolation pattern. Once defined, the subgraph does not change even if the original data changes, although this functionality is dependent both on the vendor supporting the `subgraph()` step *and* on implementing in agreement with the TinkerPop behavior.

This feature can be a powerful capability, especially when you need to do some analytical operations on a transactional system. But it should be used with caution! In many systems, the subgraph is an in-memory construct with no disk-caching capability, so creating a subgraph of the entire source graph could create memory pressure or even spur out-of-memory errors.

> ***(continued)***
>
> Also, in line with the definition of serializable isolation, any mutations that happen in the subgraph are *not* reflected in the original graph, and vice versa. Coordinating mutations between the two is up to the application developers.
>
> All of this current functionality has been verified with the Apache TinkerPop 3.4 reference implementation. But when we work with subgraphs, we should check exactly what functionality our chosen vendor supports.

Now we have our traversal source defined for our subgraph. Because of that, we can traverse our graph as we have been doing since chapter 3:

```
sg.V().has('person','person_id',2).valueMap()
==>{person_id=[2], last_name=[Perry], first_name=[Josh]}

sg.V().has('person','person_id',2).both().valueMap()
==>{person_id=[3], last_name=[Erin], first_name=[Hank]}
==>{person_id=[1], last_name=[Bech], first_name=[Dave]}
==>{person_id=[4], last_name=[Wilson], first_name=[Ted]}
```

Finds the person_id = 2 vertex and displays the property keys and values

Finds the connections to person_id 2 and displays the property keys and values

This ability to store and run additional processing on a subgraph, just as we can with any graph, is one of the things that makes subgraphs so useful. Now that we've covered how to create, extract, and work with a subgraph, let's use our personalization use case for DiningByFriends to show how we can use a subgraph.

9.2 Building a subgraph for personalization

For our personalization use case for DiningByFriends, we need to answer the question, "Based on my friends review ratings, what are the best restaurants for me?" However, it's doubtful we want restaurant recommendations for restaurants hundreds of miles away. For the sake of this example, let's assume that we are only looking for restaurants in the same area. Following the process for developing a traversal from section 8.1, we'll start by breaking the question down into its required parts. For this question, we find the following actions:

1. Locate the `person` vertex, who is the subject of the subgraph.
2. Traverse to the `friends` vertices of the subject `person`.
3. Determine the `review` vertices for each friend.
4. Find the `review_ratings`.
5. Find the restaurants with the most highly-rated reviews.

In our next step, we find the relevant vertex and edge labels in our schema based on the previous actions. Figure 9.8 highlights the parts of our schema we're interested in.

Great! We've identified the relevant schema elements . The next step in the process is to select our starting place. Looking at the previous required actions, we think the logical place to start is with the current person, because this narrows our starting

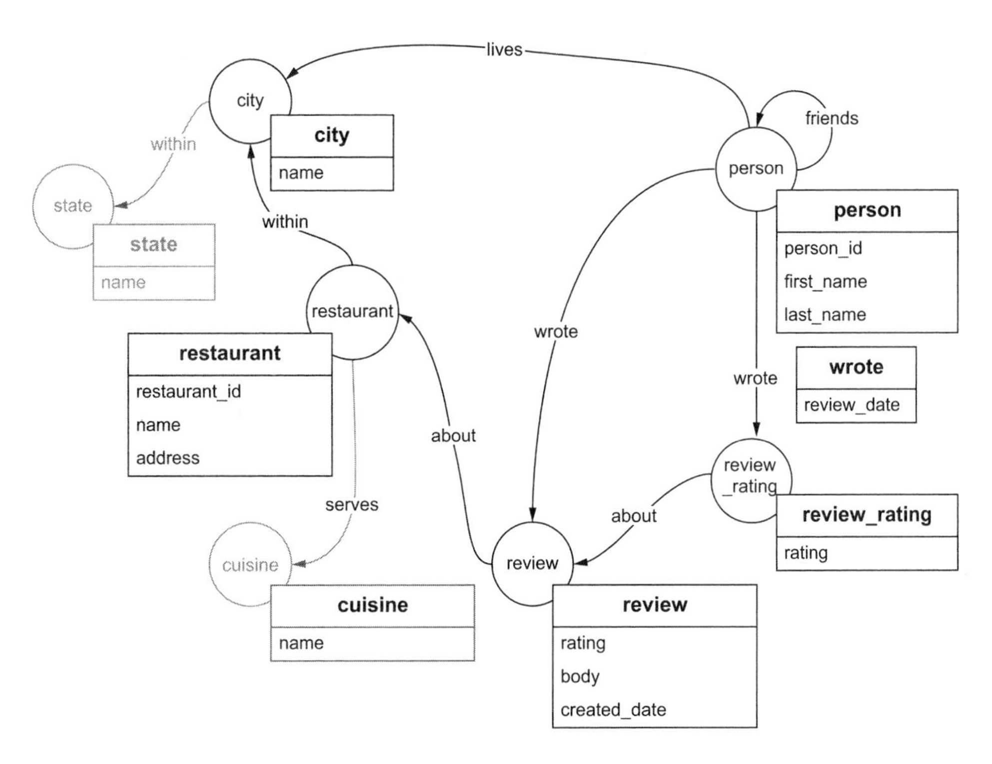

Figure 9.8 The relevant logical data model elements required for the subgraph for our personalization use case for DiningByFriends

vertices to a single vertex. Next, we need to find the end point for our traversal; in this case, we use the restaurant because that is what the question wants to have returned.

The third step in our process is to list the actions we need to take in the schema to get from our starting point to our ending point. Doing that for the required steps, we need to

1. Get the current `person`.
2. Add the `friends` edges to get the person's immediate group of friends.
3. For all of the `friends`, get their `review` and the `review_rating` vertices.
4. Add the `restaurant` vertices.
5. Include the `city` vertex for each restaurant.

Now we know the steps that our traversal needs to accomplish. Let's start developing our subgraph in an iterative manner.

First, we need the person, Josh, and his friends. For this example, we decided that the subgraph used for Josh's personalized results should include people Josh friended as well as people that friended Josh. Remember, we are using an edge-induced subgraph, so what we're really doing is collecting edges and using those edges to generate our subgraph.

NOTE There is no requirement that you must create subgraphs by traversing edges in both directions. We could have limited the direction of the edges traversed, but chose not to.

```
subgraph = g.V().has('person','person_id',2).
  bothE('friends').subgraph('sg').
  cap('sg').next()
```

Starts with the given person_id, which is the signed-in user

Traverses the friends edges to get all the friends

Next, we need to include the `review` and `review_rating` vertices as we traverse to the `restaurant` vertices. Finally, we tack on the `within` edges to identify the `city` vertices for some localization functionality. Visually, going through the data looks something like figure 9.9.

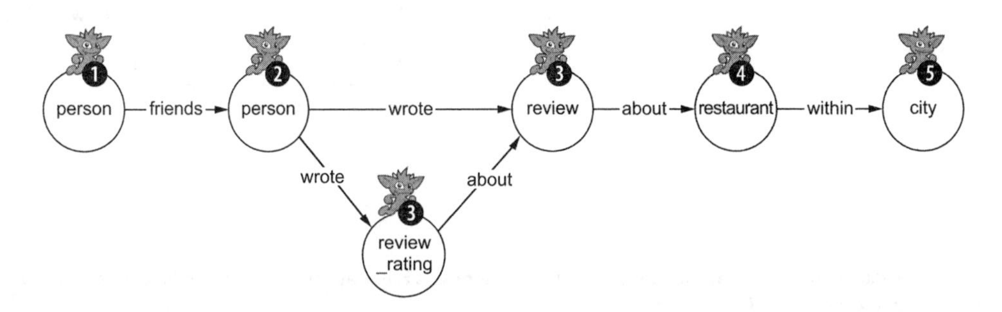

Figure 9.9 We create our subgraph by following this known-walk path through the logical data model.

Before we show the traversal, we want to take a minute to discuss a new wrinkle in our traversal—the need to optionally traverse graph elements. Looking at figure 9.9, we notice that sometimes we need to go from a `person` to a `review_rating` vertex via a `wrote` edge, and sometimes we go from a `person` to a `review` via a `wrote` edge. In the scenario, when we go to a `review_rating` vertex, we need to perform an extra step to take the `about` edge to a `review`. This means that we need to traverse our graph differently, depending on the vertex type that we are on. To handle this, we use the following Gremlin step:

- `optional(traversal)`—Attempts the traversal and, if it returns a result, then emits the result; otherwise, it issues the incoming element as with the `identity()` step

If we were to draw an example of what this additional step looks like as we move through our data, it would resemble figure 9.10.

The tricky part is that the `wrote` edge connects a `person` to either a `review` *or* a `review_rating` vertex type. This bifurcation means that if we are on a `review_rating`

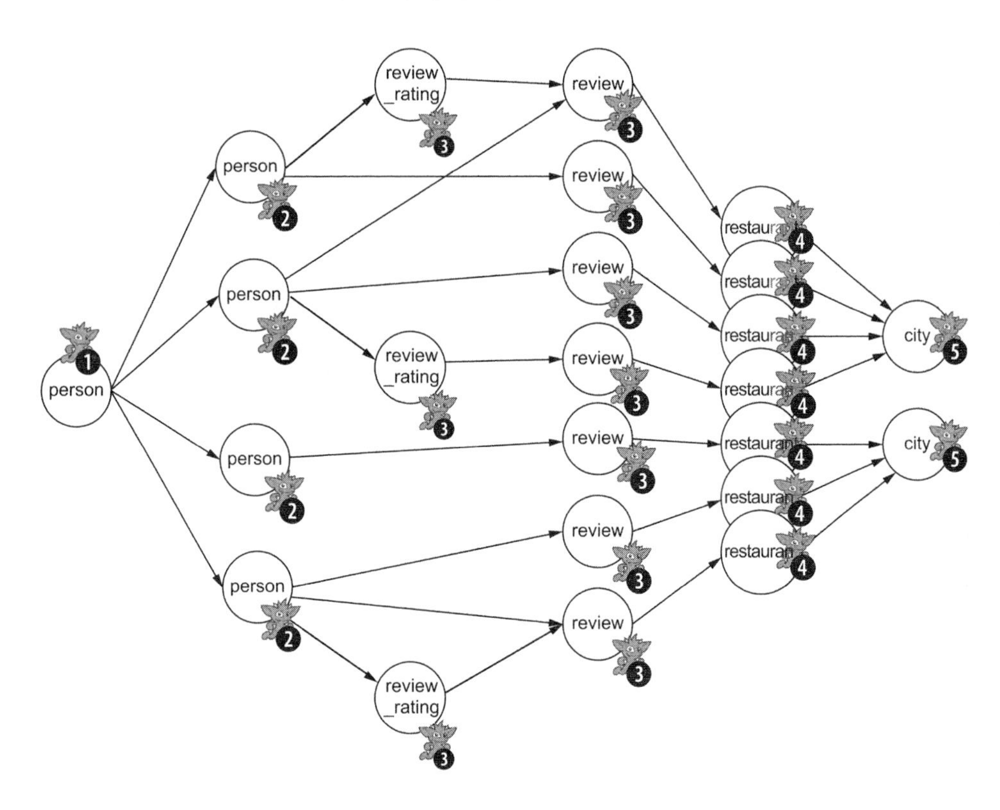

Figure 9.10 Our known-walk path from `person` to `city`, including optional steps for `review_rating` vertices through the instance data in our graph

vertex, we need to take an additional traversal step to get to our `review` vertex. After this additional step, all our traversers will be located on `review` vertices. From there, we can traverse through the rest of the graph. Extending our traversal with the additional steps we saw in figure 9.9, we construct a traversal to create our subgraph like this:

```
subgraph = g.V().has('person','person_id',2).          <-- Starts with the given person_id, which is the signed-in user
 bothE('friends').subgraph('sg').otherV().             <-- Traverses the friends edges to get all of the friends
 outE('wrote').subgraph('sg').inV().                   <-- Traverses the wrote edges to the review and the review_rating vertices
 optional(
    hasLabel('review_rating').outE('about').
 subgraph('sg').inV()                                  <-- If on a review_rating vertex, then traverses the about edge to get to a review vertex
 ).
 outE('about').subgraph('sg').inV().                   <-- Traverses the about edge to get to the restaurant vertex
 outE('within').subgraph('sg').                        <-- Traverses the within edge to get to the city vertex
 cap('sg').next()
==> tinkergraph[vertices:80 edges:121]
```

We don't know about you, but handling the same edge type (`wrote`) differently, based on the incident edge type (`review` or `review_rating`), feels a little awkward. We might not go so far as to call it a code smell; maybe it's more of a "code irksome whiff." Nevertheless, it would be nice if we could handle that case a little better. So, let's give it a try.

In attempting to rewrite this traversal, the main challenge we run into is the dual-purpose `wrote` edge. As we mentioned during our data modeling, we like to try to apply generic labels to vertices and edges; however, this comes with some tradeoffs, and this is one of those tradeoffs. After all, maybe `wrote` isn't the best term to use for connecting a `person` to a `review_rating`. What's really going on is that the user *assigns* a rating to a review. Maybe a better approach is to have different edge labels like this:

- person `wrote` review
- person `assigned` review_rating `about` review

Figure 9.11 shows how the schema looks after adding these edges. That change also changes the traversal steps. Figure 9.12 shows this change.

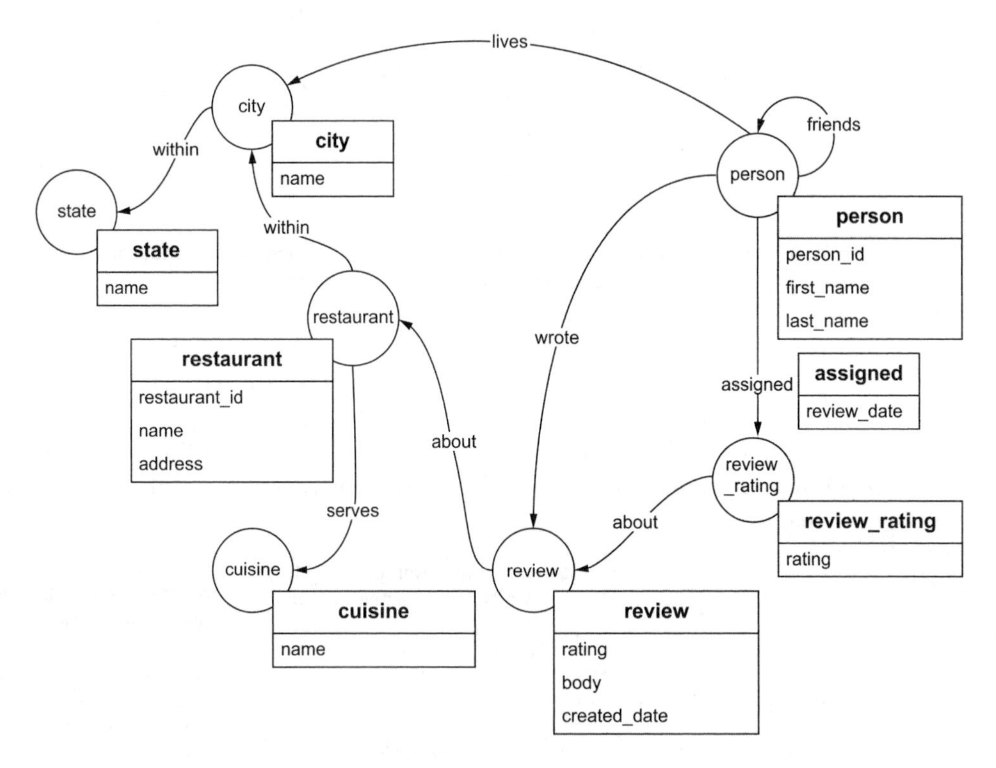

Figure 9.11 Alternate logical data model with `person assigned review_rating`

We can see that the walk has the same basic shape because the fundamental connections haven't changed. All we changed is the name of one edge, from `wrote` to `assigned`. Let's see what our traversal looks like with this data model change.

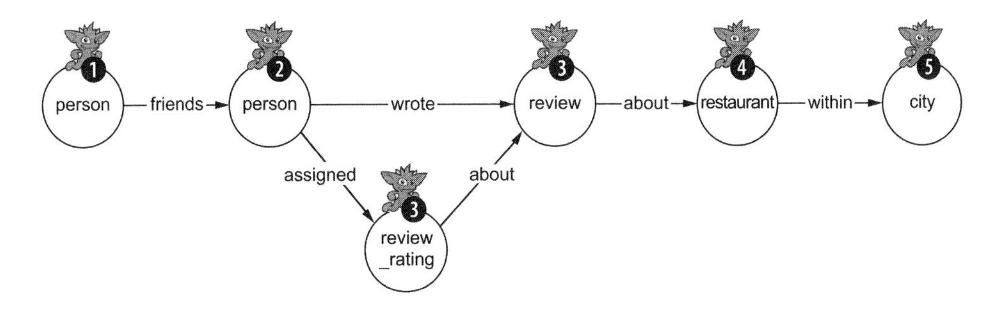

Figure 9.12 Known-walk pattern to create our subgraph for the alternate logical data model using the `assigned` edge

One difference that we notice is that in our new model, we have two different paths to take, from `person` to `restaurant`, using two different edge labels, `wrote` and `assigned`. In other words, we need to create a union of these two traversal paths, similar to how we perform a `UNION` on two queries in SQL. To handle this, we'll bring back the `union()` step introduced in chapter 7.

Modifying our previous traversal with this new step yields the following:

```
subgraph = g.V().has('person','person_id',2).
  bothE('friends').subgraph('sg').otherV().
  union(
        outE('wrote').subgraph('sg').inV(),
        outE('assigned').subgraph('sg').inV()
      ).
  outE('about').subgraph('sg').inV().
  outE('within').subgraph('sg').
  cap('sg').next()
```

Union combining the results of the inside traversals

Traverses the wrote edge to yield review vertices

Traverses the assigned edge to yield review vertices

Emits the combined results of both traversals

This might be one of those "six of one, half a dozen of the other" situations. In our first iteration, we had to use an `optional()` step; in the second iteration, we needed a `union()` step.

One thing to credit the new approach with is that it makes better semantic sense. That is, it establishes an important distinction between the `wrote` edge, which only gets us to a `review`, and the `assigned` edge, which takes us to a `review_rating`. This distinction reminds us of one of the most challenging points of data modeling—naming things. Should we use a single edge label for this use case, or two labels?

For this question, there is no "always correct" answer. In this case, we lean toward the second approach: using different labels if we think that we'll rarely traverse both to `review` and to `review_rating` in the same traversal. But the first approach does have its merits in reducing the number of labels we need to potentially traverse.

We started with that approach and that wasn't wrong. However, we don't see a compelling reason to make this change to the data model, so we'll stick with a single label for the two edges, our first approach. We're generally reluctant to change schema when there is already working code in place.

9.3 Building the traversal

Now that we have our subgraph, let's complete our work by constructing the traversal for our personalization use case. Remember, the question we want to answer is, "Based on my friends review ratings, what are the best restaurants for me in this area?" This use case is similar to some of the recommendation engine use cases. In particular, this question: "What are the ten highest-rated restaurants near me?" It is essentially the same question, except we chose to restrict it to reviews by our friends, instead of every possible review.

Given that we already solved a similar problem in the previous chapter, we can use the work we did there. However, in this case, instead of starting with the whole graph, we use our subgraph. Because our subgraph is already limited to that part of the graph personalized to the user, we can start with the reviews and proceed from there. In this case, the steps for our traversal look like this (remember each step is only being executed on the data in our subgraph):

1. Find all of the `review` vertices.
2. Traverse to the `restaurant` vertices.
3. Filter the `restaurant` vertices based on the `city_name` input.
4. Group the `restaurant` vertices by the average `review_rating`.
5. Sort in descending order by the average rating.

Did you notice that this traversal is another example of the known-walk traversal pattern? We know both the series of vertices and edges we need to traverse, and the number of times we must traverse these. The known-walk path we follow within the subgraph is quite simple, as figure 9.13 shows.

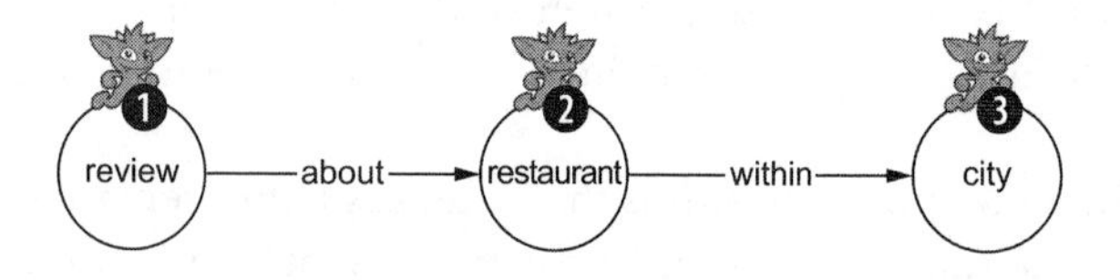

Figure 9.13 Known walk that traverses the subgraph to collect restaurants and ratings within a given city

To develop our traversal, we need to provide one input, some geographic reference for "in this area." For this example, we take the name of a city as input, and for testing purposes, we start with Houston, one of the two cities represented in our sample data. Before we start our work, let's define the city input as a variable in the Gremlin Console and create the `GraphTraversalSource` for our subgraph:

```
city_name = 'Houston'
sg = subgraph.traversal()
```

With that established, we can traverse our subgraph. Looking at the steps we need to accomplish, shown in figure 9.13, we get the following traversal:

```
sg.V().                                                  <-- Traversal starts with sg instead of g
  hasLabel('review').                                    <-- Finds all review vertices
  out('about').                                          <-- Traverses the about edge to the restaurant
  where(out('within').has('city','name',city_name))      <-- Traverses the within edge and filers on city_name
==>v[254]
...
==>v[184]
```

NOTE We start with sg, the `GraphTraversalSource` for our subgraph, instead of the usual g variable in the traversals.

The following traversal answers the question, "Based on my friends review ratings, what are the best restaurants for me in this area?" All we need to do is format our results into a logical output. Luckily, we did this in section 8.3.1, so we can reuse that code here. Combining that projection code with our traversal yields this code:

```
sg.V().hasLabel('review').                              <-- Traversal starts with sg instead of g because we are traversing the subgraph.
  out('about').
  where(out('within').has('city','name',city_name)).    <-- Where step filters the restaurants by the connected city vertex.
  where(__.in('about')).
  group().
    by(identity()).
    by(__.in('about').values('rating').mean()).         <-- An anonymous traversal (__) is always required before a subtraversal that starts with the in() step.
  unfold().
  order().
    by(values, desc).
  limit(3).
  project('restaurant_id','restaurant_name','address','rating_average').
    by(select(keys).values('restaurant_id')).
    by(select(keys).values('name')).
    by(select(keys).values('address')).
    by(select(values))
==>{restaurant_id=35, restaurant_name=Pick & Go, address=4881 Upton Falls,
     rating_average=5.0}
==>{restaurant_id=33, restaurant_name=Spicy Heat, address=4137 Hills Roads,
     rating_average=5.0}
==>{restaurant_id=9, restaurant_name=Northern Quench, address=04603
     Cartwright Stream, rating_average=4.0}
```

That looks great! We got three restaurants listed in descending order by their rating average. Before we move on, we want to show you another way to think about this traversal.

9.3.1 Reversing the traversing direction

What if we started with the geographic location, city, as our input instead of starting with all of the `review` vertices? Figure 9.14 illustrates how the known walk would look if we use the same vertices but start with the `city` vertex instead.

Figure 9.14 Known walk that traverses the subgraph starting with the city and then collecting restaurants and reviews

This known walk may look a little odd because it goes against the directions of the edges, but that isn't a problem for a graph database. Edges are designed to be traversed in either (or both!) directions. This particular example highlights an advantage of graph databases over relational ones: links can be used in either direction quite easily.

In most relational database modeling, particularly with third normal form, the foreign keys are designed to be used in only one direction. Joining in the opposite direction, if the schema supports, it is usually expensive. For most graph databases, there is no additional performance cost to go in the other direction. Simply changing directions of a relationship in the relational world is practically unheard of, but for a graph, it is a nearly trivial change. Using this approach, how do our traversal steps change?

1. Find the city based on the `city_name` input.
2. Traverse the located edges to the `restaurant` vertices.
3. Group the `restaurant` by average `rating`.
4. Sort in descending order by the average ratings.

Those steps seem pretty clear and we now have one less step than before. Let's look at the traversal for this approach:

```
sg.V().has('city','name',city_name).          <-- Traversal starts with
  in('within').                               <-- the city vertex.
  where(__.in('about')).
  group().                                        Only other change in traversal
    by(identity()).                               is here, where we traverse a
    by(__.in('about').values('rating').mean()).   different edge and in the
  unfold().                                       opposite direction.
  order().
    by(values, desc).
  limit(3).
  project('restaurant_id','restaurant_name','address','rating_average').
    by(select(keys).values('restaurant_id')).
    by(select(keys).values('name')).
    by(select(keys).values('address')).
    by(select(values))
==>{restaurant_id=35, restaurant_name=Pick & Go,
➥  address=4881 Upton Falls, rating_average=5.0}
==>{restaurant_id=33, restaurant_name=Spicy Heat,
➥  address=4137 Hills Roads, rating_average=5.0}
==>{restaurant_id=9, restaurant_name=Northern Quench,
➥  address=04603 Cartwright Stream, rating_average=4.0}
```

We get the same results with this as we did with the original traversal, so we think it is quite good, and it's (arguably) a bit more readable. When we compare the execution times, a detail we'll cover in chapter 10, the second version is about three times faster than the first version.

This increase in speed can be attributed to the fact that we filtered our traversal in the second approach earlier than we did in the first. Filtering earlier in the traversal means there are fewer traversers moving through our graph; therefore, we do less overall work. The performance of a graph traversal is directly tied to how much of the graph it must interact with. The earlier we can filter out unneeded traversers, the less work to be done, and, correspondingly, the less time the traversal will take. As we said, we'll cover performance testing in chapter 10. For now, we'll take the three-times speed improvement and go with the second version of the traversal.

This example also demonstrates that you can write many traversals differently. More than a few times, we were stuck on a certain traversal, but when we took a step back and approached it from a different starting point, the traversal came together quickly.

9.3.2 Evaluating the individualized results of the subgraph

Before we move on, let's take a quick peek at how personal our personalization approach is. Let's take a user from the other side of the graph—say, Denise—and compare her recommendations.

First, we need to create a subgraph for Denise (`person_id = 8`), using the same traversal we first developed in section 9.1. Note how all we change is the number used in the first `has()` step:

```
subgraph8 = g.V().has('person','person_id',8).      <-- Changes the input of the
  bothE('friends').subgraph('sg').otherV().              has() step to 8 for Denise
  union(
        outE('wrote').subgraph('sg').inV(),
        outE('assigned').subgraph('sg').inV()
       ).
  outE('about').subgraph('sg').inV().
  outE('within').subgraph('sg').                     The subgraph has a different
  cap('sg').next()                                   number of vertices and edges.
==> tinkergraph[vertices:72 edges:107]           <--
```

Then, we create our `GraphTraversalSource` for this new subgraph:

```
sg8 = subgraph8.traversal()
==> graphtraversalsource[tinkergraph[vertices:72 edges:107], standard]
```

Finally, we run our personalization traversal. Note that this is the same traversal that we used before, except here we use a different `GraphTraversalSource`, `sg8`, to reflect that we're traversing a different subgraph:

```
sg8.V().has('city','name',city_name).          <-- Only change: uses the
  in('within').                                    sg8 traversal source
  where(__.in('about')).
```

```
group().
  by(identity()).
  by(__.in('about').values('rating').mean()).
order(local).
  by(values, desc).
limit(local,3).
unfold().
project('restaurant_id','restaurant_name','address','rating_average').
  by(select(keys).values('restaurant_id')).
  by(select(keys).values('name')).
  by(select(keys).values('address')).
  by(select(values))

==>{restaurant_id=17, restaurant_name=With Noodles,
➥ address=50586 Keebler View, rating_average=5.0}
==>{restaurant_id=31,
➥ restaurant_name=Dave's Big Deluxe,
➥ address=490 Ivan Cape,
➥ rating_average=4.666666666666667}
==>{restaurant_id=35, restaurant_name=Pick & Go,
➥ address=4881 Upton Falls, rating_average=4.0}
```

Returns different results for Denise compared to Josh

Even in our little test data set, we can use this method of creating subgraphs to provide a truly personalized experience. When we consider the limited (and expensive) options available with relational databases, we can see how powerful the ability to define and traverse subgraphs is for the graph database engines that support this functionality. Now, having established their value, let's look at how we incorporate subgraphs in our DiningByFriends application.

9.4 Implementing a subgraph with a remote connection

There's one critical limitation with the `subgraph()` step in TinkerPop: it's not supported by the Gremlin Language Variants (GLVs), at least not as of the time of this writing. Recall that we use TinkerPop's GLV for Java, which allows us to include our Gremlin code in line with our Java code. This saves us from using string concatenations to create our traversals and then coercing the generic results into Java types such as `String` or `long`.

The issue with using the `subgraph()` step in a GLV revolves around the fact that it returns a TinkerGraph object. GLVs do not include the concept of a local graph, which means that we cannot return a subgraph and then use it for further traversing. This presents a challenge. We have a valid use case for employing the `subgraph()` step, but the Java GLV does not support it. We must switch our approach from using the GLV to using a parameterized string-based approach, similar to executing queries via JDBC.

As a result (pun not intended), our Java implementation needs to change a bit in order to submit a script to the server instead of using the GLV. To submit script-based requests, we need to take the following steps:

1. Create a `Client` object.
2. Create a string representing our traversal.

3 Submit the traversal string along with any appropriate parameters.
4 Process the results.

We'll talk about the main points for using the string-based traversal approach with the client object. But for now, see the chapter 9 source code for the `findTop3Friends-RestaurantsForCity` method in the book's GitHub repository for the full set of code.

9.4.1 Connecting with TinkerPop's Client class

We begin by connecting to our cluster. We discussed this back in chapter 6, but to refresh your memory, in order to connect to a cluster to submit a script, we need to create a `Client` object using the `connect()` method. For that, we include a string in the `connect()` method that tells the client to establish a session with the `cluster`. We chose `sgSession` for our string, but any string will do:

```
Client client = cluster.connect("sgSession");
```

Now that we have our connection to the database, we can concatenate a string that represents our traversal. This is exactly the same traversal for creating the subgraph that we developed earlier:

```
String defineSubgraph = "subgraph = g.V()." +
        "has('person','first_name', name)." +       <-- Inputs like name are not quoted.
        "bothE().subgraph('sg').otherV()." +
        "outE('wrote').subgraph('sg').inV()." +
        "optional(outE('about').subgraph('sg').inV())." +
        "outE('within').subgraph('sg')." +
        "cap('sg').next(); null";                   <-- Terminates statement with semicolon and null
```

We broke the traversal up into multiple strings, split by lines, to aid in readability. It could all be done as a single string, however, and that might make testing easier in some cases. Note that by using single quotes in our traversal, we don't have to escape the quotes within the `defineSubgraph` string. We see that `name` is not quoted, and that is intentional. This parameter is an input of our script and corresponds to a `name` key in a map of parameters that we'll submit when we send this to the server.

Also, we included the text `; null` at the end of the traversal. The semicolon (`;`) terminates the first statement, the assignment of the `subgraph` variable, and the `null` gives the whole operation something to return to the client. Technically, it gives the script a `null` to return to the calling client, and as we all know, "`null` ain't nuthin." But in this case, `null` is sufficient. (For those with aversions to `null`, an empty list (e.g., `[]`) can suffice too.) With our traversal string defined, we need to submit it to the server using the `submit()` method on the `client` object for processing:

```
client.submit(defineSubgraph, params);
```

Here, we add a `params` object, which is a map of all the parameters included in our traversal (in this case, the `name` key and its associated value). This process likely feels familiar because this is quite similar to how many SQL queries are executed with JDBC.

By default, `client.submit()` returns a `ResultSet`, which is an iterable containing one or more `Result` objects. These `Result` objects are streamed back from the server. That means that there might be a point in time where the `ResultSet` contains some but not all of the final set of `Result` objects. In this specific example, we only receive a `null` back because that is what our string-based traversal returned when we created the subgraph.

It is important to note that the `subgraph` variable we just defined does not exist in the context of our client application. The subgraph variable *only* exists on the server within our session. We can use it as much as we want, as long as we are connected to the same session. When the session goes away, the variable goes away.

When we use the server-side `subgraph` variable in a traversal, we want to handle the results. This puts us squarely in the territory of Java's `CompletableFuture` API, the details of which are beyond the scope of this book. But we can show you a quick code sample sufficient to illustrate how to process the results:

```
String findTopRests = "g.V().hasLabel('review').order()." +
    "by('rating', desc).limit(3). " +
    "out('about').values('name')";                          <- The traversal string
List<Result> results = client.submit(findTopRests,
➥ param).all().get();                                     <- Streams back all the results
results.forEach(r -> System.out.println(
➥ r.getObject().toString()));                             <- Casts to a Java object
```

In this example, the `all().get()` method ensures that all of our results are streamed back before we start the processing. Then, we use the Java `List`'s `forEach()` method. Within the `forEach()` call, we use `getObject()` to cast each individual result to a Java object, finishing with a `toString()` method. TinkerPop's `Result` class has the usual `get` methods for casting results into various types of Java objects, much like the `get` methods in JDBC's own `Result` class.

Those are all the steps we need to follow to submit string-based traversals, like subgraph traversals, to our database. We did not go through the details of how we implemented this in our sample application, but for those interested, a working version of this code is available in the code repository.

9.4.2 Adding this traversal to our application

Now that the hard work of finishing this use case is complete, the only thing left to do is to add it to our application. As in the last section, we'll follow the same process as we did in chapter 6, so you can use that as a guide.

In our example app, there is a new method called `findTop3Friends-RestaurantsForCity`. Look for it in the project's source code for chapter 9 to see the Java implementation of the subgraph functionality. If you want to test this out, we recommend running this method for Dave, Josh, and Denise in the city of Houston to see how their results are personalized.

In this chapter, we introduced the idea of a subgraph and used it to create the individualized results required for the personalization use case of DiningByFriends. Congratulations! This brings us to the end of part 2 of this book, where we extended the basic concepts and constructs we learned in the first part of the book with more complex graph traversal patterns to solve more complex use cases. In the next chapter, we'll address profiling our traversals and how to work through performance optimizations.

Summary

- Subgraphs are a subset of graph data that contain vertices and edges represented as a graph. Subgraphs are themselves graphs. This means that we can run traversals on these, but because these are constrained to a small subset of vertices and edges, subgraphs require less memory and computation power to process.
- Subgraphs can be defined in one of two ways: vertex-induced or edge-induced. Vertex-induced subgraphs are defined by specifying a set of vertices and include the incident edges. Edge-induced subgraphs are defined by specifying a set of edges and include the adjacent vertices. The database you choose determines which option is available.
- Because subgraphs return as graphs, we can traverse these and perform all the other operations you've learned to do with graphs once you've created a graph traversal source for the subgraph.
- Subgraphs can be reused and even modified, but any changes you make are done in isolation from the original graph. This means that any changes are not propagated back to the original graph data.
- When building an application that use subgraphs in Gremlin, we need to use the string-based `Client` API instead of the Gremlin Language Variants (GLVs). GLVs do not have subgraph support, so we must use a script submission method where we parameterize and concatenate strings to write our traversals.

Part 3

Moving Beyond the Basics

As we near the end of our journey through the world of graph-backed applications, our path divides. (Pun totally intended.) In one direction is the frontier of graph analytics. In the other is the familiar territory of debugging and performance-tuning when the application doesn't work quite right.

Chapter 10 explains how to troubleshoot performance and application problems via common graph database-tuning tools. We also discuss common application anti-patterns, the dreaded supernode, as well as how to alleviate or mitigate these problems. Chapter 11 closes the book with a brief look at graph analytics (complete with examples) before sharing several of our favorite resources as you look to continue your journey working with graph databases.

10 Performance, pitfalls, and anti-patterns

This chapter covers

- Diagnosing and debugging common performance problems with traversals
- Understanding, locating, and mitigating supernodes
- Identifying common application anti-patterns

Our application is built, tested, and delivered to production. We spent a lot of effort designing a system to run in a resilient and scalable manner. However, entropy is not on our side. Everything is humming along perfectly, until one day, we receive that dreaded bug ticket, "Application is slow." Knowing what's likely inside, we hesitantly click on the message, and as expected, we're presented with a vague description that says the application is slow, but otherwise gives little detail.

This chapter examines some common performance issues and techniques for mitigating those performance issues that you are likely to encounter while developing graph applications. We'll start by looking at how to diagnose common performance problems in graph traversals, including the dreaded, "Application is slow." For this, we look at a few of the most common tools available to help diagnose and debug traversal issues. Next, we will introduce you to supernodes, a common source of performance problems in graph applications. Here, we discuss what supernodes are, why these are a problem, and what to do to mitigate their effects. Finally, we'll focus on some specific pitfalls and anti-patterns that can come with building graph applications, some of which are common across databases and some are unique to graph databases. By the end of this chapter, you'll possess a solid understanding of the most common graph anti-patterns, how to detect these early on in the project, and how to prevent graph projects from going astray.

10.1 Slow-performing traversals

We have a user who's experiencing performance problems with our application and who submitted the "Application is slow" ticket. Lucky for us, the user at least told us what they were trying to do when the application slowed, which gives us a place to start looking. The ticket leads us to the problem being with this request: "Find the three friends-of-friends of Dave that have the most connections." (We know that wasn't a use case that we listed in chapters 2 or 7, but just work with us as we illustrate by example.) Digging into our application code, we locate the problematic traversal:

```
g.V().has('person', 'first_name', 'Dave').
  both('friends').
  both('friends').
  groupCount().
    by('first_name').
  unfold().
  order().
    by(values, desc).
    by(keys).
  project('name', 'count').
    by(keys).
    by(values).
  limit(3)
```

Great, we know where the problem is, but how do we diagnose this slow-performing traversal? Graph databases, like relational databases, are no stranger to slow-performing operations. And like relational databases, graphs also have tools to aid in diagnosing problems. These tools take two forms: *explaining* what a traversal will do or *profiling* what a traversal did.

10.1.1 Explaining our traversal

We should say up front that the `explain()` step is rarely our first step in troubleshooting. We usually use the profiling tool we discuss in the next section. We find that using the `explain()` step to locate issues with poor-performing traversals requires a deep knowledge of the inner workings of the databases. However, the `explain()` step is a commonly available tool across different database instances, and some people find both useful in debugging, so we'll give `explain()` a little attention.

Let's say we want to know *how* our traversal runs, but we do not want to execute it. Most graph databases perform this type of debugging step via the use of an `explain()`

step. This is similar to an estimated execution plan in a relational database, in as much as the database optimizer shows the output after it rearranges and optimizes the traversal, but *before* it actually runs the traversal on the data. (Gremlin does this through the use of strategies; see http://mng.bz/ggoR). The important part to focus on is the final traversal plan. This represents the optimized plan to be executed on the graph data. The output of the `explain()` step lists the various options that were applied in order to reach the final internal form of the traversal, the one designed to run on the physical data.

The best way to illustrate this is to run an `explain()` step and then examine the output. In the following example, the `Final Traversal` is highlighted in bold and the non-optimized options are removed for brevity. Let's run an `explain()` step on our slow traversal and see if we can draw any conclusions from the output:

```
g.V().has('person', 'first_name', 'Dave').
  both('friends').
  both('friends').
  groupCount().
    by('first_name').unfold().
  order().
    by(values, desc).
    by(keys).
  project('name', 'count').
    by(keys).
    by(values).
  limit(3).
  explain()                    <-- Performs the explain command
==>Traversal Explanation
==========================================================================
...                            <-- Output removed for brevity
Final Traversal[TinkerGraphStep(vertex,[~label.eq(person),
➥ first_name.eq(Dave)]),
VertexStep(BOTH,[friends],vertex),
VertexStep(BOTH,[friends],vertex),
GroupCountStep(value(first_name)),
UnfoldStep,
OrderGlobalStep([[values, desc], [keys, asc]]),
RangeGlobalStep(0,3),
ProjectStep([name, count],[keys, values]),
➥ ReferenceElementStep]        <-- The final traversal is what matters most.
```

As we mentioned, the important part is the line starting with `Final Traversal`. This is the optimized plan that's executed on the graph. In this example, the optimized code performed against our graph is

```
  [TinkerGraphStep(vertex,[~label.eq(person),     <-- Maps to V().has('person', 'first_name', 'Dave')
    ➥ first_name.eq(Dave)]),
VertexStep(BOTH,[friends],vertex),                <-- Maps to the first both('friends') step
VertexStep(BOTH,[friends],vertex),                <-- Maps to the second both('friends') step
```

```
GroupCountStep(value(first_name)),                    <-- Maps to groupCount().by('first_name')
UnfoldStep,                                           <-- Maps to unfold()
OrderGlobalStep([[values, desc], [keys, asc]]),       <-- Maps to order().by(values, desc).by(keys)
RangeGlobalStep(0,3),                                 <-- Maps to limit(3)
ProjectStep([name, count],[keys, values]),
  ➥ ReferenceElementStep]                             <-- Maps to project('name','count').by(keys).by(values)
```

Although it is nice to see the traversal written in optimized steps, it doesn't explicitly point out why our traversal is slow or what we can do to improve its performance. What it does do, however, is show us *how* this traversal will be executed. With enough practice and knowledge of a particular database, you can understand what needs to be done to further optimize the execution plan. But we find that simply knowing how a traversal will execute tends to lack the insight we need to fix performance problems.

This lack of insight is one reason why we rarely use the `explain()` step. Another reason is that this optimized plan is always the same, no matter what your starting vertex is. In many scenarios, we have a traversal that works well starting in one location of the graph but performs poorly for other starting points. In these scenarios, the `explain()` step won't help to diagnose performance issues.

More often than not, we want to see the actual execution of a traversal, not just the way the engine thinks it will run it. This leads us to the most commonly used performance debugging tool—profiling.

10.1.2 Profiling our traversal

Let's say our slow traversal works perfectly fine for some users, but horribly for others. Instead of looking at the planned execution, we need to profile the actual operations. This allows us to compare good runs to the bad ones and to see the differences.

In most graph databases, this type of debugging is done via the use of the `profile()` step. The `profile()` step runs the traversal and collects statistics on the performance characteristics during its execution. These statistics include details about the execution, similar to an actual execution plan in a relational database.

As with the `explain()` step, the easiest way to understand a `profile()` step is to run one and then examine the output. Let's profile our slow-performing traversal and investigate the output (shown in the table following the code input). We are looking for where the traversal spends the most time and for which step uses the most traversers. In the table, we highlighted the duration (`%Dur`) in bold within the output displayed in figure 10.1.

Upon examining the figure 10.1 output, we notice a couple of things. First, we see that the lines in the output match the bytecode from the `explain()` step shown in the previous section. This correlation makes sense because the `explain()` step tells us *how* a traversal executes, and the `profile()` step informs us *what* happens when it runs.

```
gremlin> g.V().has('person', 'first_name', 'Dave').
......1>   both('friends').
......2>   both('friends').
......3>   groupCount().
......4>     by('first_name').
......5>   unfold().
......6>   order().
......7>     by(values, desc).
......8>     by(keys).
......9>   project('name', 'count').
.....10>     by(keys).
.....11>     by(values).
.....12>   limit(3).
.....13>   profile()
==>Traversal Metrics
Step                                                               Count  Traversers       Time (ms)    % Dur
=============================================================================================================
TinkerGraphStep(vertex,[~label.eq(person), firs...                     1           1           0.836     6.49
VertexStep(BOTH,[friends],vertex)                                      4           4           0.342     2.65
VertexStep(BOTH,[friends],vertex)                                     10          10           0.248     1.92
GroupCountStep(value(first_name))                                      1           1           1.511    11.72
UnfoldStep                                                             6           6           0.093     0.73
OrderGlobalStep([[values, desc], [keys, asc]])                         4           4           9.611    74.54
RangeGlobalStep(0,3)                                                   3           3           0.111     0.86
ProjectStep([name, count],[keys, values])                              3           3           0.139     1.08
                                            >TOTAL                     -           -          12.893        -
```

Figure 10.1 The output of a `profile()` step showing associations back to the original traversal steps

Second, each line in the traversal maps to one step in our optimized traversal (shown in the output). Usually, it's straightforward to determine which step in the traversal refers to which line in the output. Unfortunately, there's no definitive documentation on how these map because they are specific to your vendor's implementation. Different vendors have different implementations, each of which has a different engine and approach toward optimization strategies. Finally, for each line of output we see

- The count of the represented traversers (`Count`)
- The count of the actual traverser (`Ts` or `Traversers`)
- The time spent on that step (`Time`)
- The percentage of the traversal's total duration spent on that step (`%Dur`)

The `Count` and `Traversers` (the `Ts` column) values won't always match; for example, when the same element is visited multiple times. In this instance, the traversers can be merged in a process in Gremlin known as *bulking*, which causes the `Count` value to be larger than the `Traversers` value.

NOTE One cautionary note is that profiling traversals requires extra resources, so the represented times may not match a non-profiled traversal. However, the proportion of time spent is the same between profiled and non-profiled traversals.

The critical questions are, where is the traversal spending most of its time and what is the count of traversers on that step? Based on the previous output, we see that more than 48% of our traversal's time is spent on the `has('person', 'first_name', 'Dave')` step. This leads us to one of two common fixes, the details of which are covered in the following section.

If we find that the longest duration steps have many traversers, we should add additional filtering criteria prior to that step to reduce the number of traversers required. In our example, however, this isn't the case. Instead, we identify that the longest duration step doesn't have many traversers associated with it; it only has one traverser. Because we know that we only have a single traverser and that our step is a filtering step, this leads us to think that we should add an index (our second common fix).

10.1.3 Indexes

Similar to relational databases, an *index* in a graph database provides a method to efficiently find data based on predefined criteria. Indexes work by allowing us to quickly and directly access the data that we're seeking, instead of scanning the entire graph to find it. Avoiding a scan of the entire graph creates massive performance improvements.

Let's say we want to search a graph to find a vertex where the `first_name` is `Dave`. Without an index, this requires us to look at every vertex to see if it has a property named `first_name` and, if it does, is the value of that property is `Dave`. While this might not be a noticeable issue in small graphs, in graphs with thousands, millions, or billions of nodes, this creates a huge performance impact.

Let's take a look at how this same scenario works if we add an index on the `first_name` property. With this option, instead of having to look at every vertex, we alternatively look at the index. The index already knows which vertices have a `first_name` property and can, with a single lookup, find the ones with the values of `Dave`. It is reasonable to expect that performing a single lookup inside an index will be significantly faster than looking at all of the thousands, millions, or billions of vertices. Here are three areas where indexes can provide the most performance improvement:

- *Properties frequently used for filtering on values or ranges.* Indexes quickly reduce the number of traversers required to execute a particular task, thereby reducing the work required of the database. This is especially helpful early on in a traversal where a minimal number of traversers is desired.
- *Properties requiring a full-text search, such as finding words that start with, end with, or contain a specific phrase.* Many databases require a particular type of index to perform a full-text search on a property because these warrant special handling to be indexed efficiently.
- *Spatial features needing to be searched if the database supports geospatial data.* Spatial properties also fall into the category of requiring special indexes to perform the appropriate queries such as, "Find all restaurants within ten miles of here."

The increased efficiency that indexes bring comes at the cost of additional storage and additional writes behind the scenes. An index makes a redundant copy of data, or at least pointers to data, optimized for retrieval by specific criteria. For these reasons, we should be prudent when adding indexes to our graph and only use these when needed to achieve the desired performance.

Every vendor's implementation has different indexing capabilities and characteristics. Some implementations, such as TinkerGraph, only offer global single value indexes. Others, like Neo4j, DataStax Graph, and JanusGraph (among many others) enable a full range of single value-based, composite value-based, range-based, and even geospatial indexes. Still others, such as Azure CosmosDB and Amazon Neptune, have no concept of user-defined indexes, preferring that the indexing details be left to the service provider to manage. We highly recommend consulting the documentation of your chosen database for the indexing capabilities, as well as the best practices for use of those indexes.

In this section, we looked at some of the diagnostic tools we can use when a particular traversal is slow. However, traversals are only one part of the application that causes performance issues. Sometimes the issue is not with the traversal, but with the data itself. Many of these data-related performance problems can be traced back to a single source—supernodes.

10.2 Dealing with supernodes

Supernodes are one of the most common data-related performance problems in graph databases. These are also particularly difficult to deal with because supernodes can't be removed. And because they are part of the data, we can only try to mitigate the problems caused by supernodes. This leads us to our first question: What is a supernode?

A *supernode* is a vertex in a graph with a disproportionally high number of incident edges. We find that supernodes are hard to define but easy to understand with an example, so let's take a look at one instance using Twitter.

When writing this book, the most followed person on Twitter is Katy Perry with 107.8 million followers (https://twitter.com/katyperry). Based on research performed in 2016, by the social media marketing company KickFactory (http://mng.bz/em5J), the average Twitter user has 707 followers. This means Katy Perry has ~152,475 times as many followers as the average Twitter user. Let's assume we stored this data using the data model in figure 10.2.

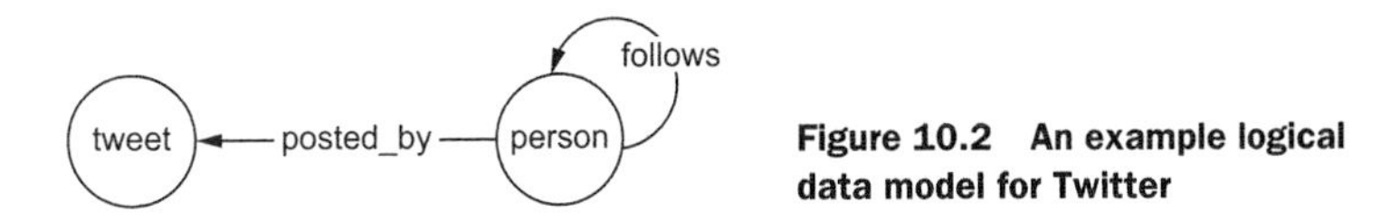

Figure 10.2 An example logical data model for Twitter

Based on this model and current research, Katy Perry has 107.8 million `follows` edges incident to her `user` vertex, and the average user has 707 `follows` edges. What happens when we notify all followers about a new tweet?

When an average user posts a tweet, we need to notify all their followers, meaning our traversal will have 707 traversers, one for each `follows` edge. In Katy Perry's case, when she posts a tweet, our traversal will have 107.8 million traversers. With this sort of follower differential, it's safe to assume that Katy Perry's tweets are more computationally intensive than the average user's post. While an extreme example, it is this sort of disparity that leads to supernodes.

The natural next question is, "What number is disproportionately high?" We wish we could give you a precise number where a vertex becomes a supernode, but it's not that easy. We need to understand two main concepts when discussing a supernode: instance data and underlying data structures. Let's look at these.

10.2.1 It's about instance data

The first concept is that a supernode is a specific vertex of a specific label in the instance data. One common misunderstanding is that a supernode refers to a vertex label, but this isn't the case. Instead, it refers to an instance of a vertex with a disproportionate number of edges compared to the other instances with the same label. Think back to our Twitter example where, because the average user and Katy Perry are both people, they have the same vertex label. In other words, it's the Katy Perry instance of the `user` vertex that is a supernode, not the generic `user` label.

10.2.2 It's about the database

The second concept to understand is that what performs like a supernode in one traversal on one database can work fine on a different database or within a different traversal in the same database. Underlying data structures and storage algorithms differ between database vendors. These differences, along with other database-specific optimizations, make it impossible to provide a generalized answer. But we do recommend that you review the documentation for your chosen database, understand the distribution of relationships in your data, and thoroughly test the chosen system based on these expected distributions.

10.2.3 What makes a supernode?

While our Twitter model is useful to demonstrate the concept of a supernode, most of us will never work with data at that scale. Instead, let's use our DiningByFriends data model, shown in figure 10.3, as a more realistic example and see if we find any potential supernodes.

> **EXERCISE** Apply what you just learned about Katy Perry and the Twitter example to our DiningByFriends model. Can you identify any potential supernodes?

When we look at our DiningByFriends data model, we see two potential opportunities for supernodes: the `city` and the `state` vertices. Why these two vertex labels? To demonstrate why `city` and `state` might potentially cause supernodes, let's use the example of two cities in the United States: New York City, New York, and Anchorage, Alaska.

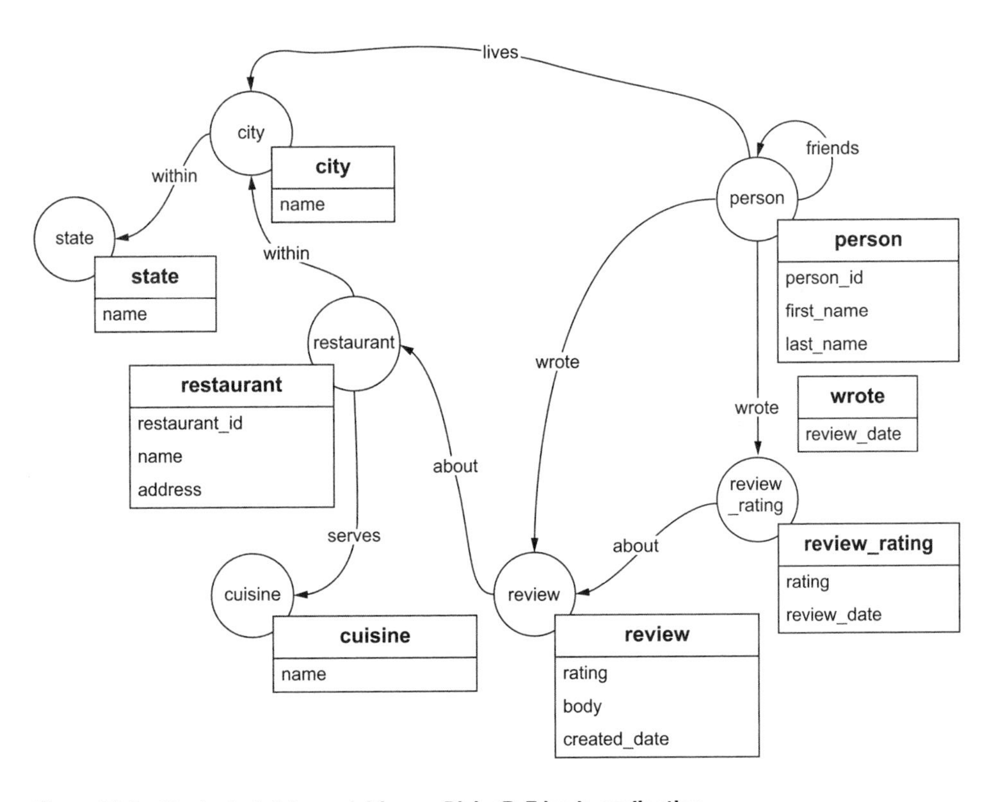

Figure 10.3 The logical data model for our DiningByFriends application

If we do a simple Google search, we find that the New York City is home to around 26,000 restaurants, while Anchorage has approximately 750 restaurants. This means that the `city` vertex for New York City has almost 35 times the number of incident `within` edges as the vertex for Anchorage, so any traversals from New York City require 35 times the work as those from Anchorage. While not quite as dramatic a difference as our Twitter example, we'd like to think that a 35-times spread in values represents a disproportionately distributed data set.

This same logic applies to our `state` vertex as well. The state of New York has approximately 1,000 cities and towns, while Alaska has around 130. This disparity represents a nearly eight times difference between the two.

While none of this disparity automatically means that we have a supernode, we'll learn how to determine that in the next section. These are but two likely supernode candidates we see in our model.

10.2.4 Monitoring for supernodes

If we can't give specific numbers of what constitutes a supernode, how do we find them? We generally employ two strategies, frequently in parallel, to detect supernodes: monitoring for growth and monitoring for outliers.

MONITORING FOR GROWTH

The first strategy is to periodically monitor the degree (number) of all the vertices in our graph and look for the top outliers. Monitoring is important because supernodes rarely exist at the beginning; these grow over time. In other words, supernodes are rarely created during the initial loading of data; instead, they tend to grow as more and more data is added to a graph. This is because many real-world networks are *scale-free networks.* Scale-free networks have many vertices with a low degree of incident edges and only a few vertices with a high degree.

Think back to our Twitter example. The majority of users have a low number of connections. There's also a small minority with a high number of connections. The same is true if you look at other networks, such as airline companies. While most airports will likely only have a few flights, there are a small number of airports, namely the hubs, that have a larger number of flights. This type of distribution of data is known as a *power-law distribution,* as figure 10.4 shows.

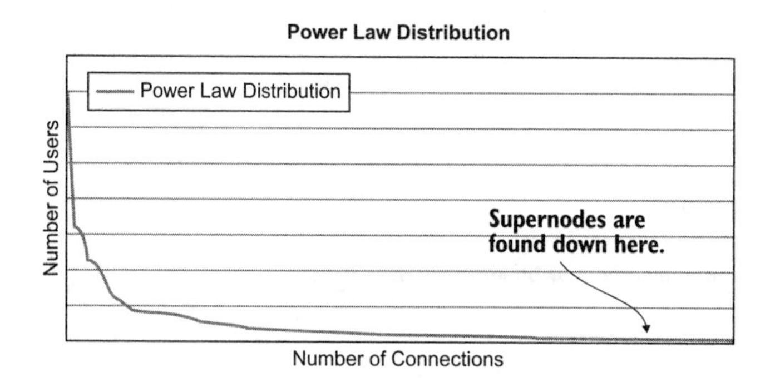

Figure 10.4 Power-law distribution with the power law having a long tail

The long tail of the distribution is where supernodes exist in scale-free networks. If we have a scale-free network and these cause supernodes, how do we check the growth of the degree of vertices? We need to monitor our data periodically to find the vertices with the highest degree. In brief, we need a traversal to take the following steps (illustrated in figure 10.5):

1. Find all vertices.
2. Calculate the degree of each vertex.
3. Order the results, descending by degree.
4. Return only the top *N* results.

Running this or similar traversals at regular intervals and tracking the results monitors the growth of potential supernodes proactively and catches these before a problem arises. While this strategy is an effective tool to find and monitor supernodes, it has a few significant drawbacks.

First, it requires us to remember to run this traversal and monitor the output on a regular basis. We're all busy, and this sort of housekeeping task is easy to delay,

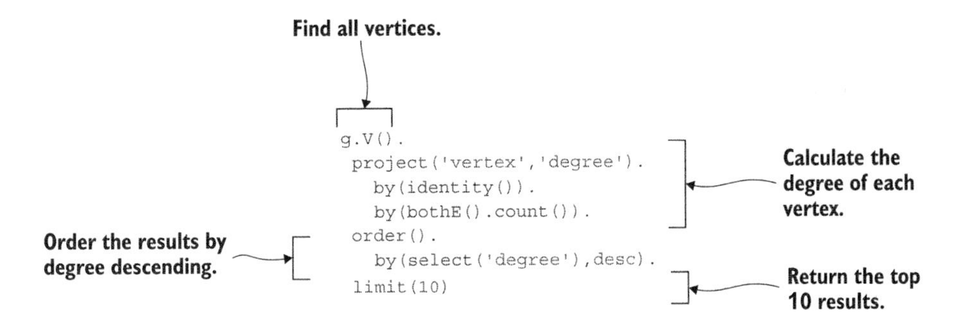

Figure 10.5 An example traversal to find the top 10 vertices with the highest degree

meaning that supernodes can creep in without proper warning. Second, the traversal in figure 10.5 is long-running because it requires visiting every vertex in the graph once (for each incident vertex) and every edge twice. As our graph grows, this traversal will take longer and longer to run and will consume increasingly more resources. But we have another possible approach for monitoring supernodes in our toolbelt. Let's look at that additional process next.

MONITORING FOR OUTLIERS

The second commonly used approach to monitoring for supernodes is to reactively monitor the performance of traversals and look for outliers. This is usually done with one of the many different application monitoring tools available on the market. When monitoring for supernodes, we look for traversals that are taking a significantly longer time to execute for one vertex than for another. Although supernodes aren't the only cause of slow performance, these are one of the common reasons why a generally well-performing traversal exhibits performance differences on specific vertices.

Although these two methods for identifying supernodes in our graph are useful, as we said, both these approaches have downsides. Due to the amount of the graph touched, each of the specified approaches tends to result in a long running query and places significant additional load on the graph database. There's no magic bullet for detecting supernodes. Still, domain knowledge, proper data modeling, and continuous monitoring are the best tools available to prevent and identify supernodes.

10.2.5 What to do if you have a supernode

If you determine that you have a supernode in your graph, the first thing to do is decide whether the supernode is actually a problem. If it is a problem, the best step is to look at ways to mitigate the supernode.

IS THE SUPERNODE A PROBLEM?

We need to consider how our traversals are traversing a supernode to determine if it causes a problem. For example, let's look at a subsection of our DiningByFriends schema containing only the `city`, `state`, and `restaurant` vertices, as figure 10.6 illustrates.

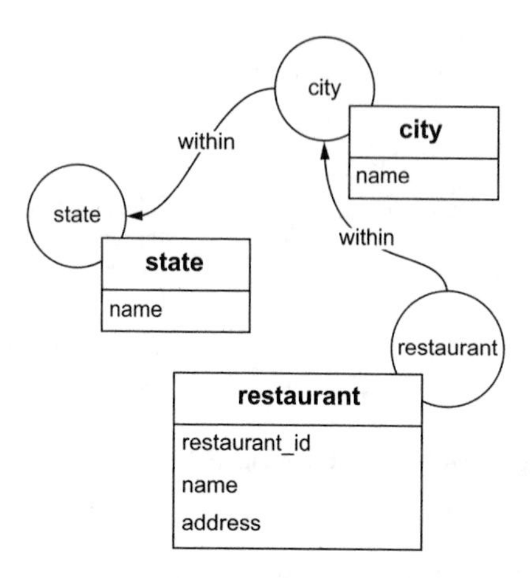

Figure 10.6 The portion of the DiningByFriends logical data model relating to restaurants, cities, and states.

As we previously mentioned, both `city` and `state` vertices are likely candidates to become supernodes within our graph. However, because these are only *likely* supernodes, we need to examin the specific traversals we run on our graph to decide if these potential supernodes will become problematic.

Let's say that the only traversal our application makes is to answer the request, "Get me the city and state for restaurant *X*." In this traversal, we only ever have one `city` and one `state` vertex associated with a specific `restaurant` vertex. This means that there is only ever a single `within` edge traversed when moving from a `restaurant` to a `city`, and only a single `within` edge traversed when moving from a `city` to a `state`. While both an instance of a `city`, such as New York City, and a `state`, such as New York, are likely to be supernodes, we traverse these in a way that minimizes the number of edges traversed, so these vertices won't cause the performance problems associated with supernodes. That is, our chosen access patterns will not encounter the worst possible branching factor, the number of successors of a given vertex, when traversing through either `city` or `state` vertices.

On the other hand, if we were to slightly change our request to, "Give me all the restaurants in New York City," we would then encounter the opposite problem. To answer this traversal, we need to traverse the roughly 26,000 `within` edges associated with New York City to find all the restaurants. This will likely cause significant performance problems because 26,000 individual traversers are required. This traversal can inflict long wait times while our database churns through these requests.

As these examples demonstrate, a change in the question can flip the same vertex from performing normally in one scenario to acting as a supernode in another. This behavior is one of the exacerbating complexities of supernodes. Not only are these highly dependent on the situation, but the negative impact is conditioned by specifics of the vendor's implementation, the hardware configuration, and the indexes configured for the graph in some cases.

In addition to the direction we traverse our supernode, there are some scenarios, especially when running analytical algorithms, that require supernodes to get the correct answer. Some algorithms rely on the connectedness of a graph as all or part of the calculation; for example, when we're in a domain such as social networking, peer-to-peer file sharing, or network asset monitoring, and we want to answer, "Who is the most connected person in my graph?" This question requires an accurate count of the degree of all the vertices in our network. With this sort of calculation, having vertices with disproportionately high edge counts is what we're after. Here, a supernode in our graph is a meaningful construct.

However, because we're traversing through the entire graph (or a large portion of it), these calculations should be treated as analytical instead of transactional. Transactional operations typically have a time allotment measured in seconds or milliseconds, but analytical operations can have a time allocation of minutes, hours, or longer.

Say that we look at our data and decide that we have a supernode. Then, we assess our questions and conclude that we need to traverse through these supernodes in a manner that's likely to cause problems. What can we do to alleviate this?

MITIGATING SUPERNODES

The most common and universally applicable approach to handling supernodes is to refactor the model to remove or minimize the impact of the supernode. This means going back to our schema and investigating potential changes to our data model to remove the need to traverse through any supernodes. To do that, we'll need to employ one or more of the data modeling strategies we have learned thus far, including

- Duplicating vertex properties on edges
- Making vertices into properties or properties into vertices
- Moving property locations
- Precalculating data
- Adding indexes

The goal of this refactoring is to minimize the number of edges that our graphs traverse. Wait! Isn't the point of a graph database to traverse edges? Isn't it better at that than any other data engine? Yes, that's true. Graph databases are optimized for this access pattern of traversing edges. However, just because graph databases are better with this operation than other engines doesn't mean that we want graphs to perform more work than required.

In the last section, we identified that the traversal, "Give me all the restaurants in New York City" is a problematic traversal. Let's figure out how to change our data model to answer this question, without touching all 26,000 `within` edges associated with New York City.

> **EXERCISE** Use the traversing techniques you learned in chapters 3, 7, and 10 to see if you can change our DiningByFriends data model to reduce the number of edges that need to be traversed.

Examining our DiningByFriends data model, we realize that what we need to do is to get the city and state properties of an address co-located to the `restaurant` vertex. If we collocate this data, then there is no need to traverse to the `city` vertex to retrieve that information.

We know how to apply data denormalization techniques to create new properties on the `restaurant` vertex for the city and state `name` properties. Because we cannot have both the city `name` and state `name` attributes on the `restaurant` vertex with the same key, `name`, we can rename these as `city` and `state`. Figure 10.7 shows this depiction.

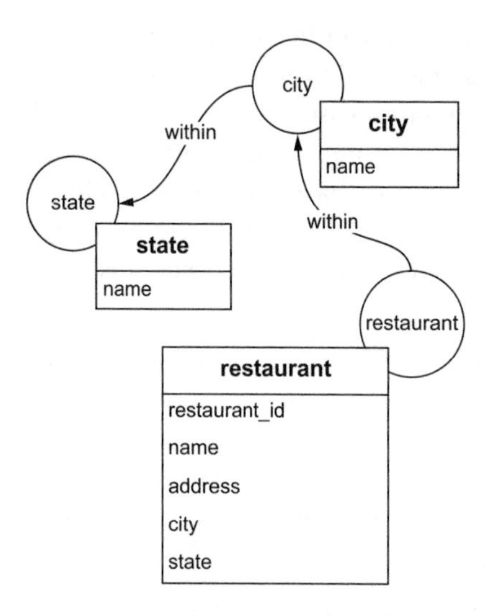

Figure 10.7 The portion of the DiningByFriends logical data model relating to restaurants, cities, and states with the `city_name` and `state_name` attributes denormalized to the `restaurant` vertex

Denormalizing the properties removes the need to traverse any edges to answer the request, "Find all the restaurants in New York City." However, it has introduced a new problem. Now we must scan every `restaurant` vertex in our system to handle this issue. To reduce the impact of scanning the entire set of `restaurant` vertices to retrieve this data, we add an index for these properties. Combining these two techniques, indexing and denormalizing, allows us to quickly and efficiently retrieve data for both requests: "Give me all the restaurants in New York City" and "Get me the city and state for restaurant *X*." We can do this because the data is now co-located on the `restaurant` vertex.

As one last method of cleanup, we can remove the `city` and `state` vertices from our data model because these are no longer being used. Figure 10.8 shows this depiction.

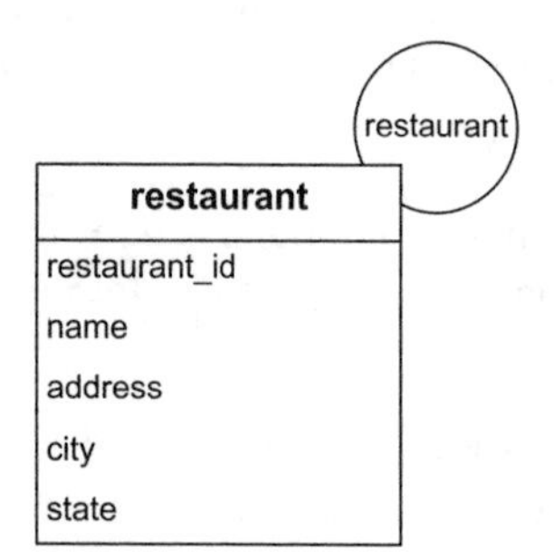

Figure 10.8 Our updated DiningByFriends logical data model with the `city` and `state` attributes stored on the `restaurant` vertex and with the `city` and `state` vertices removed

> **Vertex-centric and edge indexes**
>
> Because supernodes are a common problem within graph databases, some database vendors include specific index types to help address this issue. These index types are often referred to as edge indexes or vertex-centric indexes. We apply these types of indexes to specific Vertex-Edge-Vertex combinations; these help to index and sort the combinations to prevent the linear scan of edges, which should provide a faster graph traversal.
>
> Not all databases support these types of indexes, however, and the details on how each vendor implements these vary. If the database you select supports features like this, we highly recommend investigating vertex-centric and edge indexes as potential solutions to supernodes.

10.3 Application anti-patterns

While supernodes can become problematic as the data grows, there are other anti-patterns that you might encounter when creating graph-backed applications. In this section, we discuss

1. Using graphs for non-graph use cases
2. "Dirty" data
3. Lack of adequate testing

Each of these anti-patterns commonly appears in the design, architecture, and preparation for building an application.

10.3.1 Using graphs for non-graph use cases

"I want to use a graph database, so let's find a use case for one."

—Undisclosed graph database client

As we learned in chapter 1, while graph databases are good at several specific types of complex problems, they aren't a universal solution to all problems. It's crucial to remember both the benefits and the limitations of how graphs provide insight into our data and transform businesses in the process. A graph cannot answer a question we don't ask. Before we build a graph solution, we need a strong-enough understanding of the information to be able to model and traverse our graph.

It's also important not to overplay the flexibility of graphs. While graphs do have an amazing agility, that doesn't mean that any graph data model can answer any question or do so in an efficient manner. As we've already shown in this chapter, seemingly obvious graph implementations can hide problems, such as poor performance from a lack of indexes or unexpected highly connected parts of the data called supernodes. We think that the simplicity of graphs, their seemingly sensible way of expressing design, lulls many into believing that because graphs can do anything regarding data, they should do everything.

The answer to this problem, using graphs for non-graph use cases, is straightforward: don't let the excitement of getting to use a new technology overwhelm good software development fundamentals. Want to test if you completely understand a problem? Explain it to someone else. When we understand a problem well enough to explain it to someone, then we typically comprehend it well enough to begin working on it. If you aren't able to explain it, then spend some time refining your understanding of exactly what you want to accomplish.

10.3.2 Dirty data

The second anti-pattern involves adding *dirty* data to our graph. The dirty data we're referring to is data that contains errors, duplicate records, incomplete or outdated information, or missing data fields.

Dirty data, just like the other anti-patterns, isn't a problem unique to graph databases, but it does cause some unique issues. Although graph databases rely heavily on the connectedness between data, these also require an accurate representation of those entities to work effectively. The cleaner the data, the fewer the duplicates, the more accurately the connectedness will be represented within our graph. Consider an example of these three dirty data records in an RDBMS system representing people and their associated addresses, as this table shows.

ID	Name	Address
1	John Smith	123 Main St.
2	J Smith	123 Main Street
3	Bob Diaz	123 Main

From examining this data, we can infer that both John Smith and J Smith are likely the same person and that all three people likely live at the same address. However, if we add this dirty data to a graph, what we see is three pairs of vertices *not* connected to the same address as we would expect. Figure 10.9 shows this output.

Let's say we want to use this graph to try and determine if anyone in our graph lives at the same address. As each of the entities is represented as its own vertex, we won't find any matches. This is a problem because this sort of related link or connectedness is the basis for many graph use cases.

Thinking back to our social network, if we had multiple `person` vertices representing the same physical person, then even something as straightforward as finding a friends-of-friends link with dirty data would, at best, be hard to use. At worst, it would provide inaccurate results. What's the solution to dirty data?

The answer is simple: clean our data before we import it using a process known as *entity resolution*. Entity resolution is the process of de-duplicating, linking, or grouping records that are believed to represent the same entity together into a single canonical representation. This data cleaning process is crucial for most data sets and becomes a

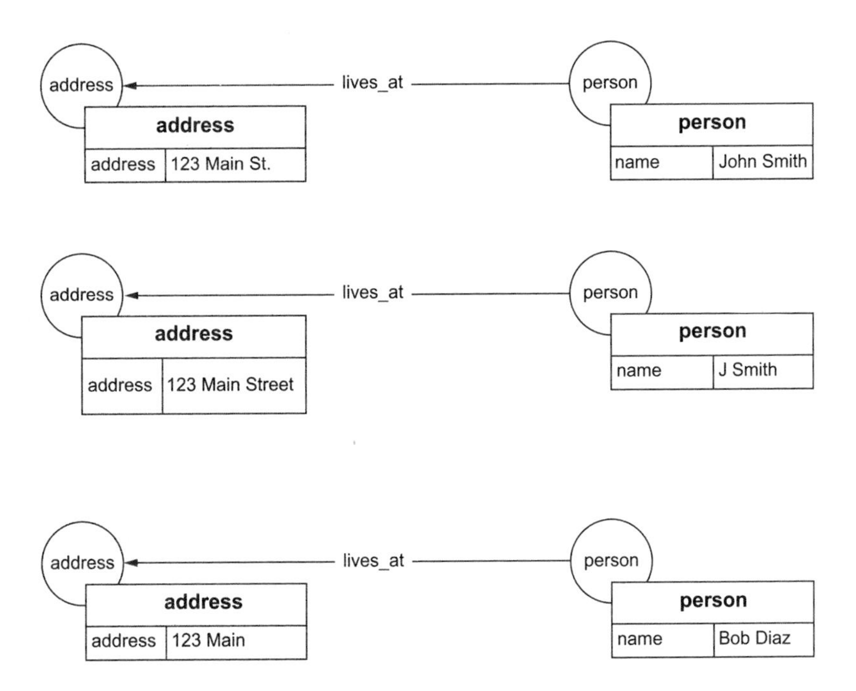

Figure 10.9 Our model showing three pairs of data not connected to the same address

greater challenge as the volume and velocity of data grows. Entity resolution is a complex process, but the crucial takeaway is that the data-cleaning process is an important part of any graph database application.

Returning to our address example, look at figure 10.10 to see what our graph looks like when we clean the name and address data to find matches before adding these to our graph. Using this graph, we can determine that Bob Diaz and John Smith share the same address, 123 Main St.

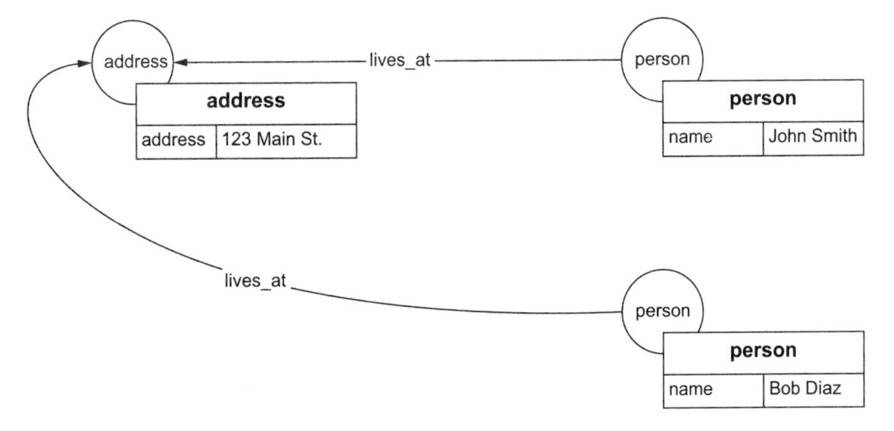

Figure 10.10 Address data graph containing our clean data connected to the same address vertex

10.3.3 *Lack of adequate testing*

The last application anti-pattern is a lack of adequate testing, which usually falls into one of two categories: non-representative test data or insufficient scale when testing.

NON-REPRESENTATIVE TEST DATA

Because of the connected nature of graphs and how we traverse these, ensuring that we test against representative samples of data is more important than with relational databases. A representative sample is especially vital when dealing with the branching factor (http://mng.bz/pzVP), the number of successive vertices in recursive queries. Unlike other database technologies, the performance of graph traversals is less dependent on the quantity of data in the graph. Instead, it's more dependent on the connectedness of the data in the graph. Testing against a representative set of data, including realistic edge cases and particularly potential supernodes, is crucial.

LACK OF SCALE

Testing at an adequate scale means using sufficiently deep and sufficiently connected data, not just sufficiently large data. Testing at scale is especially paramount with highly recursive applications or those using a distributed database. For example, if the application we build involves an unbounded recursive traversal that performs acceptably on our test graph with a depth (number of iterations) of 5, it might function horribly on a production graph with a depth of 10.

It's extremely complex to effectively simulate graph data. The scale-free nature of most graph domains represents a unique challenge. In our experience, the best approach to creating test data isn't to create it at all but, instead, to use real data whenever possible, sometimes in combination with data-masking approaches in order to protect personally identifying information in the data. But more often than not, this isn't possible. If you do have to simulate data, take extra care that the data's shape matches the expected production data.

10.4 *Traversal anti-patterns*

In this last section, we look at some of the architectural anti-patterns when building graph-backed applications. These patterns include

- Not using parameterized traversals
- Using unlabeled filtering steps

Both of these represent either a security or performance risk. Understanding how to identify and remedying these patterns is therefore essential for secure, performant, and maintainable production applications.

10.4.1 *Not using parameterized traversals*

According to the latest report (2017) on the top application security risks released by the Open Web Application Security Project (OWASP), the number one most common vulnerability in applications is injection-type attacks (http://mng.bz/gg).

Injection flaws, such as SQL, NoSQL, OS, and LDAP injection, occur when untrusted data is sent to an interpreter as part of a command or query. The attacker's hostile data can trick the interpreter into executing unintended commands or accessing data without proper authorization.

For those familiar with SQL, both the concept of injection attacks and the use of parameterization to defend against these should be well understood. What's new is that as hackers become increasingly sophisticated, they're applying the same techniques used on RDBMS databases to graph and other NoSQL databases. But what works in relational databases, parameterization, also applies to graph databases as well. Let's see what injection attacks are and then show how parameterization defends against these.

> **IMPORTANT** We've used best practices throughout this book and parameterized all the traversals we built. Gremlin Language Variant (GLV) traversals are automatically parameterized. For the subgraph traversals where we used the string API, we built those to be parameterized.

What is an injection attack?

Hackers use injection attacks when they find openings in an application that allow them to add their own malicious code to the application. This malicious code then can delete records, add users, or view unauthorized data. Missing validation logic on user input and then using that input directly in a database query is what allows this type of access. For example, let's take an application that has a REST endpoint that generates a query (shown in SQL and Gremlin) to find all of a user's friends where the `userid` is passed in from a URL.

URL	http://diningbyfriends.com/friends?userid=?
SQL query	`"SELECT * FROM friends WHERE userid = " + userid`
Gremlin	`"g.V().has('friend', 'user', " + userid + ")"`

In a normal scenario where no one is attacking the system, the URL and resulting query would look something like this.

URL	http://diningbyfriends.com/friends?userid=1
SQL query	`"SELECT * FROM friends WHERE userid = 1"`
Gremlin	`"g.V().has('friend', 'user', 1)"`

In the SQL query, values passed from the user are directly concatenated into the query. While this concatenation seems like a convenient and simple way to build an application, it poses one of the most severe types of security vulnerabilities. If a clever hacker wants to see information about other user's friends, one can, using a few simple tricks, inject SQL into the query to exploit this vulnerability.

SQL URL	http://diningbyfriends.com/friends?userid=1+OR+userid%3D2+OR+userid%3D3
SQL query	`SELECT * FROM friends WHERE userid=1 OR userid=2 OR userid=3`

To exploit this weakness via Gremlin, a hacker would need to know Gremlin and change their input. However, it's still possible as depicted here.

Gremlin URL	http://diningbyfriends.com/friends?userid=within(1,2,3)
Gremlin	`g.V().has('friend', 'user', within(1,2,3))`

With a simple change of the URL, a hacker has exposed others' private information. Using tricks similar to this URL hack, a hacker can do more damaging acts, such as deleting records, deleting the database, or hijacking our system. While we demonstrated this type of hack with Gremlin, this same technique is applicable to other graph query languages as well. In fact, any application that accepts input unfiltered or unchecked, regardless of language, is susceptible to such an attack.

PREVENTING INJECTION ATTACKS

Each graph query language and their drivers offer some type of parameterized traversal functionality. A *parameterized traversal* uses tokens to represent the input values in the query. At execution time, these tokens are replaced with values passed in from the application after being sanitized and validated. Using JDBC in Java, we accomplish parameterization using a `PreparedStatement`:

```
PreparedStatement stmt = connection.prepareStatement("SELECT * FROM person
     WHERE first_name = ?");
stmt.setString(1, "Ted");
```

If we use the Gremlin Language Variants (GLVs) that we've employed throughout this book, then we're already using parameterized queries. However, many other databases including some TinkerPop-based databases only support string-based traversals. This exposes possible string concatenation errors as shown here:

```
public static void insecureGraphTraversal(
    ➥ Cluster cluster, String userid) {
    Client client = cluster.connect();            <- Creates a client connection
    String traversal = "g.V().has(\"friend\",
    ➥ \"id\", " + userid + ")";                   <- Creates our string with concatenated parameters, which is not good
    client.submit(traversal);                      <- Submits our concatenated string
}
```

This method runs the risk of malicious code being injected into our code because we are just concatenating user inputted values to our string without validation. To protect against malicious code, we use parameters for the data being passed from the user. First, we construct our string with a token, which is just a name in the string that represents the user's input values. We then pass a map of tokens and their corresponding values:

```
public static void secureGraphTraversal(
    ➥ Cluster cluster, String userid) {
    Client client = cluster.connect();                    // Instantiates a client connection
    String traversal = "g.V().has(\"friend\",
    ➥ \"id\", userid)";                                  // Formulates our tokenized traversal (the right way to do it)
    Map<String, Integer> map =
    ➥ Collections.singletonMap("userid", 1);             // Creates a map of parameters
    client.submit(traversal , map);                       // Submits the traversal and the map of parameters
}
```

When the traversal is executed by the `client.submit()` method, the server replaces the token values with the values stored by the map. When it does this, it performs it in a way that doesn't allow malicious code to execute, just as with a `PreparedStatement` in JDBC. While this example was written using Gremlin, the same basic concept applies to other graph query languages such as Cypher.

There's an additional benefit to using parameterized queries. Most data engines, graph or otherwise, cache the execution plan. This saves us the cost of generating a new execution plan after the first time that the traversal is called with parameters. For frequently used traversals, caching execution plans can measurably improve server performance.

10.4.2 Using unlabeled filtering steps

The anti-pattern in the previous section, injection attack, is one of the most severe from a security and data integrity perspective, even if it isn't that common. In contrast, the anti-pattern for this section, using unlabeled filtering steps at the start of a traversal, doesn't pose a security risk but tends to have a massive impact on traversal performance. Let's first investigate what an unlabeled filtering step looks like in Gremlin:

```
g.V().has('first_name', 'Hank').next()
```

This seems innocent enough, but why is it an anti-pattern? If you look closely, you can see that we didn't specify the label or labels of the vertex to search. We don't give the database any hints or help on where to look to find vertices with a `first_name` of Hank. To satisfy this traversal, our database must

1. Find all vertex labels in the graph.
2. For each vertex label, find all vertices.
3. Determine if the vertex has a property called `first_name`.
4. If so, determine if the value of that `first_name` property is equal to Hank.
5. If so, return the vertex.

As we didn't provide a label to filter for in the first step, all the remaining steps are required for every vertex label in our graph, causing a huge performance impact. When transitioning to graph databases from a relational world, not filtering a traversal at the start is a common mistake, because in the RDBMS world, queries such as this aren't even possible. The key difference is that in a graph database, the starting point is all vertices (`g.V()`), while in an SQL query, the starting point is a specific table (`FROM table`). The table specification provides a natural boundary for an SQL query.

There are two different ways we can enforce filtering our traversal: either add a global index on the `first_name` property, or specify the vertex label at the start of the traversal. Let's look at both approaches.

The first approach, adding a global index on the `first_name` property, has two problems. The first is that it might not be easy to create the correct index or indexes. While TinkerGraph does allow us to add a global index on a property called `first_name`, global indexes themselves aren't common among graph database vendors. Most databases that support indexes require that the index be a combination of both *label* and *property*. Even with the proper global indexes created, we still have to search all vertex labels. Even though we can search the vertex labels faster, we still have to manipulate all of those, which is the second problem with this approach.

The second, and preferred, method to enforce filtering is to add the appropriate labels to our filtering queries and to also add indexes for the vertex and property combination. After adding the proper filtering to our traversal, it now looks like this:

```
g.V().has('person', 'first_name', 'Hank').next()
```

This pattern should be familiar because we've done this throughout the book. Although this approach makes sense, the anti-pattern of unlabeled filtering steps is probably one of the more typical ones we run across. Even in the schemaless world of graph databases, there's an implicit schema being applied to a graph by the domain; in other words, when we search for a `first_name` attribute, we know which label or labels contain the property we're targeting.

While this anti-pattern has the greatest impact at the start of a transactional traversal, it is also helpful to think about providing labels whenever possible to commonly used filtering steps, both later in a transactional traversal as well as in analytical traversals.

NOTE As a general rule, the earlier and more precise the filtering criteria we provide to a traversal, the faster that traversal runs.

Summary

- Performance issues with graph traversals can be diagnosed using one of two common methods:
  - `explain()`—Shows how a graph traversal is executed but does not run it.
  - `profile()`—Runs the traversal and collects statistics about what actually occurred. (This step is a lot more helpful.)

- Graph databases use indexes to speed up traversals by allowing quick and direct access to data, similar to relational databases. However, there are multiple types of indexes offered. The index type available is dependent on the vendor's specific database implementation.
- Supernodes are vertices in a graph that have a disproportionately high number of edges. These can cause traversal performance problems, especially when running transactional traversals.
- Supernodes can be found by monitoring the branching factor and the number of successor vertices, and by looking for outliers.
- Supernodes can be mitigated by using data modeling tips and tricks to split up the edges across multiple, different vertices or by using features such as vertex-centric indexes available in some graph databases. These indexes are designed to help alleviate the side effects of supernodes.
- Although supernodes aren't desirable when running transactional queries, these can be critically important when running analytical algorithms, such as degree centrality.
- When writing graph-backed applications, understanding the problem we're trying to solve is critical to its success. This ensures that we use the right tools and use those in the right way.
- The use of "dirty" data is a common anti-pattern that is particularly problematic in graph applications due to the highly connected nature of the data and questions. Dirty data can be addressed by properly de-duplicating and linking data to facilitate better application performance.
- You should avoid testing with unrepresentative data because the connectedness of the data dramatically impacts the performance of graph-backed applications.
- When submitting strings of Gremlin, the only way supported by some graph database vendors, you should always parameterize graph traversals to prevent injection attacks from running malicious code.
- When running transactional queries, always start a traversal with a filter traversal that specifies the vertex label (or labels) as well as the properties. Specifying a filter at the start of your traversal prevents a traversal from searching all of the vertices.

11 What's next: Graph analytics, machine learning, and resources

This chapter covers

- Graph analytics algorithms for pathfinding, centrality, and community detection
- Graphs in machine learning (ML)
- Helpful resources for graph theory, graph databases, and graph algorithms

Great! You've made it to the final chapter. It's been a journey as we've switched from thinking about problems from a relational, entity-first mindset to a graph, entity-plus-relationships mindset. Even though this is the end of the book, the next phase of your journey with graphs is just beginning. So what's next? Where do you go from here? This chapter answers these questions by providing an overview of common paths many people pursue in extending their knowledge of graphs.

Graph analytics and machine learning (ML) are two of the most common areas where exploration of graphs might take you next. This chapter introduces these two concepts and provides you with just enough information to decide if you want to explore these areas further.

We'll start with a high-level look at graph analytics and some of the unique insights that these algorithms can derive from data. We'll provide a broad overview of the graph analytics space so that you will have some understanding of what is available when you start analyzing your graph data. After we explore the world of graph analytics, we'll introduce the role of graphs in ML. This topic is a nice segue from graph analytics because graph data, graph analytics, and ML are quite complementary. At the end of this chapter, we'll close with a set of additional references and reading materials to continue your study and work with graphs.

11.1 Graph analytics

Up to this point in this book, we have worked on questions in our application that are transactional in nature; questions like, "Who are the friends of my friends?" or "What are the newest reviews for this restaurant?" These are transactional because these questions only require us to look at a small subset of our graph data. To answer questions such as, "Who is the most connected person in my graph?" or "Which person is the most centrally located?" we need to investigate most or all of the data in the graph.

Algorithms that require using most or all of the data in our graph fall into a category of problems that use algorithms known as *graph analytics.* These algorithms are useful across many domains for problems such as fraud detection, supply chain optimization, and epidemic migration prediction.

Graph analytics and graph databases

When looking at graph analytics, there are many frameworks and databases built specifically to perform these computationally intensive calculations. These sorts of specialized libraries tend to have optimizations specifically tailored to performing the long running computations that most of the algorithms require. Many transactional graph databases, such as those we have mentioned so far, have the ability to run these sorts of calculations. But if you are looking to perform these sorts of algorithms at scale, we recommend that you look at one of the analytical graph databases (such as AnzoGraph) or frameworks (such as Apache Giraph).

In this section, we'll briefly cover some of the more common algorithms and provide an overview of what types of problems each solves or the information it returns. We'll provide you with enough information to understand the types of questions each category of algorithms solves. This understanding should help you to narrow your focus as you dig deeper into these rich capabilities.

11.1.1 Pathfinding

We introduced the fundamentals of pathfinding algorithms back in chapter 4, where we used these algorithms to find friends in our social network. Although that is one way of using pathfinding in a transactional process, it can also be used analytically to explore the routes between vertices and to identify optimal paths in a graph. Each specific pathfinding algorithm works a little differently, and each has its advantages and disadvantages. In addition to the pathfinding we did in our social network, there are many other real-world use cases for pathfinding algorithms such as

- *Direction finding*—Geographic mapping tools use some variation of a pathfinding algorithm to provide directions.
- *Optimization problems*—Pathfinding algorithms can optimize various problems that deal with a large number of interdependent entities, from managing supply chains to optimizing financial trades to determining bottlenecks and points of failure in computer networks.
- *Fraud detection*—Many fraud algorithms use cycle detection, finding groups of entities that connect to themselves, to look for closely connected subgraphs as a measure of potentially fraudulent accounts.

The most common pathfinding algorithms in use are the *shortest path algorithms*, which calculate the shortest path between two vertices. There are two fundamental approaches to calculating the shortest path. The *unweighted* approach treats all paths as equal, calculating the shortest path based on the number of edges traversed. The *weighted* approach assigns relative weights to all paths, and these weights are then used in the computation. Let's see how having weighted or unweighted edges affects the results.

UNWEIGHTED SHORTEST PATH ALGORITHM

Let's say that we have a graph with three vertices, representing towns A, B, and C, which are connected by three edges that represent the roads 1, 2, and 3 as shown in figure 11.1. Now, what if we want to determine the shortest route from town A to town C? Treating all paths as equal, we see that the shortest path between town A and town C is to follow road 2, as it only requires traversing a single edge.

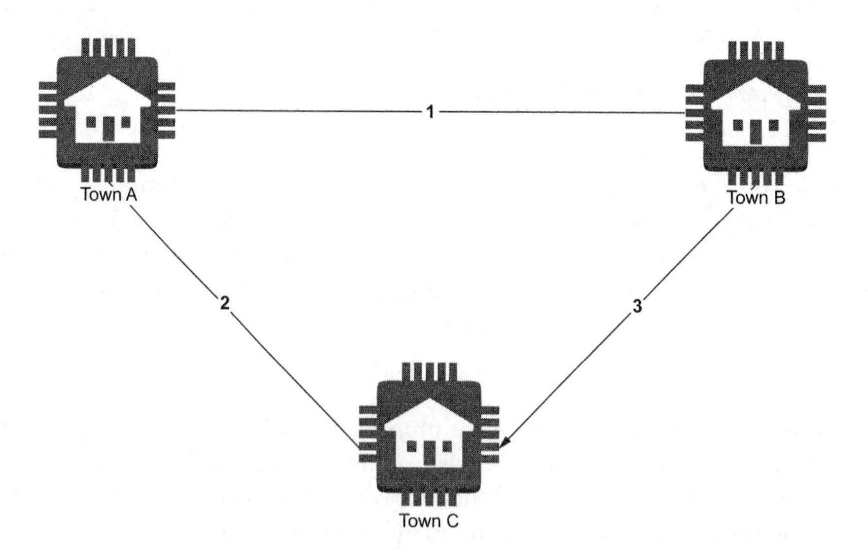

Figure 11.1 The shortest path from town A to town C is A → 2 → C.

The unweighted shortest path algorithm is a great choice when the relative cost of traversing all edges is the same or is not a concern. Social networks are a great example of when unweighted shortest path algorithms are useful. In a social network, each friend connection is equal to every other friend connection, so the relative cost of traversing these edges is equal. In fact, an unweighted shortest path is what we built in section 4.2 to find the path between Ted and Denise. However, in many scenarios, we cannot treat all the connections as equal. In these scenarios, we need to look towards a weighted shortest path algorithm.

WEIGHTED SHORTEST PATH ALGORITHM

In many scenarios, the relative cost, or weight, of moving from one vertex to another differs. To calculate the shortest path in these scenarios, we want to determine the path with the lowest overall relative cost, so we use a weighted shortest path algorithm.

Let's say that road 1 is a highway, while road 2 is a windy mountain road. Because these roads are not traveled at equal speeds, we first need to assign a relative weight to each edge. In the case of our graph, these relative weights can be compiled from multiple factors such as distance, speed, and road condition to account for the comparative travel time difference between roads 1 and 2. Figure 11.2 shows this traversal.

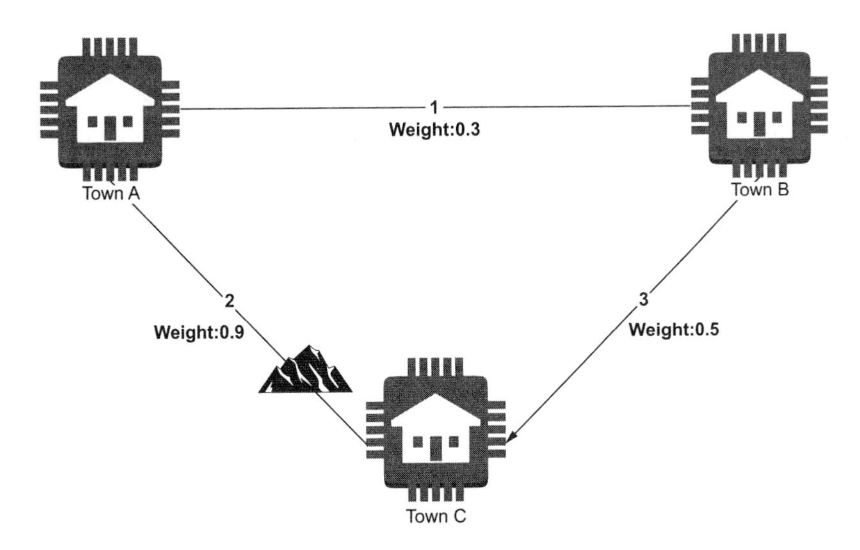

Figure 11.2 In our weighted graph, the shortest path from town A to town C is A → 1 → B → 3 → C.

To calculate the shortest path between town A and C, we no longer find only the shortest number of hops to go between the two paths; instead, we add together the relative cost of moving across the edges to find the lowest total weight. In our example, traversing edge `1` is one-third as expensive as traversing edge `2`, represented by the

`Weight` property on the edge. When we take this into account, our new, shortest path is A → 1 → B → 3 → C.

Weighted shortest path algorithms are a great choice when the relative cost of traversing edges is not equal. This is common with problems such as supply chain optimizations, where the relative cost (distance/time) of moving goods are not equal, or with network routing problems, where the time required to transfer network packets differs between connections due to hardware or other aspects such as geographic proximity.

Both unweighted shortest pair and weighted shortest pair algorithms have multiple implementations, two of the most common being Djkstra's algorithm (http://mng.bz/K5Mn) and the A* search algorithm (https://www.geeksforgeeks.org/a-search-algorithm/), both of which can be used on weighted and unweighted graphs.

11.1.2 Centrality

We use centrality algorithms to identify the importance of a vertex within a graph. *Centrality algorithms* answer questions far beyond the social networking examples for which many were invented. Some uses for centrality algorithms include

- Finding the most critical components in a computer network that can cause the most disruption if lost
- Finding the importance of a person within an organization
- Estimating the optimal timing and routing for telecommunications packets
- Finding outliers in a graph as a measurement of likely fraud

When discussing centrality, we often use the word *importance* to describe the role that a particular vertex plays in the overall structure of the graph. The specific meaning of the importance of a vertex varies, based on what is being calculated by a particular algorithm. This means that in order to interpret the results of a particular algorithm, we need to understand what type of importance is calculated by that algorithm. Let's look at an example to get a better understanding of how importance varies by algorithm.

Think about the social network we built earlier in this book. Within that social network, we could define importance by who has the most friends, or who is the most in the middle of the graph of friends, or who has the most influence over others in the network. Each of these is a perfectly valid way to define importance, but calculating each will likely yield a different result. This context-dependent definition of importance is why there are many different centrality algorithms, each of which calculates the importance of a vertex slightly differently.

Let's take a look at five common centrality algorithms and see how each measures centrality in a distinctly different way. We'll also look at the outcomes of applying these algorithms on our DiningByFriends social network.

DEGREE

Degree centrality is the simplest to understand. *Degree* is the number of incident edges associated with a vertex, so degree centrality ranks vertices based on their edge count. Degree centrality can be further broken down by measuring the *in*-degree and the *out*-

degree separately. Degree centrality is often used as a baseline for determining how connected a graph is, especially when calculating the mean, minimum, and maximum values. In the context of our social network, degree centrality shows us who has the most friends.

Betweenness

Betweenness centrality is a calculation of the number of times a vertex is used in the shortest path between all pairs of nodes in the graph. Betweenness centrality is effective at finding the critical points that connect different groups of vertices. When using this algorithm, the larger the number returned, the more important the vertex. If we run betweenness centrality on our social network, we find out who connects most to different social groups.

Closeness

Closeness centrality is a measurement of the average length of the shortest path from a vertex to all other vertices, indicating which vertices are the most centrally located with respect to all other vertices. When running closeness centrality, the smaller the return value, the more important the vertex. Running closeness centrality on our social network identifies which people are at the "heart" of the social network.

Eigenvector

Eigenvector centrality is a complex measurement of a centrality that uses the relative importance of the adjacent vertices as an input to calculate the importance of a given vertex. Just because a vertex is connected to many other vertices does not necessarily mean it is important. Instead, the importance of the vertex's neighbors is used to compute a vertex's overall significance.

If a vertex has many adjacent vertices, but those vertices are relatively unconnected, then it receives a lower score than a vertex with fewer adjacent vertices, each of which is highly connected. Running eigenvector centrality on our social network graph finds the most influential people in our social network, the ones who not only have the most connections, but who's connections are also well-connected.

PageRank

PageRank is an algorithm made famous by Larry Page and Sergey Brin of Google for its use in weighting search results. PageRank works similarly to eigenvector centrality as it uses the relative importance of the adjacent vertices to aid in determining the overall importance of a vertex. But it also includes a *dampening value* (commonly set to 0.85) to indicate a diminishing of influence as the network is traversed. The higher the PageRank return value for a vertex, the more important the vertex. As with eigenvector centrality, if we run PageRank on our social network, our results will represent the most influential people within our social network.

Centrality comparisons

Each of these centrality algorithms measures a different aspect of importance in a graph, and each of these gives us different information about our data. Let's demonstrate how some of these centrality algorithms differ by running each on our social network as the following table shows.

NOTE The code used to run each of these centrality algorithms is available in the source code repository at chapter11/centrality_algorithms.groovy. Some of the algorithms use advanced steps, which we have not covered or have mentioned only in passing.

First Name	Degree Centrality	Betweenness Centrality	Closeness Centrality	Eigenvector Centrality	Page Rank
Dave	**4**	**48**	3.33	1	0.0174
Josh	3	30	2.91	3	0.0191
Ted	1	16	2.26	1	0.0174
Hank	2	16	2.75	5	0.0197
Kelly	2	24	2.91	2	0.0183
Denise	3	26	3.56	**8**	**0.0206**
Jim	3	32	3.08	2	0.0183
Paras	2	14	**2.36**	3	0.0185

Examining these centrality measures, we see a wide variance in who is the most important. In the table, we have highlighted (in bold) the top, or most important, result for each algorithm. Given the same underlying graph, our algorithms produce different results for determining the most important vertex. Using degree and betweenness centrality, Dave is the most important. For closeness centrality, Paras is the most important. For eigenvector and PageRank, Denise is the most important. From the measurements we can conclude

- Dave has the most friends and connects with the most social groups.
- Paras is the "heart" of the social network.
- Denise is the most influential.

As we mentioned at the beginning of this section, each of these algorithms measures a different aspect of the importance (centrality) of a vertex in a graph. Understanding the meaning of each measure is vitally important in choosing the correct algorithm for your use cases.

11.1.3 Community detection

We use community detection algorithms to uncover groups or communities of vertices that are tightly connected to one another but loosely connected to other vertices within the graph. Think about friends in a social network. Does everyone know everyone else? Are there small groups of people who are close friends but with only a few friendships to other groups in the network? These are exactly the sort of groupings that community detection algorithms identify.

Community detection algorithms aren't just limited to social networks. These are used across a large number of industries and use cases such as

- Finding communities of potentially similar accounts within an e-commerce site to find distinct families within the graph
- Identifying potential fraud by looking for tightly connected components such as groups of accounts known to commit deceptive activities
- Identifying similar groups of users to provide product recommendations

As with clustering algorithms, there are a large number of potential community detection algorithms, each finding communities in a slightly different way. Let's take a look at two of the most commonly used community detection algorithms and see how these function.

TRIANGLE COUNTING

Let's say we wanted to find close-knit communities within a social network. One way to find these communities is to find the groups of people who all know each other. Let's further say that we have a social network like that shown in figure 11.3, where Dave knows Hank, Hank knows Josh, and Josh knows Dave.

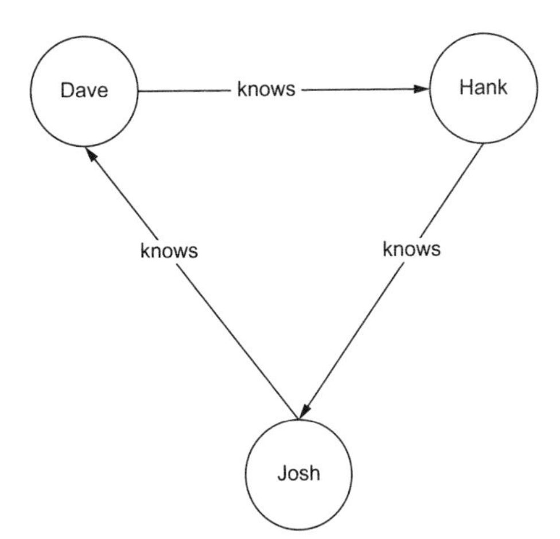

Figure 11.3 A social network where Dave, Josh, and Hank form a triangle

When we look at this figure, we see that these three people are tightly connected to one another and that this grouping forms a triangle. Counting the number of these triangles across a graph is known as *triangle counting*. Triangle counting does what the name implies: it counts the number of triangles within a given subset of nodes. If we look at figure 11.4, we see that in this graph, there are two triangles (A–B–D and E–C–F) that are highlighted.

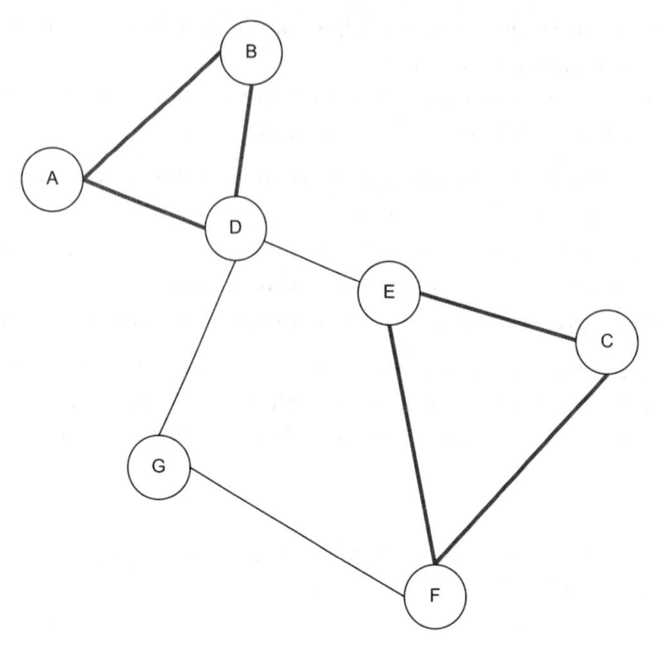

Figure 11.4 A graph showing two highlighted triangles (A–B–D and C–E–F)

Triangle counting is useful in capturing how cohesive or how closely related the network of vertices are in a graph. Graphs that contain closely associated networks or communities have a higher triangle count, and graphs with loosely connected networks have a lower triangle count.

CONNECTED COMPONENTS

Instead of triangle counting, what if we want to find groups of people who are well-connected to one another, but who are not connected to other groups? To find these communities, we use an algorithm known as *connected components.*

In graph theory, any subgraph in which every vertex has a path to every other vertex in the subgraph is known as a *component.* The connected components algorithm finds all these components within a graph. Looking at figure 11.5, we can see that our graph has two connected components within it, highlighted by the dashed lines surrounding each component.

Connected components discover clusters of related data within a global graph, which can be helpful in finding items such as families within a social graph or groups of associated or possibly duplicate accounts within an e-commerce site. The algorithm shown in figure 11.5 does not consider the direction of the edges between the vertices, so it is known as a *weakly* connected components algorithm.

However, let's say we want to find groups within our social network where our relationships between people are only one-directional, such as with Twitter. The fact that Dave follows Josh does not mean that Josh follows Dave. To find the communities within this graph, we need to take the direction of the edge into account. To accomplish

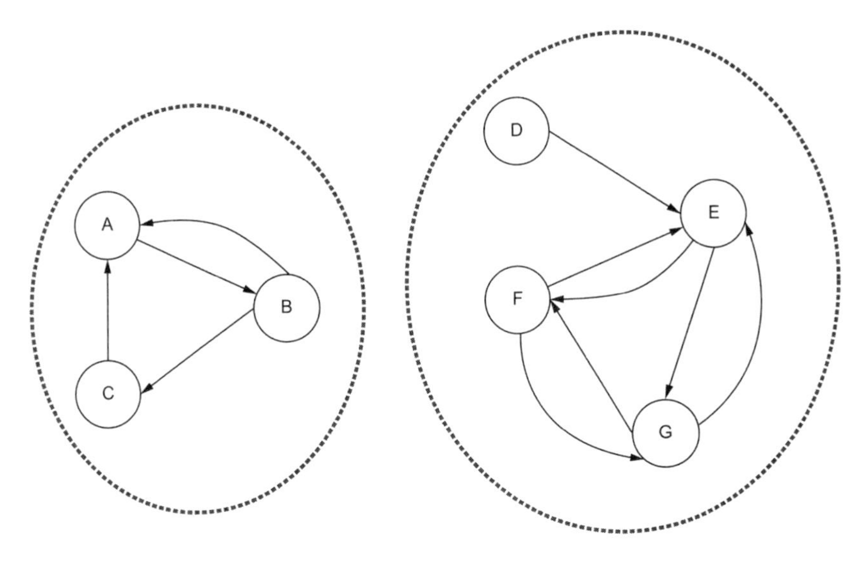

Figure 11.5 A graph containing two highlighted components identified by a connected component algorithm

this, we use a variation of the connected components algorithm known as *strongly* connected components.

Strongly connected components are essentially the same as weakly connected components, except that the former considers the direction of an edge. In a strongly connected component, a pair of edges exists between any two vertices in the subgraph with one edge in each direction. If we look at the same graph that we used for connected components, we can see that although the graph also contains two strongly connected components, the vertices included in those components are different, as figure 11.6 highlights.

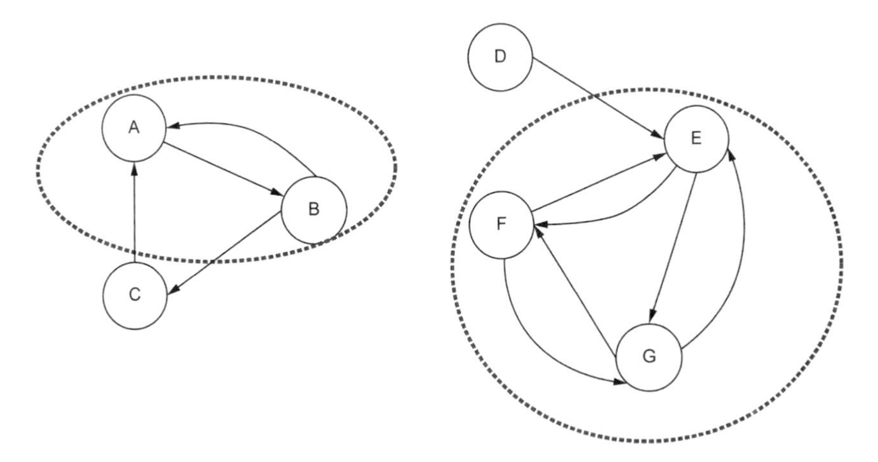

Figure 11.6 Our connected component graph highlighting the two strongly connected components, which is slightly different than with the connected component algorithm

We use the strongly connected components algorithm for detecting highly connected communities within a graph where directionality matters. Strongly connected components are frequently used in domains to find hubs of fraudulent activity or in product recommendations to find groups of similar users.

11.1.4 Graphs and machine learning

Whereas applying graphs in machine learning (ML) is not new, it is only in the past few years that this has taken off within the software industry. One irony of this is that although many ML technologies rely heavily on graphs to accomplish their learning, these neither allow for graphs as input nor provide them as output. Though this is starting to change with current research, most standard ML algorithms take as inputs fixed vectors or matrices of data. Because this is the case, how does the flexible data structure of a graph get applied to the rigid data structure of an ML model? In this section, we cover two approaches: feature extractions and graph embeddings.

FEATURE EXTRACTIONS

Often the simplest method for using graphs in ML is to extract features of a graph that provide insight into the data in the graph. Although we can use any number of graph features, here is how some of the graph analytics algorithms we learned earlier in this chapter help generate input features:

- *Shortest path*—Takes the shortest path between a person and a known bad actor as a predictive measurement for a fraudulent ML model
- *Triangle count*—Uses a triangle count in a social network to determine how social or antisocial a particular user might be
- *Degree*—Uses the degree of connections of a vertex to determine how critical a sensor is within a sensor network

These sorts of graph features are frequently beneficial when combined in ways that complement each other and provide a more holistic view of the topology and connectedness of the graph. When we combine many of these features together to create a vector or a set of vectors for ML, we now have what is known as a *graph embedding*.

GRAPH EMBEDDINGS

Graph embeddings turn sparse data into much more compact vector representations. Although much of the research in this area is driven by work done in natural language processing (NLP), it is now being applied more generically to graphs to provide inputs into tasks such as predicting new friendships and finding fraudulent activity. Graph embeddings tend to come in one of two forms:

- *Vertex embedding*—Represents each vertex as a single vector/matrix. Used to compare items on a vertex level. In the case of DiningByFriends, we might use this to compare relative social networks (a person's community of friends that we found in chapter 3).
- *Graph embedding*—Represents the entire graph/subgraph as a single vector/matrix. Used to compare entire graphs to one another. In the case of DiningByFriends,

we might want to use graph embedding to represent each personalization subgraph we found in chapter 9.

Why would we want to take the rich topology of a graph and create a vector out of it? Vector operations are simpler and faster to perform than comparable operations on a graph. Also, many of the algorithms and tooling available today are optimized for vector operations. Few are built with graph as input data in mind. So, what sort of features might we want to include in an embedding?

Well, this is the million-dollar question, isn't it? The challenge here is to ensure that whatever features we include adequately represent the topology, connectedness, and other graph attributes while minimizing the size of the vector. Larger embeddings require more time to process and more space to store, but also preserve a high level of fidelity with regards to the original graph data. Choosing the correct features to use for a given graph in a particular domain for a specific data set can be complex. Feature engineering is an entire discipline in itself, and its applicability to graph data is beyond the scope of what we are able to cover in this book. If you are interested in looking into this in more depth, we recommend that you check out some of the additional resources in the next section.

11.1.5 Additional resources

Throughout our writing and revising process for this book, we used a large number of references and resources to aid us in compiling the information condensed here. For those of you who desire to study these topics further, we include a list of our favorite, most helpful resources. We have grouped these into four areas: graph theory, graph databases, graph data sets, and graph algorithms.

GRAPH THEORY

We start with the underlying mathematical fundamentals of graph theory to provide a strong understanding of how graphs work:

- Sarada Herke, "Graph Theory Channel," http://mng.bz/9AM8—This YouTube series provides informative and entertaining videos that focus on graph theory and discrete mathematics. We found these videos to be an excellent teaching tool to quickly go from no understanding to a solid grasp of the basic concepts of graph theory.
- Richard J. Trudeau, *Introduction to Graph Theory* (Dover Books, 1975), http://mng.bz/jV49—This book provides a good foundation for understanding the mathematics of graph theory. It is written in a way that provides those of us without a mathematics background a solid conceptual understanding of the mathematics behind graph theory as long as you are willing to do a bit of work.
- Douglas B. West, *Introduction to Graph Theory* (University of Illinois, Urbana-Champaign, 2001), http://mng.bz/WqdX—This graph theory book is for those who really want to go deep into the mathematics behind graph theory. It assumes that the reader is familiar with math terms and symbology. If you do

not come from a mathematics background, you might want to familiarize yourself with math concepts in order to get the most out of this text.

GRAPH DATABASES

For those interested in taking a deeper dive into the specific tooling and database options available, we recommend the following books:

- Ian Robinson, et al., *Graph Databases,* 2nd ed. (O'Reilly Media, Inc., 2015), http://mng.bz/QxVv—This is pretty much the go-to book for building on Neo4j. The Neo4j graph database platform powers mission-critical enterprise applications such as artificial intelligence, fraud detection, and recommendations. Two of the authors are CEO and Chief Scientist at Neo4j. We highly recommend that this text be part of your library if you plan on using Neo4j.
- Denise Gosnell and Matthias Broecheler, *The Practitioner's Guide to Graph Data* (O'Reilly Media, Inc., 2020), http://mng.bz/lX1z—This book focuses on how one thinks about graph data and graph data problems, as well as some of the considerations when building large-scale graph applications. It also uses Gremlin and the Apache TinkerPop framework, but is more focused on the conceptual than language syntax.
- Kelvin R. Lawrence, *Practical Gremlin: An Apache TinkerPop Tutorial* (June, 2020), http://mng.bz/8GNg—This free online resource is the go-to place for additional help with the Gremlin language. Because it is an online book, updates are made regularly as features and syntax in the Gremlin language change.
- Corey L. Lanum, *Visualizing Graph Data* (Manning, 2016), http://mng.bz/EE2r—For anyone interested in how to approach the visualization of highly connected data, this book offers many examples and case studies. It also provides a nice, real-world perspective on how to think about visualizing graph data.

GRAPH DATA SETS

For those trying out graphs and graph databases, one of the most common frustrations is finding good data sets to work with. In this section, we provide a list of locations to find data sets ready for graph analysis:

- "Stanford Network Analysis Project (SNAP)," http://snap.stanford.edu/—This site is full of many great data sets for general graph analysis, including several extremely large data sets.
- "Kaggle," https://www.kaggle.com/—The Kaggle community provides excellent data set aggregators for all sorts of data science work. Many of the data sets are also suitable for those interested in investigating specific graph problems such as fraud or supply chain optimization.
- "Google Datasets," https://datasetsearch.research.google.com/—This is a search engine for publicly available data sets, specifically those associated with government and research projects.

- LDBC (Linked Data Bench Council), "The Social Network Benchmark (SNB)," http://ldbcouncil.org/developer/snb—A European-based, non-profit organization, LDBC provides both tools for generating social networking data sets and sample data sets of various sizes for benchmarking engine and application performance.

GRAPH ALGORITHMS

For those looking to dive deeper on graph algorithms, we have provided you with a list of useful resources on the subject. These resources focus on the algorithms and the analytics that you can use with highly connected data:

- Tushar Roy, "Coding Made Simple, Graph Algorithms Playlist," http://mng.bz/NnYX—This YouTube series in an easy-to-consume format provides a detailed overview of how the most common graph algorithms (Dijkstra's algorithm, strongly connected components, and others) work. We found this series helpful in understanding the implementation of each of the algorithms discussed.
- Algorithms Course, "Graph Theory Tutorial from a Google Engineer," http://mng.bz/X0Va—This YouTube video provides almost seven hours of detailed instruction on common graph algorithms. Each of the examples uses Java implementations to aid understanding.
- Alessandro Negro, *Graph-Powered Machine Learning* (Manning, 2018), http://mng.bz/DzR0—This is one of the few books that focuses solely on the use of graphs and machine learning. At the time of writing, this book is still in Manning's MEAP program, but we used a preview text as a reference throughout the development of this book.
- Mark Needham and Amy E. Hodler, *Graph Algorithms: Practical Examples in Apache Spark and Neo4j* (O'Reilly Media, Inc., 2019), http://mng.bz/Mo1B—This is a well-written book, providing a great overview of graph algorithms and, especially, how to use them in Python with Neo4j and Apache Spark. This is another book we highly recommend for everyone's bookshelf.

11.2 Final thoughts

Congratulations! You have reached the end of your journey into the world of graph databases. We hope that the skills you have learned inspire you to continue working with these databases. We have sought to provide you with a solid grounding and conceptual model so that you are successful as you move forward with your own highly connected data projects.

Summary

- We use pathfinding algorithms, such as unweighted or weighted shortest path, to describe the connectedness within a graph.
- We use centrality algorithms such as degree, betweenness, closeness, eigenvector, and PageRank to describe how important or influential a vertex is within a graph.
- The output of centrality algorithms can vary significantly, so understanding how each one works is important to selecting the appropriate one for your use case.
- We use community detection algorithms such as triangle counting, connected components, and strongly connected components to detect unique clusters (or communities) of highly connected vertices within a graph.
- Graph features such as shortest path, PageRank, and triangle count can be extracted from a graph to use as input into a feature set in machine learning (ML).
- Graph embeddings are a mechanism that represents the sparse multi-dimensional structure of a graph as a vector or matrix.

appendix Apache TinkerPop installation and overview

For the examples in this book, we use graph databases and tools from the Apache Software Foundation's TinkerPop project (http://tinkerpop.apache.org/). The project's software is properly called Apache TinkerPop or simply TinkerPop. This appendix delivers an overview of the TinkerPop project and explains how to install and configure the features needed to run the code examples in this book.

A.1 Overview

TinkerPop is a top-level Apache Foundation project, which offers an open source and vendor-agnostic graph computing framework with both transactional (OLTP) and analytical (OLAP) capabilities. In addition to the core libraries included as part of the project, there are a wide array of third-party libraries that are part of the TinkerPop ecosystem.

TinkerPop provides a standardized interface that is currently implemented by more than 20 separate database engines. This includes DBaaS (DataBase-as-a-Service) products (such as Amazon Neptune and Azure ComosDB), commercial offerings (such as DataStax Enterprise Graph and Neo4j), and open source software (such as TinkerGraph and JanusGraph).

NOTE A TinkerPop-enabled graph database is a database that implements at least the minimum APIs required to perform traversals via the Gremlin query language.

The TinkerPop project is made up of multiple different pieces. We have included the ones in this overview that we use throughout this book.

A.1.1 Gremlin traversal language

The Gremlin traversal language is the graph query language of the TinkerPop project and is the query language we use for the examples in this book. Gremlin supports both imperative and declarative syntaxes, but the imperative syntax is the predominant approach.

Gremlin allows for both query and mutation operations on data through the use of a series of steps that are chained together, similar to the way a functional language chains methods. This ability to chain operations enables the construction of complex traversals through our graphs. It is often useful to think of a Gremlin traversal in terms of a stream processor: data enters from the previous step, an operation is performed on it, and data is transmitted on to the next step.

A.1.2 TinkerGraph

TinkerGraph is an in-memory graph engine that supports both OLTP and OLAP workloads and is part of the TinkerPop Gremlin Server and Gremlin Console. TinkerGraph is built as a reference implementation of the TinkerPop API. It is a full-featured, open source implementation of TinkerPop. TinkerGraph is the core graph engine used in the various tools and software provided as part of TinkerPop.

Note that TinkerGraph isn't a piece of software that you download. It is the core engine that is used by the downloadable software such as Gremlin Server and Gremlin Console. Other vendors may choose to include it in their implementations.

A.1.3 Gremlin Console

The Gremlin Console is an interactive terminal application used with TinkerPop-enabled graph databases. The Gremlin Console enables users to connect to local or remote databases, load data into a graph, and interactively traverse around the graph. It can be used either as a standalone application with its own in-memory graph data or as a client to a graph database server. We use the Gremlin Console as a client throughout this book for our interactions with a separately running Gremlin Server.

A.1.4 Gremlin Language Variants (GLVs)

Gremlin Language Variants (GLVs) are like language-specific drivers that allow developers to use Gremlin as a query language, but to do so with the vernacular and idioms of their preferred development language. GLVs are exceptionally powerful and go well beyond our common understanding of database drivers.

When using a GLV for your language, be it Java, Python, C#, or JavaScript, you are using that language's tools and syntax. GLVs encourage writing Gremlin traversals in the style of the application's programming language: a Java developer uses Java syntax, a .NET developer uses .NET syntax, and so forth. In this book, we use the Gremlin-Java variant.

A.1.5 Gremlin Server

The Gremlin Server facilitates remote execution of graph commands against graph data. The Gremlin Server also allows non-JVM clients to communicate with JVM-based graph databases and provides a mechanism to communicate with databases hosted on separate machines. In this book, we use the Gremlin Server to host our graph data in a client-server architecture.

A.1.6 Documentation

The Apache TinkerPop website (http://tinkerpop.apache.org/) has a complete set of documentation including tutorials, getting started examples, and Gremlin recipes. Although we discuss some Gremlin concepts and syntax within this book, this book is not intended to serve as a replacement for the TinkerPop documentation. We strongly recommend that you take time to familiarize yourself with the available resources on the site if you choose to use a TinkerPop-enabled database.

A.2 Installation

The first step in installing the TinkerPop framework is to download the reference tools from the Apache TinkerPop site: http://tinkerpop.apache.org/downloads.html. The most recent version at the time of publication is 3.4.6, but any TinkerPop 3.4 implementation should work with the examples. For this book, you need to download and install both the Gremlin Console and the Gremlin Server.

NOTE This book utilizes the MacOS syntax for all examples, but we provide the Windows syntax for the same options as well.

A.2.1 Installing and verifying the Java Runtime

The prerequisite for running the Gremlin Console is Java version 8. If you do not have Java installed, you should download and install the latest Java Development Kit (JDK) from Oracle (http://mng.bz/ZrPj), OpenJDK (https://openjdk.java.net/), or your preferred Java distribution. To verify that Java is installed and its version number, use the command `java -version` like this:

```
$ java -version
openjdk version "1.8.0_222"
OpenJDK Runtime Environment (AdoptOpenJDK)(build 1.8.0_222-b10)
OpenJDK 64-Bit Server VM (AdoptOpenJDK)(build 25.222-b10, mixed mode)
```

This indicates that this machine is running Java version 1.8.0.222. From the response, we know that Java is properly configured and ready to use.

A.2.2 Installing Gremlin Console

Now that we have all the prerequisites installed and verified, the next step is to install and run the Gremlin Console:

1. From the TinkerPop downloads page (http://tinkerpop.apache.org/downloads), click the button for Gremlin Console.
2. We are now on a page that lists the mirrors of the sites to download from. Select a mirror and click the link to download it.
3. Once the download completes, unzip the code using either a command-line tool or a GUI editor to a directory that we refer to as GREMLIN_CONSOLE_HOME.
4. Open a command-line terminal.
5. Navigate to the GREMLIN_CONSOLE_HOME directory.
6. Start the Gremlin Console:
   a. For MacOS or Linux, type `bin/gremlin.sh`.
   b. For Windows, type `bin\gremlin.bat`.
7. Once the Gremlin Console starts, you will see it move through a loading process where any configured plugins are activated. Once the plugins are activated, you get an input dialog as shown here:

```
$ bin/gremlin.sh                    <-- Starts the Gremlin Console

         \,,,/
         (o o)
-----oOOo-(3)-oOOo-----
plugin activated: tinkerpop.server
plugin activated: tinkerpop.utilities
plugin activated: tinkerpop.tinkergraph
gremlin>                  <-- The Gremlin Console command prompt ready to accept input
```

A.2.3 Installing Gremlin Server

Now that we have installed and can run the Gremlin Console, it is time for us to install and run the Gremlin Server:

NOTE The Gremlin Server uses TCP port 8182. You may need to adjust your OS or local firewall settings in order to permit access on this port.

1. From the TinkerPop downloads page (http://tinkerpop.apache.org/downloads), click the button for Gremlin Server.
2. We are now on a page that lists the mirrors of the sites to download from. Select a mirror and click the link to download it.
3. Once the download completes, unzip the code using either a command-line tool or a GUI editor to a directory that we refer to as GREMLIN_SERVER_HOME.
4. Open a command-line terminal.
5. Navigate to the GREMLIN_ SERVER _HOME directory.

6 Start the Gremlin Server:
   a For MacOS or Linux, type `bin/gremlin-server.sh start`.
   b For Windows, type `bin\gremlin-server.bat start`.
7 You will get a message saying that the server has started, along with the process ID. The process ID is different each time you start the server. For example

```
$ bin/gremlin-server.sh start
Server started 56799.
```

A.2.4 Configuring the Gremlin Console to connect to the Gremlin Server

With both the Gremlin Server and the Gremlin Console running, it is time to connect the Gremlin Console to our Gremlin Server instance:

> **NOTE** If you have any Gremlin Console instances running, close these with the console commands `:q` or `:exit`.

1 Open a command-line terminal.
2 From the GREMLIN_CONSOLE_HOME directory, navigate to the conf directory.
3 In a text editor, open the remote.yaml file. This file contains three parameters that you might need to adjust. If you are running everything locally, then you will not need to change any of these parameters:
   - `hosts: [localhost]`—This parameter is the IP or domain name of the Gremlin Server where we want to connect.
   - `port: 8182`—This parameter is the port to connect to; it defaults to 8182.
   - `serializer: { className: org.apache.tinkerpop.gremlin .driver.ser .GryoMessageSerializerV3d0, config: { serialize-ResultToString: true }}`—This parameter is the data interchange format between the Gremlin Console and the Gremlin Server. Depending on the production database you choose, you may need to adjust it to the format provided by that database vendors documentation.
4 Save the file and close it.
5 From the GREMLIN_CONSOLE_HOME directory, start the Gremlin Console with the following commands:
   a For MacOS/Linux, type `bin\gremlin.sh`.
   b For Windows, type `bin/gremlin.bat`.
6 Once the Gremlin Console starts, execute the following command:

```
:remote connect tinkerpop.server conf/remote.yaml
```

This command uses the parameters that we just defined to connect to the Gremlin Server instance we have running.

7 A message is returned confirming that you are connected:

```
         \,,,/
         (o o)
-----oOOo-(3)-oOOo-----
plugin activated: tinkerpop.server
plugin activated: tinkerpop.utilities
plugin activated: tinkerpop.tinkergraph
gremlin> :remote connect tinkerpop.server
➥ conf/remote.yaml
==>Configured localhost/127.0.0.1:8182
gremlin>
```

Gremlin Console command connects to a Gremlin Server

Response confirms that the connection is configured

8 Next, run the command to switch from local mode to server mode:

```
:remote console
```

The Gremlin Console informs you that it has switched modes:

```
gremlin> :remote console
==>All scripts will now be sent to Gremlin Server -
➥ [localhost/127.0.0.1:8182] -
➥ type ':remote console' to return to local mode
gremlin>
```

9 Run the following command to display some basic information about the graph database hosted on the Gremlin Server:

```
gremlin> g
==>graphtraversalsource[tinkergraph[vertices:0 edges:0], standard]
```

We have now successfully connected to our Gremlin Server via the Gremlin Console. To exit the session with the Gremlin Server and close the connection, execute the following command:

```
:remote close
```

A.2.5 Gremlin Console command modes: Local versus remote

When issuing commands to a remote graph database server, you can choose either of these two modes: *local* mode and *remote* mode. The preferred method of sending commands to the Gremlin Server is to put the Gremlin Console into remote mode. This is what we did in the previous section, and it is becoming the default mode when using Gremlin Console connected to a server. Remote mode means that any commands executed in Gremlin Console will be sent to the Gremlin Server, run there, and the results will then be displayed by the Gremlin Console.

If you are going to only issue one or two commands, you can do this with local mode by prefacing each command with `:>`. This sends the command to the configured remote connection. Only the commands prefaced by these two characters (`:>`) will be executed on the Gremlin Server. Any commands that have not been prefaced with these characters run within Gremlin Console's own process. To switch between the two modes, use the `:remote console` command like this:

```
gremlin> :remote console
==>All scripts will now be sent to Gremlin Server -
➥ [localhost/127.0.0.1:8182] - type ':remote console'
➥ to return to local mode
gremlin> :remote console
==>All scripts will now be evaluated locally -
➥ type ':remote console' to return to remote mode
➥ for Gremlin Server - [localhost/127.0.0.1:8182]
gremlin>
```

A.2.6 Using the Gremlin Console

Before we fire up the Gremlin Console, there are a few additional options we should discuss. If you want to see a list of the options available for the Gremlin Console, type the following:

```
$ bin/gremlin.sh --help
Usage: gremlin.sh [-CDhlQvV] [-e=<SCRIPT ARG1 ARG2 ...>]...
 ➥ [-i=<SCRIPT ARG1 ARG2 ...>...]...
  -C, --color      Disable use of ANSI colors
  -D, --debug      Enabled debug Console output
  -e, --execute=<SCRIPT ARG1 ARG2 ...>
                   Execute the specified script (SCRIPT ARG1 ARG2 ...)
                 ➥ and close the console on completion
  -h, --help       Display this help message
  -i, --interactive=<SCRIPT ARG1 ARG2 ...>...
                   Execute the specified script and leave the console
                 ➥ open on completion
  -l               Set the logging level of components that use
                 ➥ standard logging output independent of the Console
  -Q, --quiet      Suppress superfluous Console output
  -v, --version    Display the version
  -V, --verbose    Enable verbose Console output
```

As depicted, there are a variety of options to use, but the most common one we use (`-i`) loads a script while starting the Gremlin Console. This is handy for configuring the Gremlin Console, loading data, and then leaving the Gremlin Console up and running, waiting for further input. All of the scripts provided in the book's GitHub repository (https://github.com/bechbd/graph-databases-in-action) do the following:

- Configure a remote connection to a Gremlin Server on localhost
- Set the Gremlin Console in remote mode
- Load the data, either with scripted operations or from a GraphSON import file

What follows is an example of running a simple data load script:

```
$ bin/gremlin.sh -i $BASE_DIR/path/to/
➥ data-load-script.groovy                 <-- Starts the Gremlin Console
                                               in interactive mode with a
         \,,,/                                 data load script
         (o o)
-----oOOo-(3)-oOOo-----
plugin activated: tinkerpop.server
plugin activated: tinkerpop.utilities          Uses the built-in g variable
plugin activated: tinkerpop.tinkergraph        to quickly verify data is
gremlin> g                              <--    loaded in the graph
==>graphtraversalsource[tinkergraph[vertices:4 edges:5], standard]
gremlin>        <-- The Gremlin Console
                    prompt waiting for input
```

The Gremlin Console is what is known as a REPL (Read Evaluate Print Loop) terminal. This means that the commands we type are immediately executed, and the results of that evaluation are printed to the screen. Because the Gremlin Console runs on Groovy, you can execute standard Groovy code like an addition computation inside the Gremlin Console. For example

```
gremlin> a = 1
==>1
gremlin> b = 2
==>2
gremlin> a + b
==>3
```

The ability to run Groovy code allows you to perform complex queries on your graphs and to save the results of those queries to variables that you can use later for additional calculations. This ability to write code is also extremely helpful when debugging your graph traversal code.

Inside the Gremlin Console, there are several available commands, all of which begin with a colon (`:`). To see a listing of the available commands, type `:help` and press Enter. The most common commands we use are `:exit`, `:quit`, `:x`, or `:q`, which are all functionally identical and exit the Gremlin Console:

```
gremlin> :help

For information about Groovy, visit:
    http://groovy-lang.org

Available commands:
  :help       (:h  ) Display this help message
  ?           (:?  ) Alias to: :help
  :exit       (:x  ) Exit the shell
  :quit       (:q  ) Alias to: :exit
  import      (:i  ) Import a class into the namespace
  :display    (:d  ) Display the current buffer
```

```
  :clear        (:c  ) Clear the buffer and reset the prompt counter
  :show         (:S  ) Show variables, classes or imports
  :inspect      (:n  ) Inspect a variable or the last result with the
                         GUI object browser
  :purge        (:p  ) Purge variables, classes, imports or preferences
  :edit         (:e  ) Edit the current buffer
  :load         (:l  ) Load a file or URL into the buffer
  .             (:.  ) Alias to: :load
  :save         (:s  ) Save the current buffer to a file
  :record       (:r  ) Record the current session to a file
  :history      (:H  ) Display, manage and recall edit-line history
  :alias        (:a  ) Create an alias
  :grab         (:g  ) Add a dependency to the shell environment
  :register     (:rc ) Register a new command with the shell
  :doc          (:D  ) Open a browser window displaying the doc for the
                         argument
  :set          (:=  ) Set (or list) preferences
  :uninstall    (:-  ) Uninstall a Maven library and its dependencies from
                         the Gremlin Console
  :install      (:+  ) Install a Maven library and its dependencies into
                         the Gremlin Console
  :plugin       (:pin) Manage plugins for the Console
  :remote       (:rem) Define a remote connection
  :submit       (:>  ) Send a Gremlin script to Gremlin Server
  :bytecode     (:bc ) Gremlin bytecode helper commands

For help on a specific command type:
    :help command
```

index

Symbols

A

B

C

D